Raves for City Chicks

"City Chicks is a revelation! It's time that someone expertly connected gardening to raising and keeping hens. The two practices go hand in hand. This book is filled with excellent advice so that everyone can confidently practice good earth stewardship, not to mention have a prize garden!"

— MICHAEL C. METALLO, PRESIDENT
NATIONAL GARDENING ASSOCIATION

"Another Take on Chick Lit. The chicken is still having her moment as the mascot and darling of the always-cresting locavore food movement. But as hipsters and foodies from New York to San Francisco embrace her charms and services — many people are struggling to learn how, exactly, to care for her. Enter City Chicks."

— PENELOPE GREEN
THE NEW YORK TIMES

"Far more than just another book on chickens, City Chicks opens the door to a whole new world of poultry possibilities. Keeping small flocks is good for gardens, municipalities, education, and the local food movement. City Chicks is a comprehensive information source."

— RICHARD FREUDENBERGER,
BACKHOME MAGAZINE PUBLISHER & CHICKEN OWNER

"City Chicks shows how local governments can save thousands — if not millions — of tax-payer dollars that are spent on solid waste management simply by allowing residents to keep hens to help with composting food and leaf and yard waste in their backyards."

— MIMI ELROD, PH.D.
MAYOR, LEXINGTON VIRGINIA

City Chicks might well become 'The' reference bible for the burgeoning local food movement." — CATHY TAIBBI
EXAMINAER.COM

"The focus is on how chickens fit in with so many other common needs and concerns of our century. This is a book for our times." — J.D. BELANGER, EDITOR EMERITUS
BACKYARD POULTRY MAGAZINE

"City Chicks is an outstanding book that covers it all. It is comprehensive starting with fresh eggs through to raising replacement hens...and integrating chickens into urban agriculture — with all the joy in between. Novice and expert will enjoy this book." — ANDY MARSINKO
GRANDMASTER EXHIBITOR, POULTRY JUDGE
AMERICAN POULTRY ASSOCIATION HALL OF FAME

"The best solutions today are integrated solutions. City Chicks show how to successfully produce protein along with your garden vegetables while managing waste and increasing soil fertility...a 3-for-1 benefit! — WILL RAAP
FOUNDER, GARDENER'S SUPPLY

"A 'must-have' for anyone interested in a family flock of chickens. City Chicks offers step by step instructions for raising and properly caring for chickens, with notable warnings against common mistakes that new chicken keepers make. Written in plain, no-nonsense language. Highly recommended." — MIDWEST BOOK REVIEW

"I love the lightheartedness, the humor, the fun in it, as well as the really solid information that City Chicks provides. I really love it!" — MARJORIE BENDER
RESEARCH & TECHNICAL PROGRAM MANAGER
AMERICAN LIVESTOCK BREEDS CONSERVANCY

City Chicks

Keeping Micro-flocks of Chickens

as

Garden Helpers,

Compost Creators,

Bio-recyclers,

and

Local Food Suppliers

by

Patricia Foreman

Good Earth Publications, Inc.

Copyright © 2010 Good Earth Publications, Inc.
First Edition ISBN-13: 978-0-9624648-5-0 • ISBN-10: 0-9624648-5-6
Kindle Edition ISBN-13: 978-0-9843382-2-1 • ISBN-10: 0-9843382-2-5
EPUB Edition ISBN-13: 978-0-9843382-3-8 • ISBN-10: 0-9843382-3-3

The information in this book is, to the best of the publisher's knowledge, true and correct. The author and publisher have exhaustively researched all sources to ensure the accuracy and completeness of information contained within. We assume no responsibility for errors, inaccuracies, omissions, differences of opinion, or any other inconsistency and cannot be held responsible for any loss or damages resulting from information contained within the book. Any slights against people, programs, products, companies or organizations are unintentional.

Library of Congress Cataloging-in-Publication Data
    Foreman, Patricia L.
      City chicks : keeping micro-flocks of chickens as
    garden helpers, compost creators, biorecyclers and local
    food suppliers / by Patricia L. Foreman. -- 1st ed.
      p. cm.
    Includes bibliographical references and index.
    ISBN-13: 978-0-9624648-5-0 (trade pbk.)
    ISBN-10: 0-9624648-5-6

    1. Chickens. 2. Urban agriculture. 3. Organic
    farming. 4. Sustainable agriculture. I. Title.

SF487.F67 2010   636.5
                QBI08-600316

Cover by Phil Laughlin Studio
Published by: Good Earth Publications, Inc. Printed in the USA!

About the Cover

The *City Chicks* cover is intended to show that chickens are not just dirty, stinky farm animals. In proper environments and with respectful, loving care, their personalities can shine forth including senses of humor, creative thinking, friendships and service.

As reflected by the "Hen Have-More Plan" in the theater marquee, *City Chicks* introduces chickens in roles of more than just providing meat and eggs. They have versatile skill sets and can work as urban agricultural assistants in backyard food production systems. Not only providing fertilizer, they can also be employed as organic pesticiders, herbidicers, compost creators and produce suppliers. They can also help divert food and yard waste from landfills, saving millions of tax payer dollars. These are valuable assets in any "have-more" plan.

Finally, the proud and sophisticated hen's red hat is intended as a tribute to the Red Hat Society for enriching and expanding the lives of its members, and supporting the attitude of living life to its fullest and greatest. You go girls!

About the Author

Patricia Foreman was born and raised in Indiana. She graduated from Purdue University with degrees in Pharmacy and Agriculture (Animal Science, genetics and nutrition). At Indiana University's Graduate School of Public and Environmental Affairs she earned a Masters of Public Affairs (MPA). Her majors were in Health Systems Administration and International Affairs. She completed the Virginia Master Gardener's program in 1999. Pat has kept poultry for over 20 years. Her experience includes having owned and operated a small-scale farm raising free range, organic layers, broilers and turkeys. She keeps a backyard flock of heritage chickens to help with the kitchen garden and egg supply.

Her many awards include a Fulbright Scholarship, and appointment as a Presidential Executive Management Intern. She served as a Science Officer for the United Nations in Vienna, Austria, and has worked in over 30 countries conducting workshops and providing consulting services. Agencies funding projects she has worked on include the U.S. Agency for International Development, World Bank, World Health Organization and the Pan American Health Organization.

Pat is the co-author of several alternative, sustainable agriculture books including: *Chicken Tractor, Day Range Poultry, Backyard Market Gardening* and *A Tiny Home to Call Your Own*. She loves to talk and is the co-host, along with Andy Schneider, of the Chicken Whisperer Backyard Poultry and Sustainable Lifestyles Talk Show. She has been a guest on radio talk and TV shows across America. She is available to facilitate workshops, give presentations, and provide consulting services. Contact her through Good Earth Publications.

The Differences Between *Day Range Poultry, Chicken Tractor* and *City Chicks.*

The trilogy of chicken books have overlaps in information, but each book contains unique material depending on the flock size and purpose behind the flock. Some flocks are raised for meat, others for eggs and some are breeder flocks for the next generation. Each requires different management techniques. The table below summarizes these differences. You can see the entire table of contents of each book on the Good Earth Publications Book Store at www.GoodEarthPublications.com

	City Chicks	Chicken Tractor	Day Range Poultry
Scale	Focus is on micro-flocks of urban chickens kept in backyards and gardens.	Focus is on home-stead poultry pro-duction of 25 to a few hundred broilers and layers for self-suffi-ciency, and perhaps a few customers.	Focus is on commer-cial pasture poultry production of hun-dreds to thousands of layers, broilers, and turkeys.
Where	Keeping backyard chickens as home egg producers, bio-mass recyclers and garden helpers.	Homestead perma-culture designs to improve soils with chicken tractors to fertilize garden beds.	Farm, pasture, market garden, soil improvement and multi-species grazing.
Production	Eggs, fertilizer, com-post and garden production.	Table top process-ing for the home and a few customers or farmer's markets.	Small commercial processing of hun-dreds of birds for meat and eggs.

	City Chicks	Chicken Tractor	Day Range Poultry
Shelters	Micro-flock coops and chicken tractor systems to compliment the home garden.	Smaller, bottomless shelters with popholes for free-range grazing. Movable by hand.	Details about larger shelters with and without bottoms.
Incubation	Mini-scale incubation of a few eggs, brooding and raising micro-flocks of 25 or less. Sources for mature hens.	Smaller scale egg incubation, hatchery management and brooding for flocks around 25 to 100.	Covers production scale egg incubation, hatchery management and battery brooding for flocks in the hundreds.
Fencing	Various types of fencing directing hen helpers to work different parts of a garden.	Day runs, portable fencing and panels.	Information and lessons learned about electric poultry net fencing.
Flock Types	Focus on chickens, usually as pets. No Roosters without special permits and conditions.	Main focus on layers and broilers.	Detailed information about raising broilers, layers, poultry breeder flocks and turkeys.
Feed	Bagged feed, feed supplements and growing garden crops for hens.	Bulk and bagged feed. Details about feed supplements.	Bulk organic feed ration formulas by the ton.
Marketing	Egg production for the family, and a few others.	Marketing centers around farmer's markets and direct sale customers.	Commercial markets. Covers why cooperatives benefit all growers.

City Chicks is dedicated

to my mother,

Marie Foreman.

Thanks being a role model,

and for your unwavering

love and support,

throughout all my years.

If you like *City Chicks* and the information it brings about putting family flocks to work enabling local agriculture, then please recommend it to others. We also invite you to go online at Amazon.com, Amazon.ca, Amazon.co.uk, BarnesAndNoble.com, or other book sellers to write a brief "reader review". It will take only a few minutes and help spread the word about the value of chickens — and their skill sets — that can be of service to individuals, communities and the environment.

As you read through the book, and if you find any corrections, factual or grammatical errors please let us know by contacting the publisher at info@GoodEarthPublications.com. We are always working on next editions and new titles. If you have stories and ideas to suggest that encourage "thinking outside the coop", we would deeply appreciate hearing from you.

As always, and forevermore, may the flock be with YOU!

Table of Contents

The Chicken Have-More Plan

About 60 years ago, Ed and Carolyn Robinson wrote a classic book called:

"The Have-More" Plan:
A Little Land — A Lot of Living

The Robinson's book inspired millions of people recovering from World War II, to be more self-sufficient and independent.

City Chicks was written in the same spirt as Robinson's *"The Have-More" Plan* from over a half-century ago. *City Chicks* has the ambitious intent of exploring four intertwined areas that are all parts of the chicken have-more plan.

    1. Enhancing Local Agriculture. Urban gardening and farm-yards are on the verge of a giant leap forward, ushering in a new — and necessary — era of local and home food production. People have a *right* to grow their own food and chickens have valuable skill-sets that can be employed in food production systems. Some of these "skill-sets" include producers of manure for fertilizer and compost, along with being mobile herbiciders and pesticiderers. And of course, they also provide eggs and meat. *City Chicks* shows how you can have a good meal of eggs and garden goods that only travel the short distance from your backyard.

    2. Diverting Food and Yard "Waste" Out of Landfills. Chickens can help convert biomass "wastes" into organic assets such as fertilizer, compost, garden soil and eggs. This can save BIG TIME tax payer dollars from being spent solid waste management streams. Less biomass in landfills also means that less of the global warming gas, methane, form in the land fills.

    3. Decrease Oil Dependence and Lower Carbon Footprints. Commercial food systems cannot work without oil. Over 17% of America's oil is used in agricultural production and, about 25% of this oil is used for fertilizer. The total energy input of food production, processing, packaging, transporting and storing is greater than the calories consumed. It is estimated that every person in this country requires about one gallon of oil per day just to bring food to the table. How sustainable is that? Chickens can help America kick the oil habit by decreasing the amount of oil products used in feeding ourselves ... and, at the same time, keep landfills from filling up with methane-producing organic matter.

   4. National Defense & Emergency Preparedness. Whoever controls your food supply controls you. Food supply, or lack of it, has created and destroyed civilizations since time began. In natural disasters food can become a matter of life or death. Even if grocery supply lines fail, with an abundance of local chickens you can still find food. Having wholesome, locally produced food helps keep a country safe and strong.

City Chicks ushers in a new paradigm of how to employ chickens in a variety of roles that help decrease carbon footprints, save tax payer dollars and support local food supply production. And all this is done in a way that is biologically sustainable, economically equitable, and serves us, our communities, our Earth and the future generations of all beings.

Want to be a Chicken Have-More Club member?
You already are!

**The Chicken
Have-More Plan**

1 Why Have City Chickens

There is an urban underground City Chicks Movement exploding in America. City folks are flocking to keeping chickens. Why is this happening? Most of the wind-beneath-the wings of City Chicks is the desire for locally produced foods, food supply independence, and an act of living green.

Chickens are now a symbol of local food supply, and especially local protein availability.

> *"The chicken has become the mascot
> of the local food movement."*
> – BEN BLOCK, WORLDWATCH INSTITUTE

Raising backyard chickens is part of the urban farming movement and it is gaining popularity as commercial food prices go up and its nutritional quality is questionable. Buy local, grow local, and eat local is increasingly becoming a slogan of community food self-sufficiency.

The gas crisis and the American desire to be independent from carbon based fuels are also driving forces of the City Chicks Movement. We have paid a terrible price for our oil addiction, including wars and pollution, which has caused indescribable suffering at every level of our planet's ecosystems, from the depths of the oceans to the upper stratospheres.

And there are more reasons for keeping City Chicks. Chickens are some of the easiest animals in the world to keep. Chickens are compact, easy to handle, don't eat much, and are not expensive to buy. Laying hens are generally not aggressive toward people and — unlike a dog or cat — their bites rarely draw blood. Chickens can be kept as a hobby, as pets, or as a small business. Chickens are viewed by many as an essential part of a backyard mini-farm that helps folks achieve some degree of self-sufficiency. Many chicken owners view keeping flocks as an act of green living.

There is also the charm of chickens. These cackling, cooing bipeds wearing about 8,000 feathers provide more entertainment than most cocktail parties (excuse the comparison). They have distinct personalities and are intelligently curious about exploring. The flock antics of socialization can be as brutal as a congressional debate or as loving as Mother Theresa. Hens establish deep and loyal friendships with other members of the flock. I have seen the same two hens trundling around together day after day, scratching and discussing what they find underfoot, like bookends, in constant company of each other. They even sleep side-by-side in the coop.

Chickens are more intelligent than most people realize. Research shows that chickens can find their way through a maze. They have the ability to think things through and can make choices, rather than just react to situations. This ability to think makes them trainable. My little flock's coop and their run are adjacent to my home. There is a dog ramp they could walk up to get onto the deck, but I've trained them to stay off the ramp, deck, and not to roost on the railings. I've trained them to come when I call them; they do tricks.

It's possible to have a genuine, two-way friendship with a hen, similar to that with a cat or dog. They know who you are and enjoy your company. Chickens can recognize their keepers from a distance and will come running up to greet you exclaiming (in hen talk), "What's happening dude?" or, "Where ya been?"

Chickens have a sense of self beyond themselves. I've seen a rooster charge a fox to protect his hens. He was willing to sacrifice his life to save the ladies. My flock knows the two mixed terriers that protect them, but when an unfamiliar dog shows up, or a hawk flies overhead, they sound the alarm for others to beware of potential danger. They have a sense of community and community service.

Reasons to Keep a Micro-flock of Family Chickens

Below are 13 "eggscellent" reasons why you can be "eggstatic" about keeping chickens.

1. Enhancing Local and Backyard Agriculture. Urban gardening and farmyards are on the verge of a giant leap forward of local and home food production. Chickens have valuable skill-sets that can be employed in food production systems including producers of manure for fertilizer and compost. They are also mobile herbiciders and pesticiderers. And of course, they also provide eggs and meat. *City Chicks* shows how you can have a good meal of eggs and garden goods that only travel the short distance from your backyard.

2. Diverting Food and Yard "Waste" Out of Landfills. Chickens can help convert kitchen and yard "wastes" into organic assets of fertilizer, compost, garden soil and eggs. This can save BIG TIME tax payer dollars from being spent solid waste management streams.

3. Emergency Preparedness Partners. Your local grocery has, at most, 3 to 4 days of food supply in stock; and that assumes no hoarding from a storm or disaster. The top brass at the Pentagon consider food and water supply as one of our nation's most vulnerable attack areas. Even with disasters, folks can still have a good meal of eggs or chicken soup.

4. National Defense. Whoever controls your food supply controls you. Food supply — or lack of it — has created and destroyed civilizations since time began. Communities that have family flocks of chickens help support not only national food safety and security, but also national defense! Locally produced nutritious food helps keep us strong and safe.

5. Chickens Eat Ticks, Fleas and Mosquitoes. Chickens have a keen eye for insects. In my home, even with 2 dogs and too many cats there are only the occasional tick or flea; and these buggers are probably from walks in the woods and other places. Chickens can help with tick control in Lyme disease infested areas

Chickens as National Defenders and Emergency Preparedness Partners

Local food supply is critical to our national defense, disaster preparedness and relief.

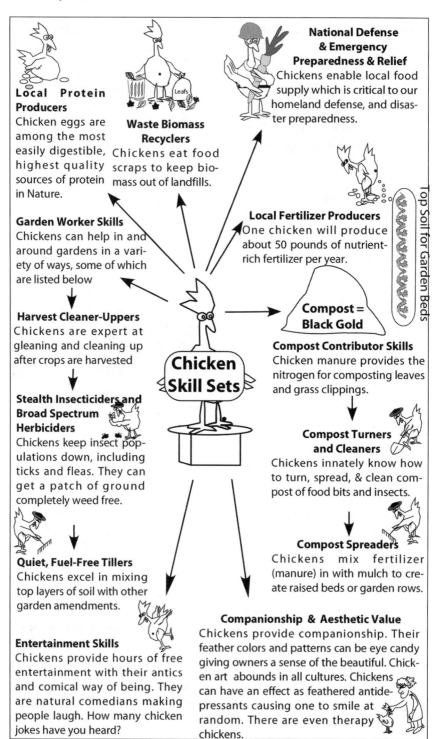

Local Protein Producers

Chicken eggs are among the most easily digestible, highest quality sources of protein in Nature.

Waste Biomass Recyclers

Chickens eat food scraps to keep biomass out of landfills.

National Defense & Emergency Preparedness & Relief

Chickens enable local food supply which is critical to our homeland defense, and disaster preparedness.

Local Fertilizer Producers

One chicken will produce about 50 pounds of nutrient-rich fertilizer per year.

Top Soil for Garden Beds

Garden Worker Skills

Chickens can help in and around gardens in a variety of ways, some of which are listed below

Harvest Cleaner-Uppers

Chickens are expert at gleaning and cleaning up after crops are harvested

Stealth Insecticiders and Broad Spectrum Herbiciders

Chickens keep insect populations down, including ticks and fleas. They can get a patch of ground completely weed free.

Quiet, Fuel-Free Tillers

Chickens excel in mixing top layers of soil with other garden amendments.

Entertainment Skills

Chickens provide hours of free entertainment with their antics and comical way of being. They are natural comedians making people laugh. How many chicken jokes have you heard?

Chicken Skill Sets

Compost = Black Gold

Compost Contributor Skills

Chicken manure provides the nitrogen for composting leaves and grass clippings.

Compost Turners and Cleaners

Chickens innately know how to turn, spread, & clean compost of food bits and insects.

Compost Spreaders

Chickens mix fertilizer (manure) in with mulch to create raised beds or garden rows.

Companionship & Aesthetic Value

Chickens provide companionship. Their feather colors and patterns can be eye candy giving owners a sense of the beautiful. Chicken art abounds in all cultures. Chickens can have an effect as feathered antidepressants causing one to smile at random. There are even therapy chickens.

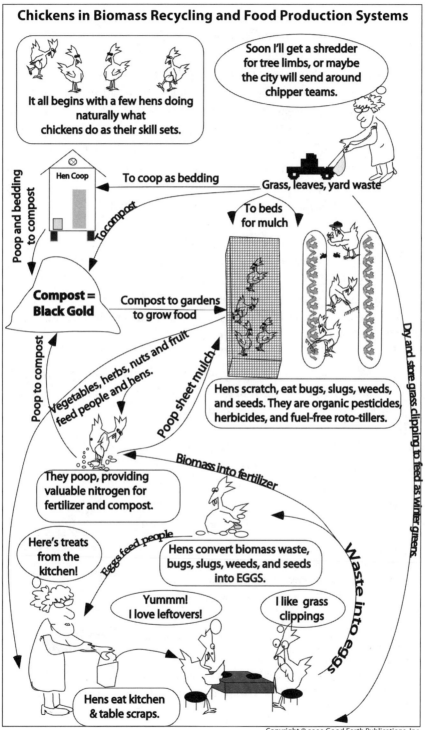

6. Easy, Quiet Pets. Chickens are easy keepers. They don't have to be house or litter trained. They don't cough up hair balls and barf on your best carpets. They don't bark at visitors or passers-by. Hens, at their loudest, have about the same decibel level as human conversation. Roosters are about the same decibel level as a barking dog. Chickens don't leave fur everywhere; you don't have to comb them. Chickens don't even want to be walked; but they do need space to range. They rarely run away. Chickens are eye-catching, fashionable, interactive lawn ornaments that are far more appealing (and tasteful) than plastic pink flamingos. Folks allergic to dogs or cats are often not allergic to chickens. Chickens are genuine "pets with benefits".

7. Personal Ecological Act. Most people who keep chickens view it as part of living "green" and treading lightly on the Earth. Eggs from a personal flock or local free range farm saves fuel, electricity, and reduces air pollution by eliminating the indoor factory farm "housing", transportation and the cold storage needed to produce and move eggs from factory farms to your home.

8. Health & Safety. There's very little risk of avian diseases, antibiotic resistant bacteria, salmonella or toxic drug metabolite tag-alongs lurking in your family flock-laid eggs.

9. Soil Producers. Chicken manure provides valuable nitrogen to mix with kitchen and yard residuals to make rich compost and garden soil. The two most valuable commodities today are topsoil and water. Already bottled water cost more than gasoline.

10. Educational. Because chickens are smaller, cleaner, cheaper, and easier to raise than cows, pigs, or goats, urban 4-H kids can participate in poultry projects to learn about animal husbandry.

A Chicken Can Biorecycle About 7 Pounds of Food Residuals per Month

11. Gene Pool Preservation. Purchasing and keeping small flocks helps preserve rare and minor chicken breeds by providing income to the heritage breeders.

12. Fashionable. No matter how you dress or what you do, keeping chickens is "in"; just ask Martha Stewart. Upon learning that you are a chicken aficionado, people regard you with new respect, and the topic of keeping chickens crops up in random conversations. Giving the hosts of a dinner party a dozen fresh eggs can be more meaningful than bottle of fine wine. Backyard eggs are becoming status symbols and have always been the source of gourmet delights.

13. Engaging Hobby. There comes a point when all you think about is chickens. You check on them at night, shining your flashlight to make sure they are sleeping. You notice your friends' eyes glaze over as you describe the antics of your incredible frolicking feathered flock. Their ungainly grasshopper-chasing athleticism becomes as addictive as football. You become eccentric even by extreme standards. Local CA (Chickens Anonymous) groups form where you can tell your stories to people who actually care, and want to hear them. "Hi, I'm Pat. I have 15 chickens".

Clucking City Workers

There are many warm-feathery reasons to keep chickens, but the most economic and politically compelling reason is to employ them to recycle food and yard waste. A chicken eats about its weight every month (6 to 8 pounds of feed). That's about 84 to 100 pounds of food/year some of which can be biomass that is kept out of the landfill.

To save you from reaching for your calculator,

Chickens as Solid Waste Management Employees

Feeding city chicks kitchen scraps and creating compost from leaves, yard waste, and chicken manure can divert biomass from landfills and save millions of taxpayer dollars in municipal solid waste management budgets.

the math is: (1 hen)(7 pounds food waste/month)(12 months) = about 84 pounds. "Big deal" you think. "That's not so much."

But what if a city had 2,000 households with three hens (or more) each? That translates to about 252 tons of biomass diverted from landfills. Here's the math: (3 hens)(84 pounds of food residuals/hen/year)(2,000 households) = 504,000 pounds = 252 tons. Now, add to that number the tons of yard waste (grass clippings and leaves) that chickens can help convert into compost and the biomass amount is as enormous as the tax-savings in not having to handle, transport and store all that biomass waste.

The chicken biomass converter and waste diversion strategy is exactly what the city of Diest in Flanders, Belgium is using to reduce their refuse management budget. By giving 3 laying hens to 2,000 households, the city of Diest is using chickens as an economical solution to the costly problem of biomass trash. Dealing with biodegradable trash costs the city about $600,000 annually. About 25 percent of the material going into landfills could be composted in backyards. Diest city officials expect to save a significant amount of trash management expense by using personal poultry power. From the city manager's point of view, the chickens' production of eggs, compost (topsoil), and fertilizer are simply spin-off added benefits to residents.

In my little corner of the world, the county landfill will be full in less than a year. Once closed, every ounce of trash must be trucked away, possibly as far as 100 miles. It's estimated that every ton trucked will cost the county (at current fuel rates) $75 or more, compared with $40/ton to dump trash in the landfill. It's simple accounting: the more residents recycle and divert tonnage away from their trash, the lower their taxes will be.

The stakes are even higher for the environment. Kitchen and yard waste sent to landfills decompose without oxygen (an anaerobic process) and generate methane. Methane not only stinks, it is 60 times more powerful than carbon dioxide (CO_2) as a greenhouse gas.

It is time to think outside the chicken coop. Policy makers can employ poultry power to lighten the trash load of local dump trucks and waste management personnel. As in the City of Diest, there could be a line item in the town budget for buying chickens to serve as low cost, non-union forming, waste management employees.

I hope there will come a time when local governments give residents compost bins and, encourage those who want them to keep chickens. There will be government sponsored classes on Chicken Keeping and Composting.

Community colleges will offer classes on family flock poultry husbandry and small scale food production. Like the Master Gardener Program, there will eventually be classes and certification for Master Composter, Master Poultry Keepers, and Master Bee Keepers.

These types of courses will help raise awareness not only of the waste management challenges and possibilities, but also of the importance of local food supply, independence from foreign oil and emergency preparedness. Yes, the humble, comical and unassuming *Gallus domesticus* might be part of the answer to the multiple crises our world is facing today.

The two sketches on the preceding pages, "Chicken Skill Sets" and "Chickens in Recycling and Food Production Systems", show how all the pieces fit together for local poultry biomass recycling, garden enhancement, and small scale food production. These odd little diagrams are simplistic in rendering, but profound in possible ramifications. Starting with the innate behaviors and characteristics of chickens, they show how these skill sets can be directed and used for multiple tasks in small-scale, local food supply systems.

In a way, these sketches are a schematic table of contents that tell what *City Chicks* is all about.

"Ask not what chickens can do for you.

Ask what you can do with chickens."

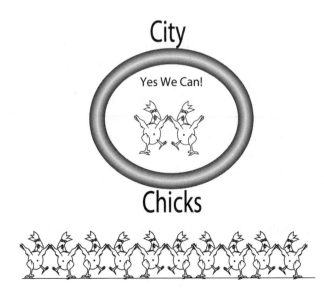

"The greatest fine art of the future will be the making of a comfortable living from a small piece of land."

—ABRAHAM LINCOLN

"Whoever controls your food supply controls you."

— ANCIENT CHICKEN PROVERB

2 Garden Chicks in Food Production Systems

This chapter introduces a new poultry profession: the garden chicken. Garden chickens' jobs are to help in (and around) food production areas. Hens do more than function as egg machines. Garden chickens often free-range in gardens with fenced-in plants and beds. They are easy to direct with portable fencing and can be assigned day jobs, such as compost aeration or cleaning up fallen fruit. Garden chickens are independent workers who wear many hats and have many food production skill sets. With garden chicks, the term "utility bird" takes on new meaning.

This chapter discusses how chickens can help grow human and hen food in backyard gardens. It explores ways that chickens can enhance crop production. Their impact is felt at all stages of a crop's growth cycle: bed preparation, weeding, pest management, and harvesting. All phases can benefit from a micro-flock of gleefully gleaning and fertilizer creating *Gallus domesticus*. The Gardening For Chickens chapter discusses the advantages of growing crops for chickens to decrease feed costs and to supplement commercially bought feed.

Garden Chicks (the book) is being researched and developed and will be published soon by Good Earth Publications.

Many testimonials compare gardening before, and after, incorporating chickens into the process. The difference is in the richness of the garden experience, both of soil and soul. The testimonials usually sound something like this:

"We had more than seven times the yield from the chicken fertilized plants than from the unfertilized ones." – PA

"Chickens have become my anti-depressants in feathers. I enjoy working in the garden so much more with my hens there. They follow me around helping me dig, pulling weeds, always looking for that opportune bug or grub. I genuinely enjoy, and look forward to their company." – ZP

"My garden flourishes with the help of my hens. I wish I had gotten some years ago." – BL

"Before chickens we had ticks, fleas, grubs, ants, grasshoppers, squash bugs, potato bugs, aphids, white grub worms, etc. After we let the poultry loose in the garden these were almost totally eradicated." – KR

"I can't imagine gardening without our chickens. They make pests, especially ticks, fleas and grasshoppers possible to manage organically." – AL

One of the more illustrative stories about keeping garden chickens is from Bailey White. She's the author of many award-winning books and sometimes reads her essays on National Public Radio's *All Things Considered*.

Bailey is an avid gardener and wrote a column for *Kitchen Garden Magazine* called, "A Gardener's Journal". One of her essays titled, "Chicken Tractor" beautifully summarizes many of the "gardening with chickens" concepts. Published by Taunton Press, *Kitchen Garden* has been discontinued and blended with Taunton's *Fine Gardening*. With permission from Taunton Press, Bailey White's "Chicken Tractor" essay follows.

Chicken Tractor By Bailey White

I came to hate my rear-tine tiller. I hated the way it would get itself into some invisible track and veer off the top of the bed down over the edge,

and skitter along the path, snatching up straw and wrapping it around the blades. I hated the way, after 100 pulls on the starter, there would still be just a little half-hearted sputter, and the muscles of my arm would begin to burn and cramp, and the mocking words "Easy-Start Engine!" would glisten brighter and brighter as the sweat dripping off my nose washed the dust off the cheerful red letters.

I hated the advertisements that came in the mail showing color pictures of a serenely frail woman, dressed in spotless clothes and standing in the middle of a beautiful garden with one hand lying limply on the handle of her trusty tiller.

I hated the smell of it, the sound of it, and the look of it. I hated its teetery little wheels and its snaggled blades like the dark, stumpy teeth of an evil leer.

And after an afternoon of wrestling with the tiller, I did not feel that deep satisfaction that comes from helping one of nature's circles complete itself. Instead, I felt that most of the energy I had put into my gardening work that day was not staying in the ground making compost rot and seeds grow, but was hovering somewhere in the air like a fetid cloud — the fumes of spent fossil fuels from the tiller and the poisonous gases of frustration and fury from me.

Then I read a book called Chicken Tractor *by Andy Lee and Patricia Foreman. The next week I sold my tiller, ordered 25 – day old chicks and built a portable chicken pen and my gardening life changed forever.*

Now, instead of dragging out the tiller, I just harness the energy of a flock of chickens by putting them in a bottomless shelter-pen that fits neatly over one garden bed. When the chickens have weeded, cultivated and fertilized that bed, I move them and their pen onto the next bed. The chickens live happy lives, with plenty of space, fresh air,

*and natural food; and as a result, my garden soil is free of
weeds and rich in nutrients and organic matter.*

Chicken Tractor *describes several systems you can use
to employ chickens in your garden. I have been using the
deep mulch chicken tractor system on my raised beds for
nearly a year now. At the end of winter, I planted lettuce
in bed 4A where the chickens had been in the fall. The lettuce grew so fast it was almost startling to watch. Yester-
day...I threw trimmings over to the chickens, who are now
working on bed 4B. I stood and watched for a while — no
sputtering road, no gasoline pall, no aching muscles, no
fits of furry, just the happy chortles of chickens joyfully
eating the very plants they themselves had nourished. And
there it was: one of nature's circles completing itself, with a
little help from me and my chicken tractor. "*

Importance of Topsoil

Many urban gardeners with the "no-green thumb" syndrome probably
don't have fertile soil to start with. They plant in thin, nutrient-poor soil
that can be measured in millimeters. There isn't enough topsoil to support
plant growth, so gardeners get lousy results.

I'd venture to say that most urban lawns and potential garden spaces
only have an inch or two of topsoil covering subsoil or clay. Digging a hole
to plant a tree can require a pick axe when subsoil is compacted. Housing
developers strip the topsoil and, when the house is finished, spread just
enough back on the lawn to grow grass. Any "extra" topsoil is usually sold
for a good price: it's valuable. Thin topsoil is why many urban lawns require
fertilizer – lots of it – year after year. The soil can't build up organic matter
because the grass clippings are collected and sent to the trash.

How can you create topsoil? Soil is formed by the decomposition of
organic matter into humus. Humus is the dark organic matter in soils

and is essential to soil fertility and moisture retention. You can think of organic matter as materials – plants or animals – that were once living. For example, in a forest leaf litter, woody material, and dead animals fall and decay. Decomposition transforms this matter into a stable humis material called humus. The decomposing is done by organisms such as earthworms, bacteria, fungi, insects, and other living beings that eat organic matter, which is the heart of the soil.

It is the humus (organic materials) in soil that gives it that rich dark brown or even black color. Humus is sometimes used to describe finished compost that has reached a state of stability where it won't break down further if conditions stay constant. Humus can remain essentially the same for centuries, if not millennia.

It's critical to understand that soil builds up ("uppens"). It doesn't deepen. Tilling often mixes subsoils in with topsoil to create not-so-good soil. I think the misconception comes from having to dig down in soil to plant. Soil rises as each year's growth dies, falls to the ground, and decays into humus. The humus contains the nutrition necessary for next year's growth. It's an ongoing cycle: the soil dwellers generate the nutrition for plants, and plants grow the biomass to feed the soil dwellers. It's a perpetual motion machine.

The valuable topsoil can be destroyed in several ways.

1. Plowing and rototilling soil causing it to loose structure, form hard-pan, and be more vulnerable to wind and water erosion.

2. Planting continuous crops in the soil and harvesting all the biomass, a practice known as mono-cropping. This depletes the soil's ability to regenerate humus. Removing all plant matter starves the topsoil dwell-ers. After years of tilling and harvesting all plant matter, soil is left with little humus. Many commercial fields consist only of remaining subsoil and clay. Hay, straw, grain, and corn are commonly mono-cropped. These farmers are harvesting their soil and selling it as a cash crop.

3. Adding astringent fertilizers that over-stimulate soil dwellers and rapidly deplete organic matter. Applying heavy fertilizers gives greater (even maximum) production in the short-term, but in the long-term it diminishes humus and soil dwellers die, leaving behind fields that can't produce without artificial inputs.

4. Using toxic chemicals (ammonia, pesticides, and herbicides) to kill soil dwellers. This disables the perpetual motion cycle in which plants provide food for soil dwellers and vice versa.

Most urban topsoil formation is impeded because the cut grass and gathered leaves are bagged and sent to landfills, so humus doesn't have a chance to form. Think about that next time you cut your lawn and export your yard waste. You are trashing locked-up sun energy in the form of grass clippings. Leave clippings on the lawn and let them reabsorb to feed the soil dwellers and build humus.

How much organic matter does soil need to be productive? It's surprising how much a little organic matter can do. A soil with only 1% organic matter is at risk of erosion unless it is continually covered with plants. Most urban lawns have approximately 2% topsoil, which is why fertilizer has to be added to maintain that putt-putt look.

A soil with 5% organic matter will have a crumb structure that can withstand wind and water erosion. This small amount of organic matter can absorb rainwater before soil begins to erode, and will hold moisture much longer. Plants grown in soil with at least 5% or more humus will be less stressed by lack of water. Some soils contain humus close to 100%, about the same percentage as peat. Soils that contain 40% or more humus block the uptake of heavy metals by plants. The heavy metals are held by the humus in a colloidal state that plants can't access, absorb, or pass up the food chain.

Organic matter acts like a living sponge. It is a giant reservoir – the soil's sippy cup – that holds water and releases it slowly for plant use.

Whether it's topsoil, compost, humus,

Rototilling Destroys Soil Structure and Life

Rototilling causes soil chaos. It disrupts soil structure, aeration and drainage. Rototilling can mix great topsoil in with sub-soil, creating not-so-good soil.

Rototilling creates compaction and hardpan. It kills worms and destroys the valuable fungus that supports soil life.

The soil looks fluffy and smooth, but it quickly looses structure and ability to hold moisture. It is much better to mulch and build soil up, rather than mix it down. It's better to mulch than till.

each serves as the growing media for plants. It holds moisture and minerals and provides drainage and aeration. It is the support base and nutrient delivery system by which most plants flourish. Topsoil provides micro-environments to support the complex soil life necessary for wholesome food production. Without topsoil, civilizations decline.

Healthy soils are living cosmos, full of life. When growers treat their soils like living entities that need care, nourishment, and protection, food production rises to optimal nutritional levels. This process requires that organic matter be replenished on a regular basis. The best way to add organic matter is with compost. Other methods include rotational livestock grazing (chicken tractors), cover-cropping green manures, and sheet mulching. All of these work but none are as immediately effective as compost.

Compost creation is such an important process that this book has an entire chapter devoted to it. For now, let's examine how your hens can serve as direct deposit fertilizers and topsoil generators.

Chickens as Direct Deposit Fertilizers

The multi-foot deep topsoil on the Great Plains was created by massive herds of hoofed herbivores, primarily buffalo. These herds would graze, fertilize, and tromp in nutrients with their hooves.

I eat chicken feed, kitchen scraps, garden goodies, insects, grubs, slugs and crawly things.

How are chicken like buffalo? They have similar effects on soil fertility. Micro-flocks of chickens grazing intensively in garden beds or lawns are like herds of ungulates grazing, fertilizing with their manure, and digging nutrients into the soil (although chickens scratch instead of tromp).

Chickens give back valuable, nitrogen-rich manure that can be mixed with leaf and yard waste to create compost and top soil.

When you use the chicken tractor system, soil fertility and organic matter content will increase overnight. Use

Chickens as Local Fertilizer and Soil Creators

the deep mulch method and you will have rich raised beds in a couple months.

> *"A standard size chicken eats about 80 to 90 pounds of feed annually, and produces about 50 pounds of valuable nutrient-rich, nitrogen-packed manure.*
>
> *Just as valuable, about 40% of chicken manure is organic matter which is necessary for creating compost and building fertile, healthy topsoil."*
>
> — ANDY LEE

Chickens cannot convert everything they eat into meat or eggs. Chickens excrete about 75% of the nitrogen (N), 80% of the phosphorus (P) and 85% of the potassium (K) they consume. The contents of commercial fertilizers are expressed in ratios of NPK, or nitrogen, phosphorous, and potassium. Chicken manure, managed and applied correctly, is home-grown fertilizer.

If you keep 10 hens in a pen about the size of a garden bed (3' to 4' wide and 10' long) they will deposit about 500 pounds of manure in a year. This much manure in one place is toxic, but if you move the chickens around your garden it becomes top-dressed fertilizer.

Here's how this system can work. Rotate the hens from bed to bed in a 30 day cycle. That means they will spend one day on each site and not return to that site for 30 days. This rotational system will cover 1,200 square feet, about the size of a 30 by 40 foot family garden.

If you don't have enough space to rotate the hens through different

Bottomless Chicken Tractor

The basic portable chicken tractor can be used as a bottomless pen. It can be used for fertilizing along rows, or left in one place to build soil in raised beds. Notice the pop hole door that lets chickens out to free-range when it's safe.

garden beds, then use the deep mulch system. In this system you add dry carbon material (such as grass, leaves, hay, or straw) daily to the pen for bedding. The advantage to adding mulch to the chicken tractor is that it stops the loss of nutrients through vaporization. If chicken manure isn't protected on the soil surface, 30 to 90% can volatilize. About 90% of that loss happens in the first 24 hours and continues as long as the nitrogen-rich manure isn't covered.

Another advantage of the deep mulch system is that it buffers the impact of nitrogen in the chicken manure that "burns" plants. Carbon in plant material links with nitrogen and releases it slowly so the next crop won't get an oversupply.

Too much nitrogen in the soil can lead to excessive foliage growth in plants. This suppresses the plant's yields because more energy goes into leaves than fruit.

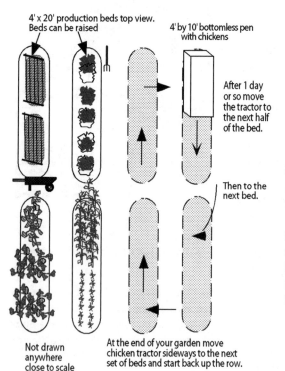

4' x 20' production beds top view.
Beds can be raised

4' by 10' bottomless pen
with chickens

After 1 day
or so move
the tractor to
the next half
of the bed.

Then to the
next bed.

Not drawn
anywhere
close to scale

At the end of your garden move
chicken tractor sideways to the next
set of beds and start back up the row.

Rotating a Bottomless Pen along Garden Beds or Rows

Direct deposit fertilization by moving the bottomless pen to a new garden section every day the hen's manure is deposited where you want fertilizer.

Put down a thin layer of mulch and/or cover crop seed on the fertilized area immediately after moving the pen to secure the nitrogen.

Build Soil with a Rotating Chicken Tractor

Suppose you are a beginning gardener and have little or no topsoil, or you already have a garden and want to improve the soil. Within one season you can add significant fertility and organic matter to your soil. Here's how.

1. Put your chickens over your current (or soon-to-be garden bed). The chickens will be confined to that space either in a bottomless pen, fenced run, or they can be hand carried from the coop to work each day.

2. Each day or so, direct the chickens to a different spot. They will leave a coating of manure and organic matter behind them.

3. As soon as you move the chickens, broadcast cover-crop seed on the old site. Then use a garden hoe to loosen the soil – about an inch deep – to cover the seeds. Apply a light mulch layer of compost, shredded leaves, grass clippings, dry hay, or straw to protect the seeds from birds and create a growing environment for the cover crop.

4. When the cover crop is about 4" to 6", put your chickens back in that space to graze. Over the course of a growing season, each bed will have three visits from the poultry fertilizers.

5. After the first and second chicken visits, apply cover crops and light mulch. After the third and final visit–or just before the growing season is over–cover the site with mulch to protect the soil over the winter. Next spring your garden bed will be easy to prepare.

By taking these simple steps you can use the chicken tractor system to mix the fresh chicken manure into the soil surface. This preserves valuable nutrients and covers new seeds to ensure good germination by keeping moisture in the soil and protecting the seeds from hungry wild birds.

The cover crop holds moisture in the soil and helps provide the seed-to-soil contact critically important for good seed generation. The root mass leftover from previous cover crops stays in the soil and decays, and as the roots decompose, they leave channels that feed and house soil life. The roots also enhance draining and moisture capillary action.

There are some disadvantages to planting a cover crop after each pen move. It takes time and seed. In most cases, the ground under the chicken tractor is moist enough for seed germination, but you will have to irrigate during dry seasons until the seed germinates and can send roots below the surface for water.

Sheet Mulch Instead of Cover Cropping

An alternative to cover cropping is to use sheet mulch. After moving the chickens, cover the ground with straw, hay, grass, or leaf clippings, and leave it there. Nutrients in the chicken manure will feed the decomposer bacteria, turn the layer of mulch into compost, and eventually produce humus (topsoil).

During decomposition, the top mulch yields other benefits. It protects the soil and soil dwellers from sunlight, wind, and, water erosion. This sponge effect results in richer soil with less work. The mulching system

Creating Raised Beds with a Stationary Chicken Tractor

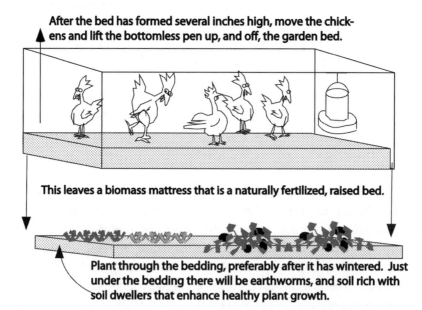

After the bed has formed several inches high, move the chickens and lift the bottomless pen up, and off, the garden bed.

This leaves a biomass mattress that is a naturally fertilized, raised bed.

Plant through the bedding, preferably after it has wintered. Just under the bedding there will be earthworms, and soil rich with soil dwellers that enhance healthy plant growth.

Put a bottomless pen over a garden bed and leave it for several weeks. Add about 1 to 2 inches of hay each day to provide fresh bedding. You can feed food residuals to the chickens, and they will mix it in with the bedding so bacteria can form humus even faster. Put the hens in the pen for day duty if they roost in a stationary coop. Make sure water and feed are available at all times.

saves money on fertilizer, tilling, and compost making. It works best with permanent beds and is ideal for the urban home gardener or small – scale market grower.

Chicken Enhanced Raised Beds

Raised beds yield more production in limited space. They are permanent, rectangular beds usually 3' to 4' wide – just wide enough to reach across to the middle. Raised beds hold soil above the ground, away from foot traffic that compacts soils.

The soil in a raised bed can be enriched and remain in the bed, where it is less susceptible to erosion. Because the soil remains in place, it can be built up for intensive gardening, which increases production. Going from rows to raised beds usually doubles the available growing area. 1' to 2' high raised beds have 12" to 20" of good soil and are home to hyper-productive gardens. Some raised beds are high enough that folks don't have to bend over. Some can be tended from a wheelchair.

To use the deep mulch method for raised beds, put a chicken tractor, or fence the area, where you want a garden bed. If there is any vegetation, mow it and leave the clippings as a source of nitrogen for future growth. Chickens love clippings and will pick through them busily, chowing down the green matter and absorbing beneficial chorophyll and trace minerals.

Every day add at least an inch of carbon material to the raised bed. Depending on the number of chickens and their age, you might need to add more. Be generous with the bedding and add enough to completely cover and mix in the manure.

Scatter some grains or feed on top and the chickens will immediately begin scratching to find the feed, mixing the bedding up in the process. After a few weeks, the once-fluffy bedding will be formed into a mattress, neatly pressed into shape and about 10" deep. Under the layers of the raised bed, earthworms, bacteria, and fungi are decomposing the mulch into compost. It will probably be warm from the fermenting action. By next growing season, the raised bed will be wonderfully fertile and ready to grow produce.

If the top layers have not decomposed into humus by spring, put about an inch of compost on top and plant your crops directly into the mattress. Salad crops will sprout within days and send their roots deep into the mulch layer, extracting nutrients and moisture.

Another way to build soil in raised garden beds is to form hen house boxes around the bed. When you clean the chicken's abode, pile bedding and droppings into the raised bed to begin composting. Add water until the mass is damp but not dripping; mulch needs moisture to start the composting process. This shallower pile will not get as hot as a full-size compost pile, but it will get warm enough to provide bacteria a comfortable environment for material breakdown. Add beneficial soil bacteria by shoveling some topsoil on top of the pile.

In a couple of weeks, the lower layers will form humus, and earthworms will migrate up the soil beneath the beds to expedite the process. They will consume the mulch and turn it into rich castings. If you are building a raised bed on hard-packed soil, difficult for earthworms to penetrate, consider importing some of the pink powerhouses. Go on an earthworm quest to gather some, then add to your future garden bed as soon as the lower levels have some humus and are an acceptable environment for the worms. Kids are especially good at this task and can collect earthworms from sidewalks after a rainy night.

Nitrogen buildup in this deep mulch method is not a problem. Some experienced growers think there will be more nitrogen than plants need, resulting in excess foliage; for example, tomatoes develop too many leaves and little fruit. We haven't experienced this problem. The high-carbon bedding acts as a buffer against nitrogen overload. As the material breaks down into humus, it traps the excess nitrogen and holds it for future crops. The plants, in their natural wisdom, extract the nutrients they need and leave the rest for future generations and other living beings.

Much of the nitrogen is used by microbes, which break down carbon material and release the nutrients for crops. Because the nitrogen is locked up with the carbon materials, there isn't nitrogen leaching into the ground water. This is a lesson in soil management. When organic matter in soil is at an adequate level of 5% or more, excess nitrogen won't leach into ground water unless it has been grossly over-applied.

These systems can be used in various combinations to build raised beds.

Chickens as Mulch Cleaners and Compost Spreaders

"Mulching" is to cover the soil surface with a thick layer of organic matter. Mulching is done for many reasons:

- Water absorption is improved under mulch, which reduces the plant's water needs.

• Weeds growing under mulch are killed due to lack of sunlight.

• Temperature ranges are buffered under mulch. The soil stays cooler on hot days and warmer during cold nights.

• Seed germination is enhanced.

• Nutrient leaching is diminished.

• Most importantly, soil dweller populations (such as earthworms) increase if they live under mulch, as it preserves soil structure and buffers soil pH levels.

One problem with organic mulches such as hay and grass clippings is that they come with weed seeds that can cause weeding nightmares for fastidious growers. The good news is that nothing excites a hen more than removing weed seeds from mulch. Give your hens access to mulch material and they will entertain themselves for hours – even days – by searching and eating seeds, chortling all the while. Chickens offer triplicate benefits when used with mulch:

• They eat weed seeds.

• They add manure and mix nitrogen with the carbon for better plant nutrition.

• They break the mulch down by scratching.

Hens Spreading Compost

Deep Mulched, Multiple Rotating Runs

Using the deep mulch system in run(s) adjoining a stationary hen house has many advantages. Adding a 6" to 12" layer of grass clippings, old hay, or leaves lets the nitrogen

Hens working on finished compost in a new terrace garden bed. Within a few days the compost will be scratched downhill, cleaned and nicely spread. More mulch will be added on top of this. Over time, a deep raised bed of rich garden soil forms. Toss scratch grains on top to encourage the hens to mix the organic matter even more.

from chicken manure mix with the carbon in non-toxic amounts. It also keeps the ground from becoming compacted. The perimeter boards, like raised beds, keep the mulch from being kicked out through the fence by scratching hens.

The chickens remove almost all the seeds, break up the pieces, and add their manure and feathers to enrich the mixture, while happily soaking up sunlight. If you have enough yard space, you can rotate runs. My hen house has three runs radiating out from the stationary coop. One of the runs is a small orchard, which the hens graze in year round.

Soon the weed-seed free, enhanced mulch can be raked up and used in garden beds. If the mulch has formed layers (which is likely in the winter) you can use a pitchfork to collect the mulch. These thick chunks make especially good mulch around fruit trees where you want thick covering so grass won't grow.

You can employ a few, or all, of the flock to range among the crops at

Fixed Coop with Rotating, Deep Mulch Runs & Garden Beds

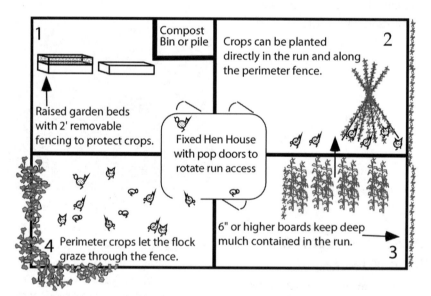

Integrated runs and garden beds can be used in rotation for different purposes. Open the pop hole , or poultry portal, to the run/garden where you want your chickens to graze, glean, clean, and fertilize. Assign only a few flock members to one spot if there is danger of too many chickens harming a particular crop.

appropriate times to work as pesticiders and herbiciders. They are efficient in taller crops like corn, asparagus, or trellised plants.

Use rotational grazing by sectioning off a run so that cover crops can grow. This can provide the hens fresh greens, especially in the winter.

Raised beds can be built inside the runs and surrounded with low fencing, like 2' chicken wire, to give you easy access to the beds and keep birds out. Put a compost bin in one of the runs for convenient composting. Let the hens help clean, mix, aerate, and spread the compost.

To mulch around trees, we put down hemp coffee bags that are free from local coffee roasters. Most of the roaster's hemp bags would be sent to the landfill if we didn't use them. The hemp bags are thick enough to block light, and they decay in a few years, adding their bulk to the humus. After pinning down the hemp bags, a thick layer of mulch is put on top.

Mulching materials can include any plant matter small enough to shovel or move with a pitchfork. Biomass from your yard can be made into mulch materials. These include grass clippings, leaves, garden residuals, and chipped tree limbs. Other mulch items can be hay, straw, wood shavings, non-shiny cardboard, newspaper, and brown paper bags.

One mulch material never to use with chickens are the cocoa-based products made from cacao bean shells.

Cocoa mulch and cacao bean by-products contain high levels of theobromine and caffeine. Both are highly toxic and even lethal to dogs, cats, and can't be good for chickens.

Chicken Condo Case Study: Joseph Martinez

Joseph Martinez is a remarkable renaissance man. He's a professor of Theater and Chair of the Washington and Lee University Theater Department. He has a lifelong interest in gardening and a large library of antique agriculture books. He has won many awards and honors. Professor Martinez and his family

Deep Mulch Run

Boards around a run, similar to a raised bed, keeps deep mulch in the run. The mulch mixes with and absorbs chicken manure. This keep the soil in the run from getting toxic from too much nitrogen. These tomatoes and marigolds are planted inside the run with a hog panel fence for protection. It also works as a tomato trellis.

were featured as a homestead family in the PBS video, "The Natural History of the Chicken".

I met Joseph when he wanted to help us process poultry at our free-range poultry farm. He wanted to experience the drama of poultry death and learn how to process his own homestead chickens and turkeys.

Joseph's chicken housing raises the bar on building and fertilizing raised beds. He calls his design "chicken condominiums" because his pens form a complex that the hens live in. Joseph designed his terraced garden beds and the pens at the same time, so the pens would fit exactly on top of the raised beds in a synergistic system. All components of the raised beds/chicken pens/poultry bridges are designed around a two foot grid. Because the measurements are standard and consistent, there is little waste of building materials and limited cutting for construction. He built 6 garden beds, 12 chicken pens, and 5 poultry foot bridges to connect the pens across human footpaths.

The 4' by 8' box pens are placed over 4' wide and 16' long garden beds. The pens are bottomless and have hinged lids that can be raised and propped up to tend and add bedding. Pop holes on the side and back of each pen connect the pens and bridges together. An external nest box is on the end of each row, so there are 3 nest boxes total.

Two of the pens have 2x4 wire fencing arched over the beds, giving the hens an outdoor play yard and access to direct sunlight.

The raised beds are constructed of railroad ties and river stones. The stones keep burrowing predators from getting into the pens. Each raised bed is 16' long, and 2 pens fit over each garden bed. There are 12 beds total in 3 rows of 4 beds each.

The pens and beds are connected using double-headed nails that are easy to remove. In the spring, when it's time to move the pens to the lawn, the nails come out and the pens are moved by hand to their new location.

To give his flock of 25 chickens access to all 12 pens, Joseph made 5 poultry foot bridges that crossed

Professor Joseph Martinez with his Chicken Condos

Chicken Condos Over Terraced Raised Beds

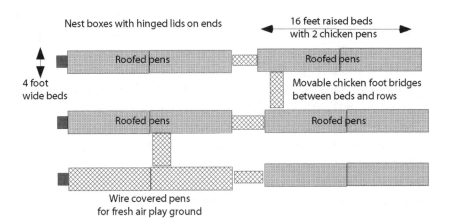

Bird's eye view of Joe Martinez's chicken condominiums. Pens were built on top of deep mulch raised beds and are connected with pop hole doors and poultry foot bridges. Joe can restrict access to certain beds that grow cover crops for rotational grazing. He scatters feed in a different pen each day so the chickens search for it in all of the raised garden beds.

This photo shows the sun room of the chicken condo and the poultry bridge that connects the pens. Notice that some pens are covered with plastic and others are open for ventilation and to allow the hens to self-regulate their body heat.

human pathways between garden beds. These bridges allow the chickens access between and across garden beds in a tunnel-like sequence. The beds lie in 3 rows along a terraced hillside, with an elevation change between them. Joseph connected the rows with poultry foot bridges, using 2 x 4 lumber, plywood, and 36" high 2" x 4" grid wire fencing. There is a 2 x 4 support beam under the plywood to keep the bridges from sagging. The longest bridge is 8' long, which is conveniently the length of a piece of pre-cut plywood and the distance between garden rows.

To secure the bridges to the beds, Joseph drilled a ½" connecting hole through the plywood bridge into the railroad tie and dropped a 4" bolt into the holes. The bolt made it easy to remove the bridges for mowing, directing the chickens to other parts of the gardens, or disassembling for storage.

All winter long, Joseph adds organic matter to the pens, including kitchen scraps, and gets old hay and leaves from neighbors. He adds about 10" of bedding to each pen, making it deep enough that hens hit their heads on the roof. It only takes a few days for the chickens to sort through the bedding and pack it down. Don't underestimate the packing power of poultry; left in the same area long enough, they can make a surface as hard as a paved road.

The beds become shallow compost piles. The soil dwellers make tunnels and air spaces so the bedding can breathe. This deep mulch system offers in-ground radiant floor heating to keep the chickens warm during the winter.

For winter housing, Joseph puts clear plastic film over the sides for wind protection. A 60 watt lamp in their roosting area provides all the heat they need in our Virginia (zone 6b) climate. If winters were severe, hay bales could be stacked around the pens (instead of plastic) with a 4" x 8' Styrofoam board on top for insulation. This system would tuck the hens in nicely while still providing plenty of ventilation and fresh air.

Joseph doesn't use feeders for his chickens. Instead, he broadcasts layer pellets on top of the bedding, forcing the chickens to scratch and dig for their dinners. Every day he puts the feed in a different pen, so the chickens won't spend too much time tilling and gleaning weed seeds from one bed. The chickens never know which pen Joseph will scatter feed in, as he chooses a different one each day. As soon as the feed is down, the chickens race to that pen, navigating the maze of tunnels and bridges at warp speed. Like NASCAR drivers, they skid around corners and careen down icy bridges to arrive first at the feeding site finish line. Rotating feeding between pens encourages the chickens to search every raised bed for leftovers. They get exercise as they explore the pens, and their wandering mixes their manure with the bedding to make garden soil.

In the fall and spring, Joseph sections off any of the pens to use as a cold frame for an extended growing season. Before planting, he takes the soil pH and adjusts it with lime or ashes from his wood stove. Joseph never adds lime to a pen if his chickens have access to it, because he's concerned the caustic powder would get in their eyes and lungs, causing irritation.

He plants directly in the chicken-fertilized mulched layers and never disturbs soil structure by tilling the beds. Because of increasing grain and feed prices, he plants cover crops in some beds and lets the chickens access a few of them at a time (rotational grazing) so they can eat the fresh greens. To herd his hens, all he has to do is open – or close – the poultry portals and/or bridge access.

In summary, Joseph's integrated poultry and garden system is a fine example of permaculture stacking. Stacking means that every element in a system provides multiple functions. Instead of simply having chickens for eggs and meat (one function), Joseph uses chickens for creating garden beds, fertilization of soils, and weed seed control in mulch. He uses a natural chicken characteristic – scratching – to enhance bedding and turn it into compost. The warmth from the fermenting bedding keeps the chickens warm in the winter. The same bedding turns into humus for food production in the summer.

His pens also serve multiple purposes. In the winter, pens stay put over the raised beds. In the summer, he uses the same pens (now portable) on grass as a rotational grazing system. The same pens that house the chickens also direct them to fertilize the yard or enhance garden soil. Sometimes he lets them free-range as pesticiders and herbiciders.

Another advantage of Joseph's system is that multiple handling of manure or compost materials is unnecessary. There isn't a toxic buildup of nitrogen, because Joseph keeps his chickens moving on grass or running around the stationary pens over the garden beds. Connecting multiple pens over raised beds for chicken housing, soil building, and food production is an elegant, simple, multipurpose system.

Garden Chicken Fencing

Fencing is the steering wheel of your chicken tractor. Fencing directs the hens to where you want them (or don't want them) in your garden system. You can get very creative with directing your hen helpers. Chickens are relatively easy to fence in and smart enough to learn a routine, such as running to the area where you've thrown food. With a micro-flock, you can carry your hens to where you want them to work that day.

There are four basic methods of fencing that can be mixed and matched to adapt to your garden design.

1. *Chickens Free-ranging.* Free-ranging for urban flocks means a flock can wander at will within the confines of a perimeter fence, but not be in a pen. Free-ranging can take place in a yard or among plants and garden beds. This works during certain plant growth phases, depending on the kinds of plants and crop characteristics. Chickens can forage and have shade, yet not bother the higher-hanging fruit. This system works with climbing vegetables, like pole beans or tomatoes, and with crops like asparagus (after harvest season) and knee-high corn.

2. *Fencing Plants in and Chickens Out.* Pens or fences can also keep chickens out of beds and away from individual plants. This keeps the chickens from harming the plants by scratching (which hurts the roots) or eating the leaves or fruit while they patrol for bugs and slugs. Comfrey cages are a good example of fencing plants in where chickens can eat some leaves but not enough to harm the plant.

Garden Chicken Fencing Systems

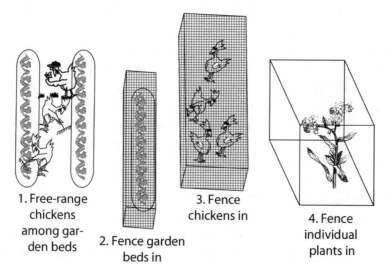

1. Free-range chickens among garden beds

2. Fence garden beds in

3. Fence chickens in

4. Fence individual plants in

3. Fencing Chickens in Pens Over Garden Beds. This is useful for seed gleaning, bug and grub eating, soil fertilizing, and tilling/mixing in biomass with topsoil.

4. Fencing Individual Plants. It can be more practical to protect a single plant from chickens pecking and scratching the roots.

Fencing Types You Can Use with Chickens

Sometimes it's best to restrict hens to a plot of ground, garden bed, or hen run. It doesn't take much of a fence to control chickens. They don't put pressure on fencing by leaning, rubbing, or charging like horses, cows, or goats.

If any flock members start flying over fences, clip the flight feathers on one wing so they can't get much lift for takeoff. Clipping wing feathers is described in the Hen Health chapter.

Here are some fencing guidelines:

• Keep perching places on fences and posts to a minimum. Your birds will look up and decide if they can snag a place to sit while they consider their next move. That's why church steeples are so steep and fence posts are sometimes so pointed. If there's no place for their little chicken feet to grip, they can't perch.

• Avoid building a wooden structure fence because of the ample perching opportunities. Chickens will view this as a landing area and observation deck.

• The fence should have bare wire (too small and wobbly for a bird to land) on its top.

• The chicken fence doesn't have to be tall. You can use a 2' or 4' high wire, depending on the size of the area to protect. The advantage of a 2' fence is you can step over it to tend the garden. You can bend over a 3' fence, and a 4' fence keeps almost everyone on their respective sides. If you have deer, fox, or dog problem, then a taller perimeter fence might be necessary.

• Welded wire fence with 2" x 4" rectangles works well. The 2"x4" is large enough for your hand to fit through, and small enough that a mature chicken can't reach more than about 7" through it.

The fence should be stiff enough to stand with metal posts or wooden stakes. For more secure fences, use 5' to 6' metal fence posts. These can be hand driven with a post pounder and easily taken up with a lever.

Temporary fencing is easy to put up and take down for rotation. Lengths of 50' are light enough that a single person can carry them. The fence can roll up so it is easy to store or move. The combination of metal fence posts and wire fence tops won't give birds a comfortable place to perch – at least not for long.

• *Chicken wire* can fence in small areas. A 2' or 3' fence of chicken wire deters feathered intruders. Give the chicken wire support and visibility with the fiberglass driveway markers. Chicken wire is flimsier than the 2 x 4 wire, but it's easier to put up and take down.

• *Snow fencing* can be used as temporary fencing to keep birds in an area. It is high enough to keep foxes from going over and sturdy enough to support itself without stakes. Wood snow fencing is heavy and hard to keep looking even.

• *Hog panels* are useful in gardens. They are 3' high and 16' long. They

Graze Reach Through a Fence

How far can a chicken reach through a fence? To find out, scatter some feed on the other side and measure. This will give you the browse line and how to space perimeter plants.

can be cut with a bolt cutter and are sturdy enough to stand without support. Wider-spaced wires make it easy to tend plants. The higher, larger 6"x 6" openings don't attract hens to fly through, although one of my hens became annoyingly adept at this. Amazingly, chickens can perch on a hog panel, but usually not for long, as the tiny grip is uncomfortable.

Other Ways to Protect Plants from Hens

The goal with young seedlings and tender plants is to create a barrier

Chicken Browse Line

Knowing the browse line is helpful when placing and spacing plants. This photo shows the summer squash browse line of chickens with 2" x 4" wire fence. The fence is inside a run, and the squash is planted directly in the fertile run soil. Some crops, like comfrey, can be positioned so that hens get part of their feed from the growing plants, but not enough to eat all the leaves and kill the plant.

that prevents the hens from pecking the leaves and scratching the roots. A plastic jug or milk carton works (use one for each plant) until the plants are about 12" high.

Stones or small logs can barricade plants so the hens can't get close enough to scratch and uproot them. Chickens don't like the taste of some plants, like tomatoes and marigolds, but these plants still aren't safe from chickens. They will want to dig in the fresh soil around the new transplant, but can be deterred by obstacles such as small rocks or wood limbs. This

no-scratch zone won't work for lettuces, kale, beetroot, or smaller plants. Hens view these as salad and will leave little behind. They will leave cabbage, cauliflower, and bok choy alone when the plants are bigger and if there is something tastier close by. However, mob grazing with a lot of hens in a small area will wipe out almost any growth.

Garden cones are one of my favorite plant protectors and double as mini-cold-frames. These cones fit over the plant and offer great protection from chickens. Adventuresome birds will hop on top of a cone but don't stay long.

Stretching 2" x 4" wire fence horizontally over raised beds protects the plants and keeps chickens off the beds. When plants are larger and established, remove the wire and let the hens roam the beds as bug patrollers or harvest helpers.

Perimeter Fencing

Perimeter fencing is usually taller and encloses the entire yard or garden area to keep deer out.

To discourage predators from digging under a permanent fence, you can lay chicken wire around the perimeter of the run. Bend the wire 90 degrees into

Fence Perching

A hog panel makes convenient garden fencing, as it can be cut to various lengths and is sturdy enough to bend and stand on its own without a post. This hen would perch on top, but not for long, as narrow wire doesn't offer a comfortable grip.

an "L" shape so about half of it lies flat on the ground. Secure with ground staples to hold it in place until grass grows over the wire. The chicken wire should reach up the fence about six inches so that it covers the lower portion. Fasten the chicken wire to the fence with staples or clips.

If birds of prey are a problem in your area, install a screen or wire roof over the run. My experience with hawks is that they will kill a pullet or bantam-size bird, but leave adult birds alone. I have rarely heard of problematic flying predators in urban areas, with the exception of owls at night. Owls generally hunt after dark when your birds will be safely tucked into their coop. However, if your birds roost in a tree instead of returning to the coop, they are likely to be eaten as an after-hours snack.

Sticks and Stones as Plant Protectors

These hens are free-ranging in a garden bed where plants are protected by small logs and stones. If the hens can't get their feet close enough to scratch up a plant's roots, they will not harm most larger plants. This bed has rhubarb, thyme, marigolds, and tiger lilies. Notice that the hens are much more interested in the mulch than in eating the plants.

Poultry Portals

Fencing isn't complete without a way to get in and out of the enclosed area. People use gates for this purpose and chickens use poultry portals, which are chicken-size passages through a fence. They are built about a foot off the ground and can be made with five-gallon plastic pots leftover from previous plant purchases. Cut the bottom off one so that 4" remains at the top. Put this through a 4" hole in the fence and use wire or string to secure it. Use the other pot as a plug to open or close the chicken's access.

Vegetables Above; Chickens Below: Trellises

Growing crops on trellises and tipi frames optimizes garden bed space. Many potential diseases can be prevented by spacing trellises far apart, so that sunlight can get in and dry damp leaves quickly (most diseases need moisture to grow). By raising the vegetables, more leaves are exposed to sunlight, resulting in better growth. Additionally, sunlight kills many problematic airborne microorganisms.

When the root base of crops is established, let chickens go under the trellis. They will use this understory for shade, dust baths, and protection. They will keep the it clear of bugs, and add their manure to the soil and mulch. Mulching under the trellis keeps moisture in the ground and helps limit the splashing of soil microbes and dirt onto the plant. You will lose the low-hanging fruit to the chickens, but the higher harvest will be safe from their pecking.

Poultry Portal

Poultry portals are chicken size passages through fences and a way to direct birds to particular area of a yard or garden. This portal is made from two plastic pots. One pot has the bottom cut off to be a frame in the fence. The other pot is a plug to seal off the passageway. A rope keeps the plug pot close; just pull the rope and the pot will snug into the frame to close the portal..

Fruit and Nut Trees

Chickens act as fruit tree insecticiders, starting in the spring with web worms. You have to break open the web for the hens to eat the small caterpillars, as they don't like the taste of the web. It helps if they are hungry when you offer them web worms.

If you let your hens free-range around trees, spread a little scratch so they will start digging where larvae emerge from the soil. The hens also offer protection from wasps and beetles.

The hens love to glean fallen fruit and will do a fine job of cleaning up under the trees. If you only have a few trees and can't free-range your hens, put chicken netting or a temporary fence around the base of a tree.

Mushrooms and Chickens

Chickens like mushrooms and readily eat the fungi growing in compost or mulch. I've yet to hear about a chicken dying from eating a poisonous mushroom; they probably instinctively solve the omnivore's dilemma of whether or not something is safe to eat. Mushrooms are high in protein – up to 30%. They are also high in sulfur, which is necessary for many metabolic reactions. Fungi can have anti-bacterial properties and immune-enhancing enzymes. For example, the Kombucha

Tipi Trellis

Chickens love trellises, especially the tipi type. The understory provides shade, privacy, protection, and the perfect place for dust baths.

Hens Gleaning Fallen Fruit

Hens cleaning up fallen fruit under an apricot tree. Gleaning the fruit as it falls keeps the wasps away. Always have fresh water available.

mushroom has health benefits revered in ancient civilizations. If you have kept a Kombucha culture you know that it multiplies and, after friends won't take anymore, you have to do something with it. Feed it to your hens either diced fresh or dried.

Mulches and Chickens

Most permaculture gardens use heavy mulch as weed control in beds around perennials, vegetables, herbs, and trees. Plants grow through the mulch. Chickens love mulch and spread it around quickly. Sometimes it's beneficial to let the chickens mess up the mulch while they feast on its abundant insect and worm populations; just rake the mulch back into place when the chickens are finished with it. Border beds with side boards a few inches higher than the mulch will keep most of the chicken-flung material within the bed. Without anything to restrain it, the chickens will spread mulch far and wide. Either way, you will have to re-mulch and tidy up the area when chickens lose interest.

Remember not to let chickens scratch through cocoa mulch made from cacao beans because of its potentially toxic quantities of theobromine, a xanthine compound similar to caffeine and theophylline.

Chickens in Woods & Shrubs

Chickens love to explore forest floors, whether the "forest" is a cluster of a few thick trees, or the mature shrubs around your home or garden. They think it is a protective overstory, covering an understory floor rife with tasty snacks. In urban forests, fencing might be necessary to keep out predators such as free-ranging, chicken-killing dogs. Don't let chickens roost in the woods at night. Owls or raccoons will eventually find and eat them.

Deep Litter in Outdoor Runs

When hens live in stationary runs, they quickly eat anything growing and the soil becomes packed, caked with manure, and toxic. The antidote to this is to treat the runs as deep litter beds. Surround each run with railroad ties or 1"x6" boards to form a deep litter bed and add 12" of biomass. The deep litter in the run will become increasingly bioactive as hens work the mulch and add their droppings. You might need to use a pitchfork to loosen the mulch. Turn the soil over with a shovel if it becomes too compacted.

Broadcast part of the chicken's feed (or scratch) on top of the deep mulch to encourage hens to dig deeper for goodies. Because an outdoor run is subject to rain and snow, this method probably won't provide enough food for your flock. Keep food available and continue to add kitchen waste or other organic material. The chickens will mix it in with lower layers.

At some point, you can harvest the deep litter—now compost—as a soil amendment to use elsewhere. Then replace it and begin the process again.

I use the deep mulch run system primarily in the winter, and plant vegetables or cover crops after rotating the chickens out. About every second rotation, I add height to the perimeter boundary of the raised run so chickens don't kick the materials out. The alternative would be to harvest the compost/topsoil and use it in the garden.

You Want Earthworms – Lots of Them

Earthworms play a major role in the decomposition of mulch and the formation of earth structure and tilth. They form channels for water to flow up and down. Their digging helps roots penetrate to deeper levels.

Earthworm castings are 5 to 11 times richer for plant nutrition than the raw materials the earthworms eat. You might think of earthworms as pre-digesters for plant nutrient absorption. The pre-digestion has to do with secretions in the worm's digestive tract and calciferous glands.

Earthworms – and most bacteria – prefer a relatively neutral soil pH, between 6.3 and 7. They also prefer a soil that has plentiful calcium. You can add calcium to your soil by adding lime. Soils that are too acidic interfere with the earthworm's ability to get the calcium.

Another benefit to having earthworms is that hens like to eat them. They provide high-quality protein and lipids.

Soil Acidity or Alkalinity (pH)

Whether your soil is acidic or alkaline is important for gardeners to know. A few reasons why are listed below:

• Many plants and soil dwellers specifically prefer either alkaline or acidic conditions.

• Some diseases tend to exist when the soil is too alkaline or acidic. For example, high alkaline conditions can cause leaves to be yellow or discolored. This happens in green leafy vegetables and is called chlorosis. The same is true for potato scab. Highly acidic soils can cause club root in brassicas.

• The pH affects the availability of nutrients in the soil.

• Soil pH is affected by environmental factors, including the type of rocks that formed the soil, rainfall, pollution and vehicle traffic, fertilizers, and the decomposition of organic matter.

Soil pH is easy to measure. There are kits available that instruct you to take a small soil sample, mix it with distilled water, and put a dab of the resulting mud on litmus paper, which is easy to find and inexpensive. If the soil is acidic, the paper turns red; if alkaline, it turns blue.

The Problems with Rotary Tilling

As a gardener, it's important to understand the physical aspects of soil management. When spring comes, most folks begin their soil preparation by rotary tilling the gardens to get ready for planting. They mistakenly think that finer soil is better and run the rotary tines over garden beds until the soil is fluffy. They must want to lay down in the dirt for a nap after all that hard work. Rotary tilling is hard work for people and even harder on the soil structure. So hard, in fact, that it carries long-term consequences that affect your plant production.

Tilling, done too deeply, mixes good soil (the surface loam) with the poor soil of subsurface sand, gravel, and clay. A mixture of fair to poor soil results, or even upside-down soil. Over-tilling puts too much oxygen in the soil, causing oxidation and over-stimulation of soil bacteria so that nutrients are lost. It also disrupts the correct seed-to-soil contact necessary for optimal germination.

Worse consequence of tilling is that hardpan forms. Hardpan is a dense layer of soil that usually forms just under the uppermost layer of topsoil. Repeated tilling, heavy traffic, compaction, and pollution can cause hardpan. Hardpan is easy to create, hard to treat, and totally disabling to garden crops. The problems hardpan cause include:

1. It prevents drainage and is impervious to water. During heavy rainfall, it holds water near the surface, causing runoff and valuable water loss. In dry times, hardpan blocks the capillary action that allows water to rise from the subsurface and water table to the surface, where plants need it.

2. It restricts roots and prevents penetration necessary for plants to access minerals in the subsoil.

3. It interferes with the migration of earthworms and other soil residents from the surface to subsoil. Hardpan can stop earthworm movement completely.

Acidic and clay soils are especially vulnerable to hardpan formation. One way to break up the barrier is with mechanical methods, such as the broad fork, which is specifically designed for hardpan. But digging is a lot of work and not the best long-term solution. Who wants to dig year after year just because the soil was not managed properly the year before? An easier way to solve the solution of hardpan is to increase the amount of organic matter (humus) in the soil. Sound familiar?

By now you understand how to employ your chickens as tillers. You can use these two-legged, feathered rotary tines instead of the gas-snarfing, fume-emitting, ear-muffling mechanical steel tillers. With chicken tractor rotational grazing and raised bed systems, the roots and top growth of the cover crops follow the chickens and act as nutrient traps. They hold the nitrogen and other nutrients in roots and leaves. The nutrients return to the chickens in their next grazing rotation and stay in the roots for soil dwellers to enjoy. If you till under the cover crop, nutrients return to the soil. If you mow the top growth, you can collect the refuse and use it in compost.

If you use these systems, you won't have to buy fertilizer for your garden crops. Fertilizer will be the natural result of residual poultry manure, green matter, and compost applications. This is a natural, elegant, and simple soil creation and enhancement system.

State universities have done a lot of research on the advantages of mowing instead of tilling or plowing under cover crops. Mowing the top growth and leaving it on the surface as a mulch replicates nature. Earthworms serve as tillers by tunneling to the surface to feed and carrying nutrients back down to their burrows. They leave channels for water and root penetration that mechanical tilling destroys.

Chickens can be a motivator to get you into the garden. They are like having a dog you have to walk. Although you may not always feel like doing it, you feel better after the experience. Once you get momentum, you'll find that gardening takes on new dimensions and meanings with chickens.

Coupled with nature and God, you and your chickens become co-creators of your food supply.

Poultry as Pesticiders

Chickens are omnivores and will eat plants and animals, but they prefer being carnivores. They are grasshopper-eating machines. They stalk grasshoppers. One hen will nail it and others will jump in, trying to snatch the hopper from the beak of the stealthy hen. The competition ends when one hen gets hold of the bug and runs away to find a spot in which she can dine in peace. Slow-moving ticks and hopping fleas don't have a chance against the dead-on aim of a hunting hen. Chickens love to search and destroy (eat) ticks and fleas. They are so efficient that my many pets are practically flea and tick free. I only find ticks on my dogs after a walk in a chicken-free forest.

The brave, feathered pest controllers patrol under trees and bushes, seeking anything bite-size that moves. They boldly go places where chemical sprays can't reach. A garden chick's "seek and eat" life goals are reliable, consistent, and effective.

Japanese Beetle Conquest

Japanese beetles invade by the millions. Hardly anything will eat them, so they have few natural predators. They don't bite or sting, so gardeners pay kids to collect their metallic green bodies from the garden, because they cause severe damage to plants. When the adults emerge from underground in the spring, they will gather in clusters and chomp on the leaves of plants, leaving only skeletal remains.

The Japanese beetle grub is a delicacy for gourmet hens, as Marlin Burkholder knows well. Marlin owns Glen Eco Farm and is a vendor at the farmers' market in Harrisonburg, Virginia. He sells organic berries – lots of them. He can make $800 in a single day with his beautiful, picture perfect berries. How does he keep the Japanese beetles off his berries and remain organic? "Easy", he says, "I put an electric poultry fence around the berry plants when the beetle grubs are just hatching, the hens patrol the berry patch and gobble the grubs down. I don't have to use pesticides."

Chickens think that Japanese beetles, in their white grub immature state, are as delectable as sushi. The white grubs are from .5" to 1.5" long, depending on the stage of development. They form a "C" shape when resting in the soil. June beetles, May beetles, and chafers are also in the white grub class and are a delicacy for chickens.

Mature Japanese beetles that invade your grape vines, rose bushes, or potato patch are high quality protein for hens. Shake the beetles off the plants or get them into a bucket with water in the bottom. Pour the bugs out to the flock and they will gobble them down like M&Ms. You can also leave the bugs in a field waterer and the hens will pick them out.

Be prepared to have back-up pest control in case the hens aren't interested in a particular pest or have more tempting dining options available. I've never seen a hen eat a ladybug, corn ear worm, or wasp; maybe they taste bitter. Regardless, you can't rely on hens to be the only answer to your bug challenges. There are some pests hens won't eat, but hens eat most of the problem insects and can definitely be part of the pest solution. This is an area ripe for research.

Tomatoes Love Chickens

Hens can be fickle in their tastes. One year my hens fought over the tomato horn worms I pulled off the tomato plants. The next year they ignored the fierce-looking creatures. I've never seen a hen find and peck a tomato horn worm off a tomato plant, but this year I planted caged tomatoes in the main run and there were wasn't any insect damage to the fruit. The chickens would eat any low-hanging, close-to-the-fence tomatoes they could reach, but a small loss of the tomato crop was well worth their organic insecticide service.

Bugs that show up usually when there is a crop to eat, like eggplant fleas, potato bugs, and squash bug cucumber beetles. One pest may or may not appeal to your hens. You'll need to cage or trellis the plants to protect the fruits until harvest.

Asparagus

Asparagus is one of my favorite perennial crops. We planted four different varieties on about a 1/3 acre patch. Like fruit trees, if asparagus beds are taken care of they can be productive for decades.

One way to take care of asparagus beds is to let poultry graze the patch. We would let the turkeys in the asparagus beds in the early fall to clean the beds and fertilize the soil. Then we did nothing to maintain the beds. Four years after we retired from commercial farming, the grasses were crowding out the asparagus so badly we would have to mow the patch. The crop yield was so low we thought the asparagus were beyond saving; it was too much work to get the grasses out and I was considering abandoning the asparagus and regularly mowing the patch.

A local farmer asked me if I wanted some rotten round hay bales, the kind that fall apart when you try to lift them. Of course I said yes; they are wonderful mulch and tons more than I needed for my small kitchen garden. Early in December (before we had heavy snow or ice) I maneuvered my garden tractor to place the bales in the asparagus patch and unrolled them, stretching the hay to completely cover the beds.

We fenced the patch and let the hens in to graze. They were in heaven.

Tomatoes Love Chickens

These tomatoes are planted inside a chicken run and protected by a section of bent hog fence. Notice how the hens have not eaten the tomato leaves or vines. Also notice how healthy the plants are because of the rich soil in the run. The hens have scratched around the edges of the plant cage, almost making a trench.

Chickens keep the slug and bug populations down, and will eat any white flies they can reach. They love tomatoes and eat as much of the low hanging fruit as they can. The run fence serves double duty; it acts as a trellis and keeps chickens away from the tomatoes. The run's deep mulch is contained by stacked 4x4 railroad ties and rebar.

Within a few weeks they had spread the old hay over the patch, leaving a heavy mulch of 4 to 6 inches. The mulch had almost totally covered the ground and killed most of the invading grasses that the hens hadn't eaten. In the spring, it was amazing to watch the asparagus penetrate the heavy, dense mulch. This spring we had the best ever crop of asparagus. My hero hens deserve all the credit; the asparagus wouldn't have made it if the chickens hadn't slowly and systematically spread heavy mulch and eaten the invading grasses, Japanese and asparagus beetle grubs.

Chickens and Honey Bees

Using the permaculture approach of stacking, hens and honey bees work together. The 1945 edition of *ABC and XYZ of Bee Culture* states that: "Bees do not bother the chickens and the work of keeping bees and raising poultry co-ordinate".

Honey bees that live close to your garden will pollinate your crops and increase production exponentially. Believe it or not, bees are also chicken food. There are tens of thousands of dead or dying bees around any hive. These bees have old, torn wings or have been thrown out of the hive; worker bees will toss dead or undesirable bees over the edge of the hive like sailors burying the dead at sea. These dead bees amount to pounds of high-protein chicken feed.

Honey bees are drawn to field waterers as a source of fresh water. Many of them will drown unless a small board (serving as a raft) is in the water for wet bees to climb onto. Chickens won't pick bees from the waterer, but they will eat any bees dumped on the ground.

Initially, chickens are instinctively cautious about honey bees and will avoid them. Apparently honey bees are an acquired poultry taste. I've watched a helpless honey bee crawling on the ground right in front of a hen, but she ignored it and pursued another similar-sized bug.

Chickens and Honey Bees

Honey bees enhance garden production through pollination. Bees can be a supplementary protein source for hens, but you don't want hens to eat healthy worker bees, as this hen is doing. Raising the hives about 18", or adding a short fence will solve this problem.

At first I thought bees were like lady bugs – not poultry fare. I assumed the bees wouldn't hurt the chickens and would let the hens free-range around the hives. Later in the season, I noticed a few of the chickens were lurking around the bee hive entrances. As it turns out, some of the hens had acquired a taste for bees and were snatching them from the landing strip at the front of the hive. These weren't the old, useless bees, but were healthy workers returning to the hive with pollen — the ones you don't want hens to eat.

The sentinel bees rarely attacked the chickens, they must have gotten used to having the birds around. If a chicken did get stung, it would shake its heads and move on. I assume the birds acquired an immunity to the stings, or perhaps the older hens were self-administering apitherapy for their arthritis and stiff joints. Eventually the guard bees totally ignored the hens, who continued to stroll through the bee yard and snack on worker bees.

The sentinel bees must have decided that the chickens were not threatening their precious honey, as the chickens never raided the hive. I think the sentinel bees got accustomed to chickens and viewed them as part of the landscape.

You want to protect the healthy honey bees from being eaten by your chickens. There are two easy ways to do this:

1. Put up 3' chicken wire with stakes to support it and fence the birds out of the hives. 3' is easy to step over to tend the hives, yet high enough that the hens won't hop over.

2. Raise the hive entrance 18 inches or more, so the hen can't easily reach the worker bees landing and taking off from the colony.

Here's a tip if you keep bee hives in a high traffic area. Position the hive so that the entry way is about 5 to 10 feet away from a tall structure. The structure could be a wall, fence, or even the side of a building. Preferably, it will be about five to six feet tall. Do not place the hive entryway up against the barrier.

Having a tall structure in front of the hive alters the bees' flyway such that they have to fly high before leaving your yard. The height is above people's heads, and few folks will ever notice bees flying overhead.

Birds and bees can be synergistic in your garden. The bees pollinate crops for higher food production and the chickens give fertilizer and garden help. Higher food production on small patches of fertile ground is what local food production is all about. Garden production can provide food for humans and chickens. Around bee hives, the dead or dying bees can be high-protein chicken food.

In summary, the only information in the literature or on the Internet are testimonials from gardeners about their experiences with gardening chickens. There is a glaring gap in the knowledge and empirical data about garden chicks' usefulness and the many roles small flocks could play in local food production.

The Gardener's Supply Company and several educational institutions are setting up to do chicken research projects in their test gardens. The focus of the research is to quantify and qualify how the chickens can be used as organic pesticiders, herbiciders, insecticiders, and fertilizers.

We believe there is tremendous, untapped, year-round potential to integrate chickens with crop production. Chickens deserve to be the mascots of local food production.

Techniques and systems will be developed to enhance the effective and humane use of hen helpers. Stay tuned. Good Earth Publications will make research results available, on the Internet, and through updated editions and new titles.

Just as milk does not "come from" cartons,

chicken feed does not "come from" bags.

— CITY CHICKS

3 Growing Food for Chickens

Feed prices are increasing at alarming rates. Recently, for the first time in my life, I've heard of grains that "won't be available until next season." Worldwide grain reserves are low. Some reasons for increasing feed prices are:

- Rising fuel costs make transportation and production more expensive.

- Farmland is being used to grow corn for ethanol instead of feed.

- Fewer acres of farmland are available due to urban sprawl, which grows the ultimate crop: houses.

- Some harmful commercial agricultural practices have depleted soil fertility.

- The human population is growing exponentially.

- Shifting weather patterns and global warming make consistent crop growth difficult.

Most folks think of "feed" as a product that comes in 50 pound bags and they drive to the feed store and pick some up when they run out. But in the old days, before cars or feed stores existed, gardeners routinely planted crops for their hens, like corn or wheat. All older agriculture books (written before 1950) discuss growing poultry feed and feeding meat scraps and dairy products to poultry as a protein sources. Urban dwellers couldn't grow grain crops to feed their chickens, but they didn't need to, because their gardens produced plenty of other plants that chickens eat. Meanwhile, their hens free ranged in large areas – not always with perimeter fencing.

Given a chance, your hens will show you that they will eat a wide variety of locally available foodstuffs. Many plants growing in your garden can be dried and stored as chicken feed, including beans, comfrey, sunflower seeds, cabbage, grass clippings, and all root crops such as potatoes, sweet potatoes, acorn squash, etc. Plants like these supplement a hen's commercial feed diet, which keeps feed costs down.

The concept of "homegrown feed" has not been widely discussed or written about for 60 years. Around 1950, after World War II, the movement away from family farms and toward modern, big-scale commercial agriculture began. Farmers traded in their horses for tractors (purchased with no money down), and animals moved out of pastures and into confinement in factory farms. Land-grant colleges were funded to teach modern industrial agriculture to the next generation of working-class farmers. Roads were built across the U.S. and mass transportation became available. Homegrown feeds slipped out of fashion as store-bought feed became more convenient, cheaper, and, (supposedly) higher quality than homegrown feed.

Well, by golly, homegrown and locally produced feed is back in fashion. Planting and growing chicken feed isn't hard or expensive. By planting edible landscapes, recycling kitchen and garden waste, and growing crops for your flock, it is possible to dramatically reduce your commercial feed bill. In some cases, feed purchases can be completely eliminated. The Vermont Compost Company has a chicken-feeding program that costs nothing – zero dollars – spent for commercial feed.

Regardless of the size of your yard, you can grow something to feed yourself and your hens. Even patio gardening can provide a remarkable amount of food. Chickens will eat almost anything humans eat, and will greedily gulp some food we would consider inedible. Some of the plants grown in your garden can be dried and stored for poultry feed, including beans, cabbage, grass clippings, root crops (potatoes and sweet potatoes), winter and acorn squash, aramanth, artichokes, comfrey, and even bamboo.

General Observations on What Chickens Eat and Do in a Garden

As the micro-flock movement grows, small production gardens will flourish. Using hens as integrated partners in gardening will inspire clever systems that grow food with, and for, chickens. New chicken gardening ideas will emerge and evolve. In the spirit of local, organic food self-sufficiency, hens will boldly go where no chicken has gone before. Their mission: to help gardeners explore unique ways to contribute to wholesome food production. They will be support staff to the executive officers of backyard production gardens.

Good Earth Publications has agreed to maintain the website: www.ChickensInTheCity.com to make available the tips, tricks, and traps of chicken gardening as they are discovered and invented by gardeners around the world. Please send your stories to info@GoodEarthPublications.com.

Meanwhile, here's a little of what we've learned through experience, books, and Internet discussions.

What Attracts Chickens to Forage?

In their search for daily snacks, chickens use their senses to determine what to eat. Foraging is a skill with a learning curve, and older, experienced birds tutor the young ones in the subjects of how to forage and what to eat. Hens must be able to forage in order to feed themselves from the landscape. I've gotten birds that had only lived in cages until they joined my flock. They didn't have a clue about how to eat anything unless it was dumped in a feeder. My older birds were role models and showed them how to forage.

I've observed that a chicken's ability to forage might depend largely on its access to free range and the amount of forage available. Larger runs with ample wild feed offer the chickens opportunities to hone their foraging skills. This is contrary to some who think that foraging is a trait that can be bred into, or out of, a chicken. I think the ability to forage is innate in any healthy chicken, but it is a skill that must be developed. It's like people throwing balls: most of us have the ability to throw, but we can learn to do it better with practice and training.

I developed this observation because of my experience with the breeder flocks we owned provide hatching eggs for our broiler production. These genetically developed chickens had not free ranged before they came to our farm, as is the case with most breeder flocks. But they turned out to be wonderful foragers once they got the chance. The same was true for our commercial, broad-breasted turkey flocks. These birds are bred for weight gain and are regarded as dumber than rocks. They are unable to breed naturally and were thought to be unable to forage for themselves. When confined, turkeys grow so fast that their legs give out beneath them. We found that, with exercise and the opportunity to free range, they could sprint after grasshoppers and other speedy insects just fine. They also learned how to scratch and glean the ground. As harvest time approached, the turkeys waddled (instead of walking) more than the heritage breed birds. We had very few leg problems, probably because they were able to exercise daily and had ample access to sunshine.

Here's a brief summary of what chickens seem to be attracted to while foraging in gardens:

• Bright colors. Chickens seem to be drawn to the primary colors and especially to reds and yellows of tomatoes and peppers, blues of berries, and the bright greens of leaves, as on lettuces.

• Reflected light. Refracted light is eye-catching and some insects, like the Japanese beetles, have wing and body surfaces that reflect light. Chickens seek these out.

• Small things that move. Chickens have keen, close-up eyesight which allows them to snatch tiny bugs and crawly, jumpy beings in the dirt, including fleas and ticks.

• Taste. The Medical Research Council's chicken genome results for taste receptors turned up few results, suggesting chickens have a poor sense of taste. This makes sense, as chickens don't have teeth to break down food with the help of saliva. They grind food in their craws (gizzards).

• Smell. Medical Research Council scientists confirmed via genome analysis that the number of smell receptors in the chicken is relatively large. Chickens have an acute sense of smell. They are not olfactory super-sniffers like dogs or mice, but their sense of smell is probably about the same as humans.

• Chicken-size food bits. The food has to be small enough to wrap a beak around, or soft enough (like bread) that it breaks apart easily.

A chicken will ignore a whole potato or beet, but if it's softened by cooking, or in cut into small bits, the hens will gobble it down.

Using Fencing as Feeders

Combining different types of fencing with certain crops can supplement the chicken's feed with fresh, raw food. For instance, hard wire fencing around zucchini allows hens to peck at low-growing vegetables, while keeping the interior and higher zucchini out of beak's reach.

Comfrey is a perennial plant that grows every year. Planting comfrey cages inside a run allows the chickens to prune the leaves without harming the plant. Comfrey is a prolific grower and can be eaten fresh or dried for storage until winter.

The Value of Vertical: Chickens with Raised Beds and Trellis

Another way to maximize production in small areas is to use the air space above garden beds, on fences, arbors, and free-standing tipi-type

Zucchini Cage

These zucchini are growing in a greenhouse at the Vermont Compost Company. The greenhouse also serves as a brooder for raising chicks. The chicks ate the zucchini as it grew against the wire.

Comfrey Cage

When a row of comfrey is planted inside the run, chickens can nibble on it and prune its leaves without harming the plant.

frames. Growing crops vertically raises crops off the ground where they are safe from pecking beaks. Hens will still snatch the low-hanging fruit, but Nature is abundant and there will be enough for everyone.

Allowing chickens access to trellised crops has multiple benefits, including pest and weed control (especially if you mulch under the trellis) and direct soil fertilization. Chickens like being under trellised plants because of the shade and low-hanging fruit. They will use the understory for dust bathing and socializing. A hen can hop as high as 3', so keep prized produce (like tomatoes for the county fair competition) at least 3' off the ground to be safe from hen harvesting.

After your crop has been trellised and the roots deeply established, let your chickens access the understory beneath the plants. You might need to put some stones or other protection around the base of the plants to protect the roots from hen scratching.

Vegetables that grow well vertically include pole beans, cucumbers, lima beans, melons, peas, winter squash, zucchini, and vining (indeterminate) tomatoes, and raspberries. Be careful with lower bush crops like peppers, eggplants, or okra, which won't survive a hen onslaught unless the plants are caged.

Also remember that while bathing, hens fluff up dirt and dust that dirties the harvest. The only way to fix this is to fence chickens out until the crops are harvested.

Lawns as Mini-pastures

Lawns – especially big backyards, can be urban mini-pastures and benefit from enrichment by rotational grazing. Lawn can be a source of fresh (or dried) green fodder for your chickens. Gather grass clippings in your lawn mower bag. Hens love tender green roughage. The chlorophyll in the greens is healthy for hens and will make egg yolks yellower. You might notice a drop in your feed bill when you feed grass clippings and garden greens. Feather plucking and egg eating tend to diminish when chickens have enough green stuff to eat.

Grass clippings are also valuable as feed-back fertilizer. The clippings contain valuable minerals including nitrogen, phosphorus and potassium, the same NPK fertilizers are made of. When clippings are removed from the lawn, 20 to 25 percent more fertilizer will be necessary to maintain comparable color and quality.

One way to improve lawns for forage is to use a mulching mower.

Mulching lawn mowers are designed to cut, then recut the grass clippings many times to create fine pieces. The smaller grass clippings can then filter more easily and quickly into the lawn canopy and not remain long on the surface. The finer clippings can also be collected for feeding to the chickens.

Mowing height is the most important factor for mowing lawn. Yard grasses, like other green plants, must manufacture sugars through photosynthesis in the leaves if they are to develop collectively into a high quality lawn. Grasses mowed at low heights have limited leaf area to sustain photosynthesis rates necessary to maintain good plant vigor.

Most people don't realize that there is a direct relationship between the height of the grass and the depth and total mass of the root system.

A lawn that is mowed too short (less than 2 inches), will have a shallow root system with little total root mass. The effect of shallow, weak roots is the grass is less tolerant to stress of any kind. For example, when soil moisture becomes limited, as in droughts, the closely mowed lawns will be stressed more, browner, and the loss of plants will be higher resulting in bare areas.

In heavy rain fall, short grass and shallow roots have less ability to absorb water and thus, less ability to decrease flooding. The deeper and more massive the root mass, the more sponge-like effect a lawn has for absorbing and retaining moisture.

Higher mowing heights (2.5 to 3 inches or more) during the summer offers several advantages:

Lawn as Mini-pastures & Grass Height

If the chickens can't get to the lawn I'll take the lawn to the chickens.

Lawn mower bags collect grass for your chickens.

Grass height not even close to scale

As grass is mowed shorter, the depth of the roots decreases. Shallower roots cause erosion and make it hard for soil to hold moisture (which means it has less drought resistance). Shallower roots can't provide as much soil nutrition to the grass and won't provide as healthy forage for your chickens.

Short lawns are high maintenance lawns.

1. Keeps soil temperatures cooler

2. Preserves soil moisture

3. Is more hospitable to earth dwellers like earthworms

4. Allows deeper, more massive rooting systems to prevent erosion and flooding

5. Helps maintain lawn consistency, integrity, and quality

6. Produces more nutritious clippings for chicken fodder

7. Decreases the amount of fuel, labor, and noise pollution involved with mowing (or weed whacking).

How often to mow a lawn? Mow it often enough so that no more than 1/3 of the top leaf blade length is removed. For example, if you usually mow your grass to a 2 inch height, then the grass height should not be allowed to grow beyond 3 inches before it is mowed back to 2 inches.

In the spring, when active grass growth is faster, some lawns require mowing more often. Mowing frequency is important with the "Don't Bag It" clipping return program. If it's too wet to mow, and the grass grows excessively tall, then set the mower at its highest level and mow the lawn. After the grass clippings are dry, lower the mower level and mow the lawn a second time in a different direction. This approach is called Double Cutting and allows the clippings to be absorbed back into the lawn, or be collected for chicken use.

Growing Sprout Gardens for Feed

A sprout garden is a way to grow chicken feed. Sprouting seeds in trays indoors provides living foods for the flock, and a living foods addition to salads.

Common Problems with Sprouting

- Seeds are soaked and start sprouting but get too dry

• Seeds are over-soaked in standing water

• Temperature is too high or low

• Insufficient rinsing and not enough fresh water

• Dirty equipment

• Stagnant air flow (add small fans in hot, humid conditions)

• Contaminated water

• Poor germination of seeds because they are old, too wet or dry, subjected to heat, or improperly stored

Growing Soil with Hens

It has been said that farmers and gardeners are really soil growers. If they tend to and take care of their soil, then the soil will produce the nutritious crops. Many growers treat the soil as just a substance to hold roots. But it's so much more than that: soil is a living process with uncountable micro-environments that house billions of living entities.

I heard an analogy: the way many growers treat the soil is like expecting to milk a cow without feeding the cow. Many growers "milk" the soil but don't "feed" the soil to keep it deep and healthy so it can produce.

Food producers must understand that they are really growing soil and, by doing this well, the soil will produce wholesome food to feed us. Another way to explain this dif-

Growing Sprouts as Chicken Food

These are sunflower sprouts growing in seedling trays inside. Chickens love these kinds of greens, especially in the winter.

ficult concept is that by growing soil, you will end up with healthy crops you can use.

Chickens can help grow soil with their nitrogen-rich manure and ability to mix that manure in with leaves, food, and yard waste to create soil. Get your neighbors to give you a steady supply of leaves and yard waste. This not only diverts it from the trash pickup and landfill, but by putting it in the chicken paddock for them to rustle and scratch about creates compost and garden beds.

Are They Getting Enough Food?

Here's one way to tell if a bird has gotten enough to eat during the day. As the birds are roosting feel their crops, located just below their throats. If the all the crops are bulging and full then the hens are getting enough to eat. Check every bird because those lower on the pecking order might not be aggressive enough to consume adequate feed. If the crops are not full, then give more feed. If some of the birds don't have full crops, then feed them separately so they can eat without competition. It's important to send your birds to bed with a "full crop" because a lot of digestion, nutritional absorption, and healing occurs during sleep.

An imbalance in nutrition can show up in many ways, including the way chickens look, how they walk, and that certain sparkle in their eyes. Thin or malformed egg shells are an indication of not enough calcium and Vitamin D.

Hen Heated Cold Frame and Green House

In commercial factory farms, a massive number of chickens produce massive amounts of carbon dioxide, manure, feathers, and heat. These are pollutants that the factory must deal with. In the small-scale permaculture systems, these same pollutants can be harnessed as yields (assets). The key is the relationship between all the elements in the design. Pollutants in a larger system can become valuable assets in a smaller system because of their placement and scale. This has meaning for backyard mini-farms.

Here's an example. In the summer, place your hen house on the shady side of a greenhouse. This lets you harvest the chicken's body heat at night, and their carbon dioxide for the plants. The early morning heat from the glazed sides will warm the hens up to lay.

In the winter put the hen house on the south side of the green house to get solar gain for the coop and as protection from the cold northern winds.

Some producers keep a micro-flock of chickens in their green house, or next to a cold frame. Be aware that with too many chickens, ammonia build up can be a problem, especially for tomatoes which are ammonia sensitive. Combining a chicken run with a greenhouse can help keep ammonia levels down.

Placing the coop against the southern side of a house or garage offers even more winter protection and solar gain.

The biggest problem with combining chickens with green houses and cold frames is the danger that the flock can get overheated. Heat buildup in a greenhouse can be rapid and even fatal to chickens because of the extreme temperatures. Additionally, without proper and adequate ventilation, high humidity can build up and have detrimental health effects. Temperature swings and high humidity can cause early molting and disruption in egg production.

In summary, there is a lot to be developed and learned about growing food for small flocks of chickens on small plots of land. Some of the areas to be researched are planting crops and forage systems directly in the run.

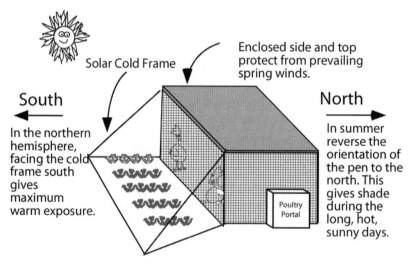

Solar Cold Frame

Enclosed side and top protect from prevailing spring winds.

South ←

In the northern hemisphere, facing the cold frame south gives maximum warm exposure.

Poultry Portal

North →

In summer reverse the orientation of the pen to the north. This gives shade during the long, hot, sunny days.

Hen Housing Combined with a Cold Frame

Combining hen housing with cold frames can benefit both hens and plants. Be careful to NOT let the cold frame get too hot during the day and overheat the chickens. Adequate ventilation at all times is critical for the flock's health.

This allows rotating chickens through a series of runs and garden beds in orchestration with different crops.

Crops for Chickens

Crop	Timing of chicken access	Feeding or Storage	Comments & Observations
Asparagus	Let chickens graze beds between harvest seasons.	Chickens glean asparagus beds. They enjoy being in the understory of the tall plants.	In the fall, mow and mulch deeply. Let chickens spread and fertilize the mulch.
Beans, bush	Cage plants in to keep hens away from roots and beans until they are harvested.	Dried beans can be stored at room temperature.	Beans are easy to harvest. If they are large, soak and/or grind. You can feed beans still in pods.
Beans, Pole	Once plants are established, hens can graze underneath.	Dried beans can be stored at room temperature.	Protect the roots. Chickens will eat low-hanging beans..
Beets	Keep hens out until after harvest. They will eat leaves.	Feed raw or cooked. Chopped peelings are a hit even with week-old chicks.	Chickens love beets in small pieces or cooked. Beets will turn their droppings red. For chicks, shred in a food processor to get tiny, chick-size bits. Many health benefits.
Brassicas, broccoli, brussels sprouts, cabbage	Keep hens away until heads have formed. Will eat cabbage worms.	Cold storage. Best fed shredded.	Chickens will eat any of the brassicas, and prefer smaller pieces. Fresh and dry leaves store well for winter forage.
Comfrey	Prolific perennial. Can grow in cages or along fences for hens to prune.	Feed fresh, wilted, or dry. Stores well for winter use.	Deep roots enhance soil fertility. High in protein. Supports pollinators like honey bees. Has medical uses.

Crop	Timing of chicken access	Feeding or Storage	Comments & Observations
Melons, cantaloupe, watermelon	Hens will eat flowers and peck tender fruit.	Stores well in cool storage. Will clean rinds.	Keep them out until after harvest.
Lettuce, salad greens, Swiss chard, kale, bok choy	Before planting or after harvest.	Dried.	Inexpensive and easy to grow. Plant a row on the outside of their run so they can graze through the wire.
Pumpkins, squash, gourds, zucchini	Easy to grow. Seeds are high in protein.	Easy to store.	
Potatoes of all kinds, beets, other root crops	After thick foliage is established.	Stores well as a root crop. Inexpensive and easy to grow.	Chickens like potatoes, preferably cooked. They eat raw small pieces but tend to ignore whole potatoes.
Fruit and nut trees	Provides forage all year. Let hens in during spring for pest control and fall to clean fallen fruit.	Fruit keeps best in cool conditions. Nuts can be stored shelled.	Chickens gleaning fallen fruit keep hornets and wasps from gathering. Hens enjoy the shade and protection trees offer.
Grains	When reseeding, add grain seeds to grass mixture.	Rotationally graze and let hens do the harvesting.	Hard to grind by hand. Mow down and let the birds graze and glean the seeds.
Amaranth	Annual, easy to grow, high protein grain.	Seed heads can be cut and thrown to runs, or kept for winter feeding.	Can grow as a summer bedding plant. Very attractive. Put in a corner or on the borders of the garden.
Sun-flowers	After plant is about 1 foot tall.	Seed heads can be cut and thrown to runs, or kept for winter feeding.	Colorful and adds diversity to a garden. When stalks are dry, run over with the lawn mower to grind for bedding.
Rape, kale and mustard seeds	These are good fodder crops and grow easily.	Let hens glean, or harvest and dry for winter use.	

The nation that destroys its soil destroys itself.

— FRANKLIN D. ROOSEVELT

Even as humankind is ratchetting up its demand on soil,
we are destroying it faster than ever before.

CHARLES MANN

NATIONAL GEOGRAPHIC

SEPTEMBER 2008

4 Creating Compost and Garden Soil

This chapter goes into detail about how to collect and transform food and yard "waste" into black gold (top soil) and chicken feed. It shows how valuable food scraps, garden and yard collections, and chicken manure can enhance soil fertility for food production. It also describes how to work with local restaurants to divert their leftovers from the trash into your compost system.

Zero waste resolutions are being declared around the globe. In a zero waste program everything is reused, recycled, or transformed into something usable. With zero waste, the entire concept of waste doesn't have any meaning.

The ecological term for such "waste" materials as kitchen and yard waste is biomass. Biomass can be living or dead organic biological material. It is a renewable energy source and can be used to make compost.

In the new "green" language of conservation, biomass of food scraps and yard materials are called residuals. The word "residual" comes from Latin *residuus,* which means the remainder of something after removal of a part or the completion of a process; it has more positive connotations than the word "waste," which implies worthlessness.

Residuals are a form of organic equity. That corn cob, watermelon rind, and orange peel all contain energy — in the form of solidified sunlight — and nutrients. Composting facilitates nature's most necessary cycle. By transmuting kitchen scraps and yard residues into topsoil, you create a small and elegant recycling loop without making a single trip to the recycling center.

When you don't compost you are throwing out valuable energy with the rest of your trash. All that potentially useful energy gets buried, entombed, and wasted in an already stuffed landfill. Kitchen and yard waste sent to landfills decomposes without oxygen (an anaerobic process) and generates methane gas. Methane not only stinks, it warms the atmosphere 21 times more than the carbon dioxide generated by composting.

Composting, done properly, is not smelly, gross, or time intensive. Compost happens naturally all around us. In forests, the decay of fallen leaves and dead plants provides the nutrients necessary for growth of other plants. Decaying matter is the raw material for new life.

One household using 5 pounds of kitchen waste each week to feed chickens or grow their backyard compost pile would keep a quarter-ton of material out of landfills each year! About 25% of the material in landfills could be composted in backyards. Even in the concrete jungles of big cities, people can use composters the size of 5 gallon buckets. The resulting compost can be used in patio container gardens to grow plants like basil, beans, cucumbers, and tomatoes.

There are numerous composting methods and practices. The scale of composting can be as large as windrows in fields or as small as a hole dug in a backyard. There are many kinds of compost bins, from trash cans with holes to specially designed composters available at garden stores. There is no "right" way to compost. It might seem daunting, but in reality, compost makes itself.

Properly prepared and cured compost provides nutrients and stabilizing humus to the soil. It improves the soil's water-holding capacity, decreases runoff, and makes more moisture available to plants. Compost encourages better drainage and aeration of the soil and improves produce quality, plant hardiness, and quality of life for soil dwellers. It affects the entire food chain's ability to resist diseases and facilitates the formation of protective soil organisms, natural antibiotics, auxins (plant hormones), and other biological substances.

The soil chemistry of compost improves the availability of soil-generated plant nutrients in the life cycle, but compost is more than a nutrient carrier; it is the foundation for building healthy soil. A garden is a living organism, home to a universe of beings. Topsoil is a living entity in its own right and contains intertwined, synergistic processes. It contains soil dwellers that are co- and inter-dependent. With the oversight of a good gardener, a garden is a life-affirming co-creation.

Types of Compost

There are four main classes of compost, and many variations on a theme exist within the compost classes. The four main compost types are:

1. Garden compost

2. Animal manure compost

3. Mineralized compost

4. Biodynamic compost

5. Worm compost

Each of these methods has had volumes of books and articles written about them. If you want more information, there are compost sources and references listed in the resource section of this book.

1. Garden Compost is the type most gardeners and homeowners make. It is made from weeds, grass clippings, tree trimmings, leaves, spent mulch, vegetable trimmings, kitchen waste, pet hair, and anything else that will decompose. Usually garden compost has topsoil added to inoculate the pile with beneficial compost-making organisms. Garden compost takes a long time to break down — usually a year or longer.

2. Animal Manure Compost is made from droppings mixed with bedding such as straw, sawdust, chopped corn stalks, etc. This is a well-established way to use farm residuals effectively. It is much better than spreading raw waste in fields (which, unfortunately, is often done) because the high levels of nitrogen in manure can "burn" and kill crops. Excess nitrogen run-off can also contaminate the water of nearby streams and lakes.

Why is commercial animal manure composting important? Because 20% to 75% of antibiotics given to factory farm animals is excreted via urine or feces. According to the Agricultural Research Service, composting animal manure reduces antibiotics in the residues by more than 95%. Large-scale commercial composting is the most sanitary and safest method of manure disposal.

3. Mineralized Composts are commercially produced. This compost process involves rock phosphate and some kind of potash products, which can come from a fireplace or wood stove.

4. Biodynamic Composts are specially formulated from specific ingredients using procedures described by Rudolf Steiner. Biodynamic agriculture is a branch of organic agriculture that views food production as a system dependent on the interrelationship of soil, plants, and animals. It does not use artificial chemicals on soil or plants and is based on the use of manures and composts. In this system, compost is made using fermented herbal and mineral preparations as compost energizers. Different compost preparations are formulated for special decomposition processes in the compost pile.

5. Worm Composting or Vermicompost is usually done indoors in bins, where the worms produce liquid fertilizer and worm castings. Worm castings are a solid, odorless by-product of worm digestion and are incredibly rich as a soil additive, conditioner, or light mulch. Raising worms for food is a branch of agriculture called micro-livestock.

Backyard Garden Composting with Chicken Manure

The manure produced by a micro-flock of laying hens is great for garden compost. Chicken droppings (collected from the bedding under roosts) produces better, faster compost than that made with garden and kitchen waste alone.

A few hens eating food waste and adding their nitrogen-rich manure to composts in hundreds – or thousands – of backyards would keep tons of residuals out of landfills and create tons of locally available fertilizer and topsoil. It's a double-prize activity with a multitude of unintended bonuses.

The use of chickens in urban waste management is not new. The city of Diest in Flanders, Belgium gave three laying hens to each of 2,000 households. The focus of this project was not to generate eggs, meat, or manure for fertilizer. The goal was to reduce garbage. Diest was encouraging residents to use chickens to divert food and yard waste from the landfill.

An article in *The Guardian* entitled, "From Mountain to Molehill," written by Sophie Unwin, explains the Flanders goal. Sophie interviewed a 71 year old poultry owner. While showing off his flock of bantams and leghorns, he says, "They eat everything – grass cuttings too wet for the compost, and they even love bones." The waste management program in Diest proves that local authorities can use a variety of waste prevention measures – including compost bins and chickens.

Why chickens? Because each chicken can bio-recycle approximately seven pounds of food residuals a month. Do the math: 2,000 households with a micro-flock of three hens – each eating seven pounds of food waste a month – equals 504,000 pounds or 252 tons of refuse kept out of waste management programs every year. That refuse is transmuted into eggs, compost, and fertilizer; this is chicken alchemy at its finest. It promotes self-sufficiency in community food supply, enriches soils to grow food, and flock owners enjoy more than fresh eggs. Now that you know the possibilities for urban composting, here is how you can do it, in the privacy of your backyard, in your spare time, with items you have lying around the yard, including hens.

Compost 101

It's hard to mess up composting because Nature does most of the work, employing billions of organisms to digest and ferment the biomass. Just follow a few simple and easy guidelines. The process and results of composting are dependent upon:

- The right amount of carbon to nitrogen. Compost is made mostly of carbon, with a little nitrogen added in. About 25 to 30 portions of carbon (plant or brown) to 1 portion of nitrogen (manure or green) is optimal. This ratio is usually expressed as 1 to 25, or just 25.

- Particle size of the solids. Smaller bits decompose faster. Particle size should be about ½ inch or smaller. Larger particles will take longer to decompose. Shredding, chopping, and even bruising help organic materials decay faster.

- Size and shape of the compost pile. There are many different sizes and styles of composters and they all work.

- The moisture content. Moisture is critical for the bacteria and compost dwellers to thrive. However, too much moisture will kill the good, aerobic critters, and encourage the bad, anaerobic critters.

- Aeration. This can be done by turning, perforated tubes, or chickens.

- Temperature. Hot piles can have temperatures up to 170 degrees.

What are Carbon Residuals?

Old and mature plant material consists mainly of carbonaceous compounds like cellulose and lignans. Wood chips, cardboard, and newspaper qualify as old plant material. Older plant materials contain little nitrogen and water to support bacterial growth, so they break down very slowly.

Young green plant material has more water and nitrogen, which breaks down quickly. Composters have to estimate the ratio of carbon to nitrogen in their pile.

The following is a list of high-carbon biomass materials used for composting.

High Carbon Content Materials (Highest Levels First)

- Wood shavings and chips

- Sawdust

- Chaff from crops such as rice hulls, corncobs, corn stalks

- Straw

- Hay from legumes

- Old mulch

- Weeds, grass clippings

- Leaves plant stalks, vines, twigs, branches

- Vegetable trimmings (corn husks, cabbage leaves, etc)

- Kitchen waste such as coffee grounds, tea bags, citrus rinds, pasta, greens, grains, breads, nut shells, ripe refrigerator left-overs, crushed egg shells

- Wood ashes (a source of lime) from the barbecue, fireplace, or wood stove

- Other compostable materials such as paper towels and bags, hair clippings, feathers, and cotton or wool materials shredded into small bits

• Shredded paper (not shiny or colored) and cardboard, like tubes from toilet paper or paper towels. Larger quantities of paper and cardboard should be recycled into future paper products; this also saves national forests

NPK Fertilizer Basics

According to modern agriculture, there are only three basic fertilizer ingredients: nitrogen (N), phosphorus (P) and potassium (K), or NPK. Where does the NPK come from?

Nitrogen (N) is refined from petroleum products (oil) and natural gas into ammonia (NH_3).

Phosphorus (P) is obtained from mining phosphate rock deposits from quarries that are located primarily in North America, North Africa, and China.

Potassium (K) is mostly from global reserves of potash that were created from sea water from ancient oceans. As the seawater evaporated, potassium salts crystallized into beds of potash ore. Over time, as the surface of the earth changed, these deposits were covered by thousands of feet of soil. Most potash mines are deep, as much as 3,300 feet (.63 miles) underground. Some potash is strip mined, which is horribly destructive to the land, wildlife, and waterways.

Potash is also formed from the ashes of plants, which gives you a clue about the fertilizer value of your yard wastes.

Whereas most of the bagged fertilizers (NPK) you buy in the store have been shipped from half a world away, your humble backyard micro-flock, garden, yard, and kitchen residues can provide the basic ingredients for high quality, homemade fertilizer that contains far more than just NPK. Below is a brief summary of NPK and fertilizer values of various compost ingredients.

• Chicken Manure: NPK ranges: N (4-8), P (1.1-2), K (0.8-1.6).Poultry poop is very high in both nitrogen and phosphorus.

• Comfrey leaves are an excellent source of potassium (K). A nutritional mulch for plants is to put comfrey leaves under grass clippings. Comfrey is also an excellent compost activator.

• Wood ash is an excellent source of potassium and calcium. The NPK can vary but an average is: N (less than 1), P (1), K (3). Wood ash contains a lot of trace minerals that the tree needed to grow. Ash does the same job as lime, raising the pH and "sweetening" acidic composts piles.

• Finished compost can have a wide NPK range, from: N (1.4-3.5), P (0.3-1), K (0.4-2). Compost is safe to put abundantly around plants. Most commercial NPK fertilizers warn that plants can be burned with too much, or direct-contact application.

Properly made compost contains far more than just NPK. Compost usually consists of a number of ingredients from different sources. It will have adequate amounts of most essential major and trace elements for healthy plant growth. Compost, combined with animal manures, produce the most complete, safe, and environmentally friendly fertilizer.

As you can begin to understand, locally produced fertilizer and fertile soil to grow food is a big deal, and it's a good deal. By using local ingredients to create fertilizer not only do you decrease the dependence on oil and foreign minerals, you increase your ability to feed yourself. Whoever controls your food supply controls you.

Carbon and Nitrogen Ratios

Your primary source of nitrogen for fertilizer will be chicken manure, although horse, cow, and goat manure are also nitrogen sources. All the carbonaceous items have a low nitrogen content. Chicken manure establishes the carbon/nitrogen balance easily.

Chicken manure is so high in nitrogen that it can burn plants if not composted or moved daily (like the direct manuring of bottomless portable pens).

Too little nitrogen causes the compost to break down slowly, resulting in a poor finished product. Too much nitrogen causes odor problems because offensive ammonia is formed. A smelly compost pile is a sign that nitrogen is escaping into the atmosphere.

Chicken droppings (with bedding) are about a 2 parts nitrogen to 1 part carbon. About 50% of a hen's droppings are deposited at night, so it's helpful to install a dropping-collection pit below roosts for easy pickup.

Finished compost of any kind ranges from 14 to 20 parts carbon to 1 part nitrogen. Stable humus in fertile soil is 9 to 14 parts carbon to nitrogen.

Compost piles should be about 25 parts of carbon (brown) to 1 part nitrogen (green or manure), which means they have about 25 times more carbon than nitrogen. This is expressed by a ratio of 25:1, or simply 25.

What Not to Put in Compost

It's also important to know what materials not to include in composting. There are toxic products and materials that break down too slowly to be useful.

• Plastic of any kind, even those little stickers on fruit. In order to break down, compostable plastic must be sent to an industrial composting facility, not backyard piles or even municipal composting centers. Whether plastics bio-accumulate in the food chain is unknown and studies are needed to determine this crucial fact. Furthermore, chickens can ingest plastic bits and suffer from impacted crops.

• Too much animal fats, meat, bones. Some animal products are OK, but too much can rot and smell. Chickens love meat.

• Sawdust or wood chips from pressure treated lumber. These contain arsenic and other toxic chemicals.

• Plastics, metal, rubber, glass. These materials do not break down quickly and contaminate soil.

• Disposable diapers. They contain plastic.

• Glossy magazines. The chemicals that make them glossy are not good for the soil.

No Plastic in Your Compost

Don't put plastic of any kind in your compost piles, even the produce stickers. Toxic tag-alongs can get into your soil from plastics breaking down.

• Anything that might be toxic. It can pass back through the soil to you.

Compost Mixture Guidelines

• A mixture of high-nitrogen and high-carbon substances will ferment better than carbon materials alone.

• Any amount of manure mixed with bedding material enhances garden compost by increasing its nitrogen content.

• Raw chicken manure can be up to 1/3 of the total mixture.

• Adding soil to your compost bin helps the fermentation process. Soil contains bacteria and other soil dwellers that are conditioned for digestion. Weeds usually contain about 5% soil.

• Wood ash and lime. If you have acidic materials like sawdust, leaves, pine needles, or pine bark – and not much chicken manure – add a little lime or wood ash to your compost pile. It takes only about 1 pound per 50 pounds of organic matter to get the job done.

Composting Pet Manure

Doggie doo is safely compostable. But cat feces possibly contain a parasite called *Toxoplasma*. The parasite is very common; over 60 million Americans carry it. Very few people carrying toxoplasmosis experience symptoms because their immune system keeps the parasite from causing illness. You can get exposed to toxoplasmosis by eating under cooked foods, drinking untreated water, and (to a much lesser extent), gardening.

When otherwise healthy people are newly infected with *Toxoplasma gondii,* they can get mild "flu-like" symptoms that last for several weeks. The parasite remains in their body in an inactive state, but can reactivate if the person's immune system is suppressed.

If a woman is infected before becoming pregnant, she will have already established antibodies that will protect her unborn child by passive immunity. However, if a pregnant woman has not previously been infected, she has not established antibodies and, if infected subsequent to conception, can pass the infection to her unborn baby (congenital transmission). This can result in a miscarriage, a stillbirth, or problems later in life.

Experts advise that it's okay to compost kitty litter, but they recommend removing the feces first if the resulting compost will be used in food production.

There are some brands of kitty litter that advertise their compostability. Two of these are Feline Pine® (made from pine tree waste) and Swheat Scoop® (made from wheat waste).

Most clay-based kitty litters are not recommended for composting. Many of these are "scoopable clumping litter". They consist mostly of sodium bentonite clay that forms a hard lump. Clay may not be what your compost needs. The specific instruction for disposal is: "landfill" or "dispose in the trash". These brand names include Super Scoop® and Fresh Step®.

Clumping clay litter has some potential health concerns in animals. The bentonite forms a hard ball when it gets wet. When a cat (or chicken) digs or scratches the litter it stirs up clay dust and the animal breaths it in. Once it gets into their lungs, the bentonite expands from the moisture, and in time can build up, causing lung problems.

Some clumping litters post a warning on the bag; "Do Not Let Cat Ingest Litter". But how can an animal not breathe while visiting the litter box? Some of the litter attached to feet, fur or feathers could be ingested, potentially forming a hard mass in the intestines over a period of time. This could be fatal.

Bedding from the cages of critters like gerbils, guinea pigs, ferrets, parrots, or other birds is usually wood shavings or compressed wood based pellets. These are compostable. Just dump the dirty bedding (as long as it doesn't include glossy paper) into your compost pile.

How Compost Kills Pathogens and Self-sanitizes

Pathogenic organisms are killed in composts by high temperatures, competition, and the antibiotic effects of other microorganisms. At 140 degrees Fahrenheit, the following organisms will be killed in less than an hour: poliomyelitis and hepatitis viruses, *mocrococcus aureus* and *streptococcus pyogenes, mycobact, tuberculosis, brucella abortus, and salmonella spec*. The eggs of round worms (*safaris lubricities*) will also be killed. When a compost pile reaches 140 degrees or higher, the heat will last for days and kill most pathogenic germs.

Compost Moisture, Aeration and Fermentation

Compost needs air for fermentation. Aeration also keeps the pile from smelling bad. Living soils are about 50% solid matter; the rest of the volume is filled with water and air space. Living soil breathes just like other living beings. The soil dwellers take in oxygen and give off carbon dioxide and other vapors. Your compost pile requires a composition similar to that of soil for the compost dwellers to thrive and cause fermentation.

About 50 to 65% moisture is optimal for fermentation. This means the matter is moist, but not dripping or dusty and dry. No moisture should drip out when a handful of compost is squeezed.

You'll have controlled aerobic fermentation when your pile:

• Has a temperature curve, heating to 150° to 170° F, then cooling down. It will reach the peak temperature within 3 days and the pile will stay hot for 3 days to a few weeks, depending on its size. Composts with a lot of soil won't get very warm. Composts with a lot of manure, young plant matter, or garbage can get very hot.

• Doesn't dry out or form a gray, moldy layer 3 to 10 inches below the surface. If you see this layer then the pile is too dry. Add more water and turn the pile.

• Doesn't form a black, foul-smelling zone in the center. Such piles are too moist and not generating enough heat. Turn the pile, putting the outer layer in the center and the center on the outside. If water is needed pour it down the holes that reach the middle.

Another way to hydrate a compost pile is to sprinkle the pile as you turn it. Try using a 4" to 6" PVC pipe that is a little higher than your compost pile, drilled with holes or slits along the sides. The pipe allows aeration and is an easy way to get water to the center of the pile.

• Has many earthworms and soil dwellers living in it. Bacteria, fungi, and other soil dwellers decompose raw materials into finished compost. There are compost inoculation mixtures and biodynamic treatments designed to increase the vigor and diversity of the compost pile community.

Turning Compost Piles

Compost piles are turned to encourage aeration. Turning also mixes in outer materials so they can ferment. If the particle size, moisture content, shape, and size of the pile are correct, your compost may not need turning at all.

 Employ your hens to help turn the compost pile. Chickens are miniature front-end loaders. They patiently bring down and spread out piles and can level a beautifully built mound in two to three days. By giving chickens periodic access to your pile they will do the turning for you and, in the meantime, feast on insects, weed seeds, and kitchen leftovers. To get them started, turn over a few shovelfuls of the pile and toss in some scratch. When the chickens are finished, rebuild the mound and keep the hens out until it's time to aerate it again.

Compost Pile Shape & Size

There are numerous ways to build a compost pile. Many garden supply catalogues feature composters that would work for a garden micro-flock composting system. A layer of straw or dirt on top of the pile helps prevent it from drying out or getting soaked by heavy rainfall. Properly built and covered piles will seldom be saturated by rain. If building a pile in the winter, use a plastic cover that extends from the top (but not quite to the ground) to enhance fermentation and keep the pile from freezing.

Compost pile size does matter. A bin 3' x 3' x 3' is about the smallest volume that will ferment well. Conveniently, 3' x 3' is the size of most pallets.

Compost Sites

Usually compost piles are tucked away in a corner of the garden. Ideally, the site is slightly sloped and has good drainage. Wet spots or standing water impede fermentation. If compost is in an area that gets shade and doesn't dry out too much, earthworms and other useful critters will stay in the ground and help with subsequent fermenting.

Inoculate new compost sites with ripened compost from older sites to encourage fermentation. Position the compost bin over future garden sites or places where you want to enrich the soil. Plant creeping plants (such as beans or peas) around the bin like a trellis and they will cover and shade it.

When is Compost Done & What is It?

The materials in your pile will become brittle and the fibers will gradually break down into a fine, spongy mass. The color will turn from gray or

Tumbler Composter

Katie Letcher Lyle uses a tumbler composter. This keeps the compost confined and above ground. It's easy to turn for aeration.

Earth Machine Composter

Cathy Wells is proud of her new composter and plans to grow more of her own food.

Pallet & Plywood Composters

Wire Composter with Drain Pipe for Ventilation

A super easy, cheap DIY composter made of wire and drain pipe.

Do-it-yourself plywood and pallet composters. They are easy to take apart and move. These two are sitting over future garden beds.

light brown to a dark chocolate brown, thanks to the valuable earthworm castings. When compost is finished, the earthworm buffet is closed; they have eaten all their food and will leave to search for the next restaurant. Finished compost has a pleasant smell.

If it contains chicken manure, garden compost will ferment from 1 to 6 months. It can be piled up and stored if you don't need it yet. Usually compost is applied to garden beds in the spring and fall.

Finished compost has many ingredients. It has sand, silt, and clay from the original mixture. It contains anywhere from 0.5% to 2% of the major fertilizer ingredients: nitrogen, phosphorus, potassium (NPK), and almost all of the trace elements. It also has several amino acids, vitamins, auxins, biotin, and many enzymes. Trace minerals in compost include iron, zinc, iodine, cobalt, boron, copper, manganese, and molybdenum. Aerobically-made composts contain a number of antibiotics to protect plants from diseases.

Compost Equipment

Backyard composting doesn't take much equipment. The basics are:

• A pitch fork or shovel to load, unload, and turn compost.

• A shredder to chop, grind, and make small pieces.

• Lawn tractors with front-end loaders are wonderful for managing free-standing compost piles and moving materials around.

• A lawn mower grass-clipping collector bag for yard residues.

• Chickens (feathered front-end loaders and two-legged turners that help aerate and spread compost).

My hens often hop on the compost piles and play "Queen of the Mountain". They get excited when I shovel out parts of the pile so they can access the inner layers. When it's time to turn the compost (to aerate the middle layer), I take the front boards off and the chickens spend most of their active time pulling the pile down and searching for bits of food, seeds, and insects.

Hens also work as tillers to mix the compost and other soil amendments into the garden beds before planting. Put the compost on top and spread some scratch grains, or dig up a spade or two of dirt, and they will spend hours tilling the top layer. Restraining them with a portable fence helps focus their attention on your desired of plot of ground or compost pile.

Commercial Composters

After paying what seemed like too much for a homegrown heritage tomato at the farmer's market, my friend Katie Letcher Lyle decided to start a kitchen garden and raise more of her own food. Katie's uptown, old-style home has that Southern hospitality feel and look. Its beautifully maintained

Wire Compost Bin over Garden Bed, With Chickens on Day Duty

Michelle Patterson built a temporary compost bin using stakes and wire fence directly over her raised garden bed. The chickens in the compost bin are borrowed from her neighbor to work day duty. Note that the waterer is sitting on a bucket to keep the water clean. A board leaning against the inside of the fence was placed to give the hens shelter from the sun. At dusk the hens will be let out of the composting area and taken back to their flock to roost safely in the coop.

yard and flower gardens could be featured in magazines. Her home is on the Lexington Home and Garden Tour.

Katie wanted to make topsoil for raised garden beds, so she bought a tumbler composter, which makes turning compost and aeration easy. Tumbler composters keep the yard tidy and are almost pest proof. The drum has air vents and interior fins that mix materials as the drum is turned. A tumbler can produce finished compost in about five weeks. Some people have two tumblers so that one can ferment and mature while they add fresh ingredients to the other.

Do It Yourself Composters

Composters can be very inexpensive. You can make a free-standing pile without any support at all or make an inexpensive bin (or series of them) using wire and stakes, pallets, plywood, concrete blocks, and even raised beds.

Several Do It Yourself composters are described below.

Wire Composters

Simple and inexpensive composters can be made with 3' to 5' high, 2x4 wire fence. Create a circular one from about 8' of fence with the end wires overlapping. A 6" perforated drain tube stuck in the middle will aerate the pile.

Most of this compost pile is grass and yard clippings, but kitchen waste is included that adds nitrogen or "green" matter. If you look closely, you can see egg shells, which contribute valuable calcium to the soil.

This temporary compost site is directly over Michelle DuBois Patterson's raised garden bed. She plans to expand the raspberry patch, located to the right of the pile, and the raspberries will naturally spread to the composted part of her garden next spring.

Chickens are assigned day duty to aerate, clean, and enrich the compost materials. They are moved to the fenced compost area in late morning, after laying their eggs. Wherever hens are assigned day duty they must have access to water and shelter. A 5 gallon waterer sits on top of a bucket so that the hens' scratching doesn't fling dirt into the water. An old piece of plywood serves as a lean-to shelter for protection from the sun and rain.

Michelle made this compost bin by pounding 5' wooden stakes in the ground and wrapping 3' high 2" x 4" wire fence around them. As she weeds and clears beds, she tosses the biomass into the compost pile, which is not more than a few feet away from her garden.

Pallet Composters

The composter on the left is made from scrap plywood and has a pallet base. The composter on the right is made from 5 pallets wired together. The front is held together with a bungee cord, which is easy to remove when hens need to come in. After they leave, shovel the semi-finished compost back in the bin for more fermenting. These composters are easy to take apart and move. The two shown in the photograph are sitting on a future comfrey garden bed.

Freestanding Compost Piles

Compost doesn't need a container. Freestanding backyard piles are usually 2 to 3' high. Higher ones will need more space at the bottom, like a pyramid. A 3' pile will have a conical base of about 7' to 10'. Like all compost piles, free standing ones will compress and spread to a wider area than the original mound.

Trench Composting

An ancient and simple way of composting is to bury food residues. Dig a hole or trench about 18" deep, add residues, and cover with dirt. The biomass will compost *in situ* with no more handling necessary. This works best for small amounts of kitchen and table scraps.

Trenches can be dug between rows of a growing crop or in an unused bed. Don't plant directly over a trenched area for about 6 weeks so the leftovers have time to decompose. Cover crops are the exception to this rule of thumb.

Naturally, it's difficult to use the trench compost method in winter or whenever the ground freezes.

Bucket Composting

This composting method doesn't require a yard. It does require some chopping and turning that takes only a few minutes. You need two (or more) waterproof containers, a stirring stick, and a bucket of rich, non-sterilized garden soil. It's important that the starter soil be non-sterilized because this supplies the live (good) bacteria for your bucket compost.

Here's How to Bucket Compost

1. Depending on how much food residue you have, get two or more waterproof containers. Five gallon buckets with lids work well. Round containers make stirring easier and more effective. Drainage holes in the bottom of the buckets can help keep the compost from getting soggy.

2. Finely chop kitchen waste or put it in a food blender. Have the mash moist but not dripping.

3. Add the waste to the bucket with a thick layer of garden dirt.

4. Stir the mixture to aerate it thoroughly. This step is essential; it keeps air in the mixture and prevents anaerobic bacteria from growing and causing foul odors.

5. Drain excess water with holes in the bottom of the bucket; try putting a bucket in a bigger bucket to catch any drainage. Certain food residues, like vegetables and soups, contain a lot of water and need more topsoil added to absorb excess moisture.

6. After about two weeks, or when your bucket is too full to stir easily, stop adding fresh kitchen scraps. Start the same process with the next bucket, adding residues until it gets too full to stir completely.

7. Every day, stir the fermenting compost in all the buckets. Be sure to go deep and mix up the bottom layers. In about two to three weeks you will have rich, sweet-smelling compost ready for your patio garden or potted plants.

8. If fruit flies become a problem, sprinkle diatomaceous earth (DE) on top of the compost just after stirring. The DE will keep their larvae from hatching.

9. A foul-smelling compost bucket indicates that the wrong bacteria are growing. Bad smells are also caused by waterlogged ingredients that stop air flow. Drain any excess liquid and thoroughly stir the mixture to get it aerated. It might be useful to transfer the mixture to a bucket

with holes drilled in the bottom for drainage. Set the mixture outside to drain or inside another bucket without drainage holes. Add dry topsoil or wood shavings if it is still too wet. If the smell is unbearable, throw it out and start over.

When your first compost is complete, you can use it as starter for future batches, adding a layer with each residue addition. Your bucket-made compost contains good bacteria for fermentation so you won't need more topsoil to inoculate future batches.

Summary of Steps for Successful Composting

1. Mix about 2 parts garden and kitchen waste with about 1 part chicken manure and bedding; the ratios don't have to be perfect.

2. Add water until the mixture feels like a damp, but not dripping, sponge.

3. Stack your pile or put the mixture in a compost bin or bucket.

4. The compost will begin cooking. After it has cooled, stir it for aeration. Move the material on the edge to the center. The chickens can help you with this phase, but you will have to rebuild the pile for the second cooking.

5. Let the compost cure until it is a dark, crumbly, earthy, sweet smelling material. This can take from weeks to months.

7. Spread a thin layer of compost on your garden or flower beds and work it into the soil. Chickens can help with this part as miniature scratch-and-turn tillers.

Trouble shooting Compost

Problem: The compost stinks.
Solutions: A stinking compost pile can have several problems:

1. The wrong bacteria are fermenting. Add good garden soil to inoculate with beneficial bacteria.

The Four Seasons of Composting with Chickens

 **Spring**

• Dig out any finished compost from Winter composting and spread on gardens or lawn. Use chickens to spread, shift and clean the finished compost.

• If using movable compost bins, relocate to new sites over garden beds.

• Employ chickens as quiet, fuel-free tillers to mix top layers of soil with other garden amendments.

Summer

• Employ chickens as compost turners and cleaners. Chickens innately know how to turn, spread, and clean compost piles of food residuals and insects.

• Collect grass clippings for bedding, yard hay and composting.

• Collect garden waste and add into compost or deep mulch runs.

 **Winter**

• Keep filling composters. In colder temperatures the biomass decomposes slower. With spring it will speed up.

• Feed dried grass (yard hay) to flock as supplemental.

• Mix dried grass and leaves with with wood shavings to use as coop bedding.

Fall

• Clean out composters and spread finished compost into your garden soil.

• Collect leaves for bedding and composting.

• Get city wood chippers to give you the chips for garden mulch and compost materials. This saves them transportation.

All Year Round

• Feed kitchen and table scraps to flock.

• Collect chicken manure for compost. Chicken manure provides the valuable nitrogen.

• Collect compostable biomass from your yard and neighbor's yards to stockpile for bedding and carbon material.

2. The compost doesn't have enough aeration. Turn the pile to get it aerated and mixed.

3. The compost is too wet. Add more dry carbon material and mix thoroughly.

Problem: Flies and vermin in the compost.

Solution: Poorly managed compost piles and garbage not only attract flies, but act as their breeding ground. A fly lays about 200 eggs at a time, which can hatch within a day and emerge as flies within a week. Rotting masses of garbage and manure stink, attract flies, and become breeding beds.

The beauty of compost is that when fermentation is aerobic, few (if any) flies will breed in it. Flies don't like compost. You might still see them in small puddles around the manure piles – or on waste not in the piles – but they won't be in the properly constructed and fermenting compost pile itself. Aerobic fermentation and sanitary housekeeping at the compost site will avoid the problems of flies and rodents.

Problem: The pile is not decomposing:

Solution: There can be several reasons for a pile not decomposing:

- Not enough aeration. Turn the pile.

- The particle size might be too big; shred or remove larger pieces.

- There might not be enough nitrogen; add more green material or chicken manure.

- The moisture level might be too low; try adding water.

A well built pile can yield compost in a few weeks to a couple of months, but an unattended "toss and forget" pile made with non-shredded materials can take a year or longer to decompose.

Problem: My pile shrinks and I don't get much compost.

Solution: Compost piles can seem to melt and disappear. This means the materials are being fermented, digested, and taken underground by the soil dwellers. Build a bigger pile. Consider asking your neighbors to let you have their compostable materials.

Problem: The compost pile is attracting pests.

Solution: Don't include meat or fat products in the compost. Put in a rodent screen to prevent pests from burrowing under the pile.

Problem: The finished compost is lumpy, stringy, and has bits of twigs and eggshell still in it. It isn't fine and crumbly.

Solution: Don't worry; your compost is usable as is and will still improve soil fertility and tilth. You might let it sit and "mature" a little longer, then pick out the bigger pieces. Try sifting the compost through a hard wire screen to remove the chunks.

Problem: My compost has a lot of weed seeds that I don't want to introduce into my yard or garden.

Solution: Spread the compost on a garden bed and let your chickens clean up the weed seeds. If you don't have chickens, let the compost ferment longer, until the seeds germinate or decompose.

Composting is easy and can be done in almost any backyard. It turns "waste" into a rich soil additive – black gold for growing food. Chickens aid composting by providing nitrogen-rich manure and by sorting through and turning compost piles. Composting can dramatically reduce the amount

Restaurant Food Residuals Collection

During lunchtime, the Healthy Foods Co-op features a vegetarian Cafe. Employees collect food residuals in 5 gallon buckets and Mitch Wapner collects them every day. They divert about 1.5 tons of waste from the landfill each year. Mitch's chickens glean, clean and turn the compost piles enhancing the composting process.

of biomass deposited in landfills and waste management systems, helping local governments minimize "waste" management spending.

Master Composter Program

Local yard waste bans and brimming landfills are stimulating nationwide interest in Master Composter classes. The University of Delaware Cooperative Extension has created a Master Composter program that compliments the Master Gardener program. A Master Composter is an expert in the art and science of composting.

Two of the first Master Composters are Hetty Francke and Gail Hermenau. They developed workshops that educate public school students. Their classes teach respect for our planet and emphasize that everyone – children and adults – have to be responsible and protect it.

Composting for Restaurants and Others

This section examines three case studies about composting more than one household's kitchen scraps and yard waste. Serious gardeners are willing to collaborate with anyone interested in contributing to the green movement, especially when the green moves to their compost bins. Additionally, various entrepreneurial opportunities exist in the world of composting. These case studies explore such possibilities:

1. How a retiree cooperates with a local food co-op and eatery by collecting biomass waste to feed his chickens and makes compost to use in his production gardens.

2. The 1950s efforts of E. E. Pfeiffer and his urban compost operation in San Francisco.

3. The operations of a contemporary urban compost company and the forces that oppose its successful and serviceable operations.

Each of these case studies show how little efforts can equal big results.

Case Study 1: Restaurant Residue Recycling in a Market Garden

Mitch Wapner is a retired veterinarian. He sold his equine practice several years ago and moved to Rockbridge County, Virginia with his wife, Cindy, and their three kids. Wanting to live lightly on the planet, they bought Paradox Farm and built a straw bale home, complete with solar hot water and passive solar gain to heat the house. They grow almost all their food in the garden and have a root cellar for winter storage. They keep bees for crop pollination and honey, and chickens for eggs and meat. Mitch sells honey, eggs, and produce at the local farmer's market.

He collaborates with the local Healthy Foods Co-op, which sells local produce and has a small lunch café featuring vegetarian dishes. The staff collects kitchen food scraps, expired items from their produce section, and food waste in 5 gallon buckets. This saves about 60 pounds of residuals a week, or 1.5 tons/year from going to the landfill.

Every day or so, Mitch takes the kitchen, café, and produce residuals back to his farm where his chickens greet him with enthusiasm. He lets the flock glean through the pickings from the compost piles. This compost is spread on his gardens to grow food for his family.

Case Study 2: Recycle for Restaurants & Groceries

Inspired by Mitch's cafe/chicken/compost system, we decided to do our own research project. We started small, collecting the residuals from the Masamoto Sushi restaurant owned by Isamu and Sandy Masamoto. As you might guess, sushi restaurants have few leftovers. My flock of 25 could handle more.

We approached McCoy's Deli, owned by Chris and Christy McCoy, about collecting their residuals. McCoy's Deli is green – and not just because they are Irish – and they were pleased to collect the food residuals for us.

This plan had 5 simple steps:

1. Give the restaurants collection buckets with lids for residual collection. We got about 30 kitty litter containers from the Humane Society. These containers hold up to 28 pounds each.

2. Weigh the containers to keep track of the amount of food residuals recycled.

3. Deposit the residuals on top of a future raised garden bed in my backyard where the chickens can have access. The future garden bed

has a layer of deep mulch below. We added wood chips and old hay as needed to supply the residuals with enough carbon.

4. Track how the consumption of food residuals affected the chickens' intake of commercial feed.

5. Track differences in egg production and quality.

The chickens loved the carry-out meals. They would come running every time they saw a white bucket. They delighted in the variety of cuisines, as chicken feed must get boring after a few years. My mixed-terrier guard dogs became more attentive to the chickens and food scraps as well.

In one month we collected about 200 pounds of food residuals, which equals over a ton of residuals kept out of the trash management stream every year. The chickens' commercial feed consumption dropped about 10%.

The food residuals melted away every day as they were eaten and scratched into smaller pieces, disappearing into the deep mulch. The key to deep mulch composts is to have lots of carbon (from straw, hay, grass clippings, etc.) so the residuals get re-covered daily. Without bountiful carbon, food scraps would remain on top and are fly-attracting eyesores. The other option is to rake up what hens don't eat and move it to the compost bin. In our deep mulch system, pests and scavengers were not a problem, as there was never waste residuals uncovered for very long.

Unintended consequences were positive and numerous. Local service organizations and municipal committees requested that I do a presentation to their groups about composting. The city of Lexington and the county are looking into providing composters to residents.

Commercial Urban Composting as an Alternative to Landfills

Urban composting of biomass residuals as a waste disposal method has been successful in the past and will hopefully gain favor in the future. A fascinating book, *Secrets of the Soil*, by Peter Tompkins and Christopher Bird, tells the story of Dr. E. E. Pfeiffer, who identified and isolated strains of microorganisms that could digest every component of biodegradable garbage in a city dump or slaughterhouse waste pile. Dr. Pfeiffer's bacterial compost super-starter was capable of converting tons of city garbage into usable, clean compost.

In 1950, a plant was set up in Oakland to compost as much as 400 tons of city garbage each day. Dr. E.E. Pfeiffer convinced capital investors to invest on the basis that:

> *"It costs Americans, as taxpayers, billions of dollars a*
> *year to cart way as garbage precious minerals and organic*
> *material taken out of the soil in the form of food, while it*
> *was costing farmers billions a year to put chemical fertil-*
> *izers back into the ground."* — DR. E.E. PFEIFFER

Once the compost plant was operating the process was simple. Workers would bulldoze tons of raw garbage onto conveyer belts. Then giant suction fans sucked up most of the paper waste. Next, huge magnets collected metal objects. Finally, gloved workers picked out glass and big wooden objects. The remaining goop was fed into a giant hopper with rotating blades that shredded the mass into a gunky, slurry material, while spraying it with water containing Pfeiffer's bacterial concoction. Dr. Pfeiffer referred to his brew as biodynamite because of its decomposition power.

Bulldozers formed the gunky mass into windrows. Within two to four days, the raw materials would heat to over 150 degrees as the super-bacteria decomposed and digested the garbage. In less than a week the process was complete. The piles cooled and shrank into finished compost. Like a phoenix rising, the city garbage was transformed from stinking gunk to valuable compost and plant food. Like all properly made compost, it smelled sweet. Plants thrived in the city compost, producing superior yields and increased protein levels.

Think of it. Household food waste could leave the dinner table, be transformed into compost and shipped out as rich, healthy black soil back to farms and nurseries. The entire cycle would only take three weeks. Pfeiffer stated:

> *"If all the U.S. garbage were processed each year we would*
> *have (tons) of compost, enough to fertilize millions of*
> *acres. Garbage dumps would just about disappear."*
> — DR. E. E. PFEIFFER, CIRCA 1950

Despite the huge success of the garbage compost plant, it closed within two years because of pressures from chemical fertilizer producers, worried about losing business. It is an example of the common American tragedy in which private interests are deemed more important than public welfare.

Case Study 3: The Vermont Compost Company

History is repeating itself. A similar drama is unfolding in which political interests and powerful economic forces are attempting to shut down the operations of two shining, compost-creating stars: Vermont Compost Company and Burlington's Intervale Compost Center. These compost projects have been in operation for about 15 years. Their story is an excellent illustration of the possibilities, challenges, and rewards that exist for local compost companies.

The Vermont Compost Company is a pioneering, paradigm-shifting company located within the Montpelier, Vermont, city limits. It was founded by organic farmers to produce quality composts and compost-based live soil mixtures for certified organic plant production. Karl Hammer is the owner and president of the VCC.

The VCC is a viable model for small-scale, community-integrated compost operations. Its operating site is about 5 acres and includes the home of the owner, a building in which potting soil and compost is packaged, and a large garden.

This District helps its 22 member communities work toward Zero Waste. Their mandate is to:

> "Work toward Zero Waste means that we will strive to capture all of the resources inherent in trash so they can be reused and recycled in this region, instead of wasted by burying them in a landfill or burning them in an incinerator. Reducing waste up front through good product choices is also a critical component of a Zero Waste effort."

The District has two food waste diversion programs, one with schools and the other with businesses. The School Food Waste Diversion

Vermont Compost Company's Logo

Program collects food residuals from 4,000 students in 9 schools who sort their food waste for composting. This has diverted 65.4 tons of food waste from landfills as of 10/11/06. The schools have competitions and awards for the best residual collectors. They have science projects to raise salad greens in potting soil made by the VCC, and they give awards for the best growers. Classes take annual field trips to the landfill and the VCC.

Business Food Waste Diversion Program works with 42 businesses that produce food residuals. Among these are restaurants, workplace cafeterias, and other institutions. They have diverted 833 tons of food waste to composting since the program began in April 2004.

The VCC mixes food residuals with old or late harvest hay, bedding, and manure from local dairies. The mixture is piled about 20' high and left to ferment.

What's exciting and unique about the VCC compost process is the use of chicken workers. The biomass mixture is put in windrows and, while

Food Residuals Delivered to a Commercial Compost Site

Food residuals being delivered to the Vermont Compost Company. Each participating school or business is given special containers in which to collect the food scraps. Chicken workers are front and center for the delivery.

fermenting, VCC's 1,200 free-range chickens swarm over the steaming piles. The compost chicken workers aid the compost process in four ways:

1. *Sanitation.* Daily and consistently, chickens gobble up the bits of food, insects, and other edible delights in (and around) the windrows of future compost. As the piles become more biologically active with earthworms, crickets, beetles, and slugs, the hens get even higher quality protein food. At the same time, they make food less available for vermin, thus keeping the rodent and insect populations down.

2. *Odor Control and Aeration.* The hens' scratching keeps the piles aerated. Aeration keeps food scraps cleaned up, which in turn keeps offensive odors down. The scratching of 1,200 hens is a formidable force. At the same time, the chickens add their nutrient-rich, nitrogen-charged manure to the compost.

3. *Superior Compost.* The synergy that chickens bring to the compost process helps produce quality organic compost and potting soil.

4. *Egg Production.* 1,200 hens equals a lot of eggs. VCC's hens produce an average of 1,000 dozen eggs per month, which provides an additional

Chickens Moving Compost Mountains

You can see how chickens can move mountains. 1,200 chickens are the equivalent of 4 tons of heavy equipment. They move and mix tons of compost without using any fuel.

income stream to VCC and high-quality, locally-produced protein for the community.

VCC uses heritage breeds known for their foraging prowess, including Black Australorps, Rhode Island Reds, and hybrid splashes from several generations of VCC chicken breeding.

VCC has no feed bills for its adult chickens. The only food the chickens get is what they salvage from the compost piles. They are in scratch heaven; VCC is a chicken's dream come true. They have all that food and all day long to scratch, peck, and dust bathe in the sun. It is the alchemical transformation of "residuals" into eggs and garden soil.

VCC doesn't have fences keeping the chickens in or predators out. These biddies are truly free-range. Of course, without fences to stop them, predators would dine on daily fresh chicken dinners, were it not for the two poultry-protecting German Shepherds on 24 hour guard duty. There is also a buffer pasture surrounding the VCC with guard mules to chase off foxes and wandering dogs.

The chickens are locked up at night. The doors open in the morning, and they are free to travel anywhere their hearts desire. Karl Hammer claims that his chickens are "free to leave" and work on the compost piles totally at will. The chickens spend most of the day in the composting windrows, returning to the barn periodically to lay eggs and drink.

The Central Vermont Solid Waste District estimates that VCC keeps an average of 13 tons of food scraps out of landfills each week. That's 1,352,000 pounds per year of potential landfill stuffing converted to compost. VCC produces superior quality compost and sells it to local farmers and gardeners for organic fertilization of crops or potted plants. The compost is certified organic.

Here's where history repeats itself. The VCC was served

Eggs from Food Residuals

Vermont Compost Company's 1,200 hens lay about 1,000 dozen eggs each month. They aren't fed a single grain of commercial feed. Instead, they eat the food residuals, insects, and other critters from the compost piles.

administrative orders by the state's Natural Resources Board to "cease and desist" its composting operations, to remove all compost from the premises, and to pay an $18,000 fine for violating ACT 250 – even though there have been no court hearings or judgments on the pending appeal. A similar order was issued to the Intervale Compost Project in Burlington, Vermont.

The power struggles behind these orders have evidence of being riddled with high-level political nepotism and favoritism. Although the stakes are high for the VCC and the Intervale Compost Project, they are even higher for the environment, as biomass waste management practices will have a significant effect on the long term health – or disease – of the planet.

What's happening to some of Vermont's compost businesses is similar to what happened to Dr. Pfeiffer's compost plant almost 60 years ago.

Composting waste biomass is cheaper, more efficient, and environmentally friendly than landfill stuffing. But instead, the government routinely spends millions of tax dollars on "waste management" to build new landfills, buy more dump trucks, contract bigger transfer stations, and pay employees to haul trash. Furthermore, landfills generate methane and leak toxins.

The Vermont Natural Resources Board is trying to close down remarkably creative and environmentally-friendly companies that solve a host of disposal problems, empower local food production, provide employment, and divert valuable resources away from landfills and into the food chain.

Because of exhaustive negotiations and huge public outrage, the Cease and Desist order has a two year moratorium for the VCC.

I believe this is a critical time in our planet's history. With global warming and weather patterns changing we must, as a world culture, seek alternative ways to manage resources and decrease our collective carbon footprint. The composting issue is at the core of our daily activities and lifestyles. The cycle starts at our dinner tables with the food we don't eat, and includes landscape waste and non-polluting farm manure.

The beauty of local composting companies is that they cycle topsoil back to our gardens and farms. If chickens are used in the compost process, the valuable, high-quality protein of free-range eggs is an added bonus.

Compost projects are precious. They are life-affirming and ought to be expanded, funded, protected, and developed in every town and city across the planet. They must not be ordered to close because they damage short-term corporate profits. Narcissistic thinking that encourages private gain at the expense of the public good must end.

Karl Hammer, President of the Vermont Compost Company, says it all in his essay: "Suicidal Obsession":

> *In the food-insecure, grain-short world of now, it is*
> *suicidal when a community that allows a farm opera-*
> *tion that produces eggs without buying grain, that pro-*
> *duces composts and potting soils, manages pastures, and*
> *now hosts a new market garden is ordered to "cease and*
> *desist". The farm is the victim of a suicidal obsession with*
> *definitions in law that don't respect or understand whole*
> *biosystems. We are in an emergency about food and self*
> *sufficiency, folks. We need more chickens, more compost*
> *and more crops and more people working on the land.*
> —KARL HAMMER, VERMONT COMPOST COMPANY

In summary, this chapter is about transforming biomass "waste" into an asset. This transformation can be done backyards. On a larger scale, it can be done as municipal service that collect residuals from food serving institutions such as restaurants, school cafeterias, colleges, prisons, processing plants, and any business that serves food. Urban compost companies can recycle the residuals into compost and potting soil.

Compost happens, just give it a chance. Compost happens even better with nitrogen-rich chicken manure; give it a chance as well. You will be providing a service to yourself, your community and our planet.

"*There is magic in that little world, home; it is a mystic circle that surrounds comfort and virtues never known beyond its hallowed limits.*"

— ROBERT SOUTHEY

"*Be it ever so humble,
        there's no place like home.*"

— JOHN HOWARD PAYNE

5 Chicken Housing, Furniture & Interior Design

In urban areas, hen housing should ideally be small, upscale, stylish, and even have whimsical, artistic touches that invoke smiles and magical thoughts of Hobbits or other little critters. Cute coops will help eliminate resistance from neighbors who view chickens as smelly, noisy, vermin-laden farm animals. When we proposed allowing hens within the city of Lexington, Virginia the 2 the biggest concerns were:

1. Property values would drop

2. Smells and flies would be a nuisance

This chapter is about housing hens in such a way that property values will be enhanced by having chickens next door. Hens are remarkably easy to shelter, but challenging to protect from predators. They don't need space to store stuff or display memorabilia, and they share one bedroom and bath. All they want is a clean, draft-free, well-ventilated place to sleep, a semi-private place to lay eggs, good vittles, and enough room to run, take dust baths, squabble, and roam in the sunshine.

There is a trend in home construction called "housing for all ages". This concept also applies to chicken housing and is important to keep in mind while constructing your coop. It's not unusual for hens to live 8 years or longer. Arthritis can set into the tender joints of older hens, and they need easier access to roosts and nest boxes than the "spring" chickens. Over the years I've watched hundreds of birds coming home to roost and have observed how my older hens walk stiffly and slowly, especially in the mornings. I dearly identify with them. I don't mind keeping my geriatric hens. They have served us well with years of egg production. They don't lay eggs as often as they once did, but the eggs of older hens are larger than those of their younger colleagues, and mature flock members are valuable as experienced garden workers. Older hens act like grandmothers and tutor younger ones in flock culture and the wise ways of the world. They deserve respect.

There are many styles, shapes, and sizes of hen houses. Summer housing is often portable, bottomless, and open air for use in the gardens. Winter coops are usually stationary and provide stable shelter against the wind and cold. Some shelters are simple, open front sheds that don't close but provide a place to roost out of the wind.

Micro-flock coop and chicken tractor designs have become remarkably creative and popular. There are books written specifically about coop design. Vendors advertise upscale, ready-made shelters in poultry magazines such as *Backyard Poultry* and *Practical Poultry*. Many micro-flock building plans are available online. Manufacturing chicken tractors, arcs, chalets, huts, and coops could be a good local cottage industry, much like building storage and garden sheds. I predict that builder supply stores like Lowe's and Home Depot will start selling micro-flock shelter kits.

There are annual coop tours in major cities that feature urban hen housing. Portland, Oregon, hosts a "Tour de Coops" and Madison, Wisconsin's "Mad City Chickens" club hosts an annual "Backyard Chicken Coop Tour". Austin Texas hosts an annual "Funky Chicken Coop Tour". Other cities hosting coop tours are Atlanta, Raleigh, Tucson, and Salt Lake City.

And the movement is growing. About 600 chicks were given to people in Roswell, Georgia as part of the Chicken Stimulus Package Giveaway.

Some of the coops featured in these tours are fanciful delights. The ever-upscale and fashionable Martha Stewart hired an architect to rehab an old chicken coop at her Turkey Hill Farm in Westport, Connecticut. Martha keeps Araucana hens who lay stylish, bluish eggs. Auracanas were first bred by the Araucanian Indians of Chile, which adds to their exotic appeal.

There isn't any one right way to design and build hen housing. I don't doubt that soon there will be books on coop feng shui. Modifying existing structures to accommodate chickens is easy and requires only basic carpentry skills and tools. Permanent structures that can be modified for hen housing include:

- Converted storage, garden, or tool sheds

- Re-modeled sections of a barn or outbuilding

- An area inside a garage

Portable shelters (chicken tractors, arks and small coops) are popular for yard and garden use. These structures are usually bottomless pens that keep hens restricted to a small area of grass or over a garden bed. Bottomless pens are intended to be moved, or have fresh bedding added every day.

Designs for creative hybrid huts combining coops with chicken tractors have become increasingly available and are remarkably versatile. Some are double-decker units with enclosed upper loft-living quarters, accessible by a ramp from an open area below. Some have detachable runs that double as cold frames.

These hybrid coops provide hens secure sleeping quarters and a small enclosed area to range about. Some of them have pop hole doors in the runs to let chickens free range, and regular doors to allow people inside access.

One hen condo sold by Stephen Keel, owner of HenSpa.com looks like a trash container designed by Frank Lloyd Wright. It camouflages the coop by blending into urban backyards. It might allow contraband hens to live in forbidden city limits without getting busted by the police.

Some movable chicken tractors are like mini-barns or playhouses. They have solid floors so litter and manure can be collected for composting. The hens free range during the day and are protected by electric poultry netting or movable, temporary fencing. Some mini-coop condos are put over fenced, raised beds and the hens are directed with poultry portals and tunnels to work different beds in different parts of a garden.

However you build your hen housing – no matter what materials you use – keep in mind the following features.

Photo Credit: www.HenSpa.com

Coop Disguised as Trash Can

City Chicken Urban Coop Photo Tour

The octagon coop on the left is a gazebo-like hen housing. On the right is a raised, A-frame ark. These coops show how urban coops for micro-flocks can blend in with million dollar, in-town homes, as well as upscale urban neighborhoods.

Brent Stavig's passive solar hen house has a sun room and play yard. Facing the coop East gets the hens up and going early. The trellis over the coop and run provides shade from the hot afternoon "dragon" sun.

Dennis Harrison-Noonan owns a home repair and remodeling business. He specializes in the smaller projects that require a high degree of careful attention to details and preferences, like this portable 4'x8' "Play House" coop he designed. Dennis sells both the plans and finished coops. His website is: www.isthmushandyman.com.

Designing and building City Chick coops for micro-flocks might be a profitable cottage industry in some areas.

Mural Georger owns this artistic, portable coop. The semi-permanent run has a 2" x 4" wire fencing with metal fence posts. This type of movable fence allows the chickens to be directed where fertilizer is needed.

PHOTO COURTESY OF SEATTLE TILTH

Scott Wallace owns this stylish coop attractively tucked along a fence. Hens are gardeners and love a flower box gracing their coop. Notice the pitch of the coop's roof matches the neighbor's roof to blend in with the neighborhood architecture.

Ingela Wanerstrand clearly has the green permaculture thumb with plants growing on the coop roof and the compost pile next to the coop for easy composting. The window box has the potential to grow food that the chickens could harvest through a window screen.

Terrie Abrahamson's coop fits the whimsical, hobbit style. This permanent, roundish hen hut is made of stone and stucco. Deep windowsills let in light. The run is arched PVC pipe secured by metal posts. Poultry netting keeps the chickens in and flying predators out. A high perimeter wood fence provides additional safety. The kids play set next to the hen hut give the hens entertainment and a sense of excitement.

Julie Metzger owns this coop with stylish cedar shake external nest boxes. External nest boxes give more room inside small coops and make egg collection easy.

Kathy Pelish built her "3 boxes" raised ranch coop with the highest box off the ground and containing the nest boxes. This segmented, multi-level coop design could be used on steep hillsides.

PHOTO COURTESY OF SEATTLE TILTH

Protection from Predators On, Under, and Above Ground

Almost every critter loves a chicken dinner. It's a good rule of thumb to believe that where there's a hen there will be a predator above, on, or below the ground. In urban areas, dogs will probably be the most prevalent chicken chaser. But raccoons, hawks, feral cats, or an escaped ferret can wipe out your entire flock quickly if the hens are not protected.

Hen House Location, Location, Location

Before you build your hen shelter, ask yourself if it will fit through, over, or around structures or obstacles in your yard. These obstacles include garden gates, garage doors, tree stumps, or anything that might restrict its movement. Will you be able to get a wheel barrel or garden tractor to the shelter when it's time to clean the bedding out?

Easy access to feed and water is important for the hens and their keepers. Carrying buckets of water and 50 pound bags of feed isn't a chore most of us want to do often. There are housing designs, equipment, and daily care systems that minimize the effort of keeping hens. These systems, tools, and techniques are woven throughout this chapter.

Coop Requirements

Coops, arks, chicken tractors, whatever your flock lives in has basic requirements that apply to all structures. These requirements are discussed below.

Coop Space

It's better to give your hens more, rather than less, space. Animal scientists recommend as little as 2 square feet per bird. Dang! A hen can hardly turn around in an area that small, and any scratching would put litter in other hens' faces. By all humane standards, this is too small an area, even for bantams. For micro-flocks, 6 to ten square feet per bird is reasonable to give each lady enough real estate to be a decent chicken. Squeeze even the most mild mannered hen in with too many others and she will morph into a PMS moody bitch. Not a pretty sight.

If your coop is primarily for roosting (sleeping quarters) and the hens have access to a run, you can get by with less coop space. Hens tend to sleep well close together. But during the day chickens need room to flap, scratch, dust bathe, nap, eat, drink, lay their eggs, and debate poultry politics.

My hens have a small coop with access to a large fenced run and garden

Example of Coop Furniture Arrangement

The above photo is a 6' by 8' tool shed converted into a coop. The metal nest boxs are high enough for the hens to get under for more floor space. They use the ladder to access the roosts. The metal can holds feed and the upper shelf stores bedding. There is chicken wire attached to the side of the shelf that keeps chickens from roosting. Lattice keeps

the hens out of the droppings pit. A 6" threshold keeps litter in the pit, and shelves on the doors make tools and medicine easy to reach.

Inside the coop (photo on left) the automatic pop hole is on the left, the yellow plastic baby pig creep feeder is in the middle, and a metal creep feeder on the right holds oyster shell. The open studs are good places to store pitch forks, shovels and arm-extending egg retrievers. There is no place for rodents to hide.

Not showing is a lamp to light the coop and the extension cord that runs from the house.

beds. The coop is only 6' by 8' and has just enough space for 6 metal nest boxes, roosts, and a large wall feeder and oyster shell dispenser. A field waterer is outside in warm weather. In the winter, a metal waterer hangs inside the coop just above a heated platform to keep the water from freezing. There is a black, 4" deep flexible rubber waterer outside. The black walls absorb heat from the sunlight and water semi-thaws on all but the coldest days.

Coop Ventilation

Hens must have good ventilation and fresh air to be healthy. Ventilation is the best way to manage moisture. There are 4 sources of moisture:

1. Atmosphere (air)

2. Condensation caused by moisture condensing from the air and dripping from the rafters

3. Ground moisture wicking up from wet soil below and dampening the litter

4. The hens' substantial body heat and moisture production which causes condensation

Condensation, over time, can have harmful consequences. Tight construction can result in moisture collection on interior walls and damp litter on the floor. Damp conditions encourage mold, bacteria, and ammonia production, making the indoor air quality toxic and a conduit for disease in the flock. If there is moisture on the walls or in the bedding, then the housing does not have enough ventilation.

Ventilation in the summer is necessary to keep your birds cool. This can be accomplished by an open-front housing design, or operable windows or screened openings to provide cross-ventilation. Operable windows need to have a latch that a raccoon can't open.

A design flaw often found in smaller coops, chicken tractors, hen huts and arks, is that they don't provide nearly enough ventilation.

Coop Heating

How much protection do mature hens need from freezing weather?

This question has been researched and debated for over a century. Mature chickens, with their thick coats, have substantial protection from getting chilled. As long as the birds are dry and out of direct wind they can withstand cold relatively well down to very low temperatures. However, in subzero temperatures, tender appendages (including feet and toes, large combs and wattles) are vulnerable to frostbite.

Generally, artificial heat or tightly enclosed quarters are not required in coops that house mature, healthy, well-feathered birds.

However, unfeathered young birds, or birds that are in heavy molt with bare skin showing, need warmer quarters. A cold draft can quickly cause hypothermia (loss of body heat) and kill a baby chick or unfeathered bird.

Add artificial heat or arrange the housing so that the heat given off by feathered birds keeps the unfeathered ones warm. It's not unusual to see a poorly feathered bird tucking herself under the bosom of a fully feathered friend for warmth, even during the day.

The general rule on heating and cold protection is to construct or seasonally manage chicken quarters so that the ambient air temperature will not get low enough to cause frostbite or harm the hens. Sensitivity to cold varies depending on the following factors:

1. The breed of bird. Breeds with large combs and wattles need more protection and warmer houses than smaller, pea-comb and cheek-feathered breeds.

2. Health and vigor of the birds. Birds with poor circulation and low vitality, or poor feathering, will be more sensitive to adverse affects of cold.

3. The amount of exercise and movement. Moving increases circulation and decreases frostbite tendency.

4. Humidity (moisture) levels. Cold, damp housing is harder for birds to cope with than housing in a cold, dry atmosphere. Wet birds can chill quickly by losing valuable body heat and become vulnerable to frostbite.

5. Wind chill factor. Cold wind blowing directly over a flock adds

to the cold and can cause body heat loss, increasing the birds' susceptibility to frostbite.

Severe Weather Protection

Plan your shelter to provide protection from the worst possible weather in your area. Consider extreme temperatures, hurricane winds, or floods, and ask yourself, "What will it take to safely house hens in the worst conditions?" The answer will depend on which climate zone you live in.

In temperate climates, insulating the walls won't be necessary, as most hens can tolerate temperatures down to zero. In northern climates, where nighttime temperatures can drop to 30 below, most hen houses have extra insulation in roosting areas.

High winds can topple coops and turn chicken tractors into kites. Hur-

PHOTO COURTESY OF GARDENER'S SUPPLY COMPANY

Winterized Ark

This micro-flock hut has been modified to keep hens comfortable during a Vermont winter. Foam insulation is installed on both sides, and the screened open front provides ventilation; the chicken-keeper lifts it up to let the hens out to free range. During severe storms and sub-zero temperatures a plexiglass storm window slides between rails on the sides of the ark, and the tarp can be let down at night for extra protection from high winds. The roosting perches are in the top of the hut. They are ventilated so moisture can't condense. Deep mulch covers the bare ground below. With this system, be careful to avoid severe temperature swings inside the coop that will stress the flock.

ricane strapping or methods to tie down pens might have to be part of the housing design.

Hen's don't swim well, if at all. Flooding can drown a flock quickly. If you live in a flood plane, or have seasonally wet areas, arrange hen housing away from flooding or standing water.

Whatever the environmental challenges, your coop should be constructed tightly enough to prevent cold drafts (especially on the roosts), open enough for ample fresh air, and secure enough to keep predators out.

Hen House Flooring

There are 3 flooring options for hen houses: dirt, wood, or concrete. Each has pros and cons.

A *dirt floor* is cheap, easy, and can enrich the soil beneath your coop, but an earthen floor in a non-portable coop can become toxic from too many high-nitrogen manure deposits. Dirt floors become muddy if not well drained. The biggest downside is that predators can dig through the dirt for chicken dinners.

A *wood floor* has the advantage of keeping birds off the ground and thereby dryer. Wood floors provide protection from predators and deter rodents. The coop can have doors or removable sides for cleaning. However, a wood floor (even one made of pressure treated lumber) can eventually rot if not kept dry.

Concrete floors are permanent and easy to clean. Concrete is rodent proof, although rats can burrow under concrete slabs and raise families there. Concrete floors are more expensive and elaborate to construct. Concrete tends to be colder and moister because it wicks moisture up from the ground.

No matter which flooring material you choose, a thick layer of litter, 3 inches or deeper makes the coop easier to clean. In the deep mulch system, with daily management and the right ratio of litter to manure, the birds can stir their poop with the bedding and it will compost underneath them.

Coop Interior Design & Furniture Essentials

There are certain items that are essential parts of hen households:
1. Waterers
2. Feeders
3. Nest boxes
4. Nest box bedding & liners
5. Floor bedding

6. Roosts & perches
7. Pop holes
8. Roost ladders and pop hole ramps
9. Dust box
10. Flat surface storage
11. Lighting
12. Heating
13. Runs, fencing & backyard day ranging
14. Tools and supplies for chicken care

The placement and use of the furniture has important social and functional considerations for the flock's health and well-being. It also affects the ease of tending the flock.

1. Waterers

Chickens need to drink a lot of water. In hot weather, 6 hens can consume

Double Walled Hanging Waterer

Double wall founts can sometimes stick together and be hard to separate when you need to refill them. To get more leverage, use your feet to secure the base while you twist the top to release the catch, then lift up. All hanging founts have handles you can use to suspend the waterer with a chain, but the chain can slide to one side, causing water to spill. Several layers of tape on the handle will keep the chain in one place. Use double-ended snaps to make the waterer easy to remove for refilling or height adjustment.

about a gallon per day. To drink, chickens dip their beaks into water and then lift their heads high so the water to rolls down their throats. They only get a few drops with each dip of their beak, so it can take a long time for a hen to satisfy her thirst, especially if she is dehydrated. Knowing how a hen drinks will help you select and position waterers.

Chickens often drink at the same time in groups. They form a rhythm, dipping their beaks in cadence with each other. Group drinking can encourage a dehydrated, sick, or wounded bird to drink. Bring one of the ailing hen's friends to the waterer and the healthy bird will start to drink, enticing the sick biddy to drink as well. Be mindful of the pecking order if you use this technique; a higher-ranked bird will whack your sick one, ending the rehydration session abruptly. But buddy drinking can encourage an ill hen to sip if she is dehydrated and disinterested in water. This technique seems to boost morale and interest in food as well.

The best choice of waterer depends on many factors, including how it gets refilled, ambient temperatures, and flock size. Water weighs 8.3 pounds per gallon (2.2 pounds per liter). A small, 3 gallon waterer weighs almost 25 pounds. A 14 gallon waterer weighs over 120 pounds. In warm temperatures, backyard waterers can be refilled from a hose. But in freezing temperatures water usually has to be carried by hand.

The problem with the use of large waterers for micro-flocks is that the water gets stale after a few days. Your hen's water is fresh enough if you are willing to drink it. If you are not willing to drink the same water you are giving your chickens, it's not good enough and you are creating health problems for your flock.

Your choice of waterer styles are hanging waterers, ground waterers, and automatic waterers.

Hanging Waterers

Hanging waterers hold 2 to 8 gallons. They hang high enough to keep dirt and litter out of the water. If they are hung correctly, the water is always level and won't spill unless made to swing by frolicking hens.

Hang the waterer as high as the chest of your smallest hen so that she can easily submerge her beak to drink. For bantams, this might be only a few inches off the ground. For heavy breeds, the waterer can be hung much higher. If you keep bantams and heavy breeds together, position a stepping stool so the smaller ones can reach the waterer.

Ground Waterers come in 3 to 14 gallon sizes and are usually made of plastic. Most electric waterer heaters are not approved for use in ground waterers because they will melt the plastic. Most ground waterers have a handled top that unscrews, and they must be somewhat level to operate properly.

Automatic Waterers. When there is no danger of freezing, the easiest way to assure a constant water supply is to use a waterer with a float connected to a water source (such as a garden hose or gravity fed tank). Several kinds of small-scale automatic waterers are available. If used with a garden hose, they might require a pressure reducer. Usually they are used outside the coop.

Trough Waterers

A trough waterer uses a float to maintain a consistent water level. They are usually about 3.5 inches wide and shallow (2 or more inches deep). They come in different lengths, the better ones have brass fittings, and most have

Trough Waterer with Float

Trough waterers are long and shallow so many birds can drink at once. The float (on the right of the trough) is attached to a hose and keeps the water level constant. Trough waterers are usually attached to a heavy board for stability. They are easy to move to different locations to disrupt traffic patterns. The rotating board on top discourages perching.

a rotating reel to prevent birds from perching. These long-bodied waterers need to be attached to a board for stability and elevated when used in bedding or mulch. Some can be hung from a fence. When hens are working in a garden, a trough waterer provides a constant supply of water that doesn't have to be hand carried.

An easy do-it-yourself automatic field waterer can be made from a heavy, sturdy rubber tub with a stock tank float connected to a garden hose. The tub should about 4 inches deep, shallow enough that young pullets can get out and won't drown in it. Do not use this waterer with baby chicks; they could drown. 5 gallon buckets are too deep and not safe to use as waterers because of the drowning factor.

The water float attaches to the side of the tub with screws. If the tub is double-sided like the one below, drill through the plastic to secure the holding screws on the float.

Putting the waterer on a stand makes it easy to level and keeps the water cleaner. Put the float on the downhill side so that it shuts off sooner when the tub is full and won't slowly overflow.

The parts to the DIY automatic waterer are:
- Rubber tub
- Stock tank float
- Elbow connector (preferably metal)
- Water flow regulator, preferably with a "Y" connector

DIY Automatic Waterer

This do-it-yourself automatic waterer provides a continuous water supply. It can be connected to a garden hose or rain water harvest barrel. Putting it on a stand keeps the water clean and the waterer level. The wooden float in the water acts as a raft to rescue honey bees that fall in; they would drown if not for the float.

- Garden hoses – preferably contractor's grade
- Plumber's Teflon tape to seal the connections so they don't drip

The plastic "Y" and elbow connectors are cheaper but tend to break and leak easier. The metal or brass connectors are more expensive but should serve you several seasons. The "Y" connector acts as an extra outlet for water that irrigation can attach to, such as a soaker hose.

Don't let this system freeze. Even though it won't harm the rubber tub, the tank float and connecting parts can crack.

While tending your chickens, quickly rinse out the tub to remove algae, wasted feed, and dirt. It is convenient to keep a scrub brush at the waterer. Periodically moving the waterer around the pen or coop changes the traffic pattern and prevents the ground from being worn bare.

Rain Water Harvesting

Rain water harvesting is easy to set up with a field waterer. Position the waterer so it is downhill and run a hose from the collection container to the waterer. Gravity pressure will keep the waterer full.

The formula to calculate how much rain water you can collect off a roof is as follows:

Collection Area (sq. ft) x Rainfall (in/yr.)/12 (in/ft)
  = Cubic Feet of Water/Year

Cubic Feet/Year x 7.43 (Gallons/Cubic Foot) = Gallons/Year

Rain Water Harvesting

A simple rain water harvesting system like this rain barrel under a downspout can be attached to an automatic waterer. Rainwater harvesting provides non-chlorinated water for your hens. Photo taken at the Gardener's Supply test gardens.

For example, if you have a 500 square foot roof and your area gets 36 inches of rain fall per year the calculation is:

500 square feet of roof x 36 inches of rain/year = 1,800/12 (in/foot).

This gives you 1,500 cubic feet of rain water/year.

Multiply (1,500 cubic feet of rain water) * (7.43 gallons/cubic foot) = 11,145 gallons of water per year.

Winter Water Systems

Micro-flock winter water systems are usually different from warmer weather watering because of the possibility of freezing. Plastic waterers can crack when the water freezes inside them. You will probably have to carry fresh water to your hens in the winter.

The best choice for winter waterers are metal, double wall founts that come in 2, 3, and 5 gallon sizes. These are made of heavy gauge galvanized steel that won't crack when the water freezes inside. For micro-flocks, 2 gallon waterers will provide enough water, assuming it's changed every day or so.

Here's a simple winter water system you can use without electricity. Get 2 metal waterers that you exchange every day. In the morning, fill the first waterer with warm (not hot) water. Hot water will burn

Waterer Heater Base

The photo above shows a hanging waterer on a heated base. The fount could sit on top of the base without support from the chain, but the chain keeps the waterer from sliding off. A grout brush for cleaning the waterer is conveniently hanging on the chain. The photo below shows the underside of a fount heater, where the heat source is located. The heater works best when the waterer sits directly in the middle of the heater.

the tender tongues and throats of thirsty hens. Usually it won't freeze until the hens get a few good drinks.

The second waterer in the coop will have frozen overnight, so bring it inside to thaw for the next day's use. The hens soon learn to drink before the water freezes. Unfortunately, unless you have a base heater, the water in the trough (or lip) of the waterer freezes first, making any water remaining in the container unavailable.

Waterer Heaters

Waterer heaters (also called fount heaters) keep waterers thawed, but require electricity. Most smaller hen houses are not wired, so run an outdoor, heavy-duty, contractor-grade extension cord to your hens' shelter. The electricity powers the electric heated platform to keep water in a metal waterer from freezing. Here are some things to remember about fount heaters:

- Many are not approved for use with plastic waterers.

- They are approved for indoor, or sheltered outdoor use.

- Put the heater on a solid surface or on a stand for stability. This keeps air, bedding, moisture, and rodents out of the heater base and away from the heating element.

- Hang the waterer barely above the heater to level the waterer, keep it in place, and prevent spillages.

- Water exposed to wind is likely to freeze. Position a barrier (like a bale of straw or plywood) to keep wind off the waterer.

- Heaters with covered bottoms are preferable to those with open bottoms.

- Fount heaters are usually 3" high and 16" top around.

- The heat source is in the center of the heater top.

- Many have an automatic thermostat that turns on at 34 degrees

Fahrenheit (just above freezing). and can be effective down to 10 degrees.

• Unplug the base when not in use.

• If you don't have a place to store it, the base can be unplugged and used to keep the waterer elevated in warmer temperatures, but the cord and plug will wear out faster.

Heat lamps hung directly over waterers are used by some poultry keepers. Whether or not this method works depends on how cold the ambient air temperature gets. A heat lamp can cause problems; it can start a fire if it is dropped onto hay or shavings. If hens get scared and flail about, they can slam into it hard. Make sure it can't hit anything and break.

A heat lamp almost caused a fire in one of our large brooders. One of the heat lamps wasn't secure and slid onto the bedding. Luckily, I smelled the smoke and was there to take care of it before a flame broke out. A fire in poultry bedding spreads fast.

The second problem is that heat lamps can overheat the hens. Many hybrid chicken tractor/coops have small living spaces. A heat lamp in a confined area could easily overheat the hens and induce unseasonable molting. In small roost spaces, the birds' body heat will keep them warm on cold nights.

A 60 watt light bulb in an aluminum reflector can work, depending on how cold your climate is. Position the reflector and bulb so the light shines directly on the metal waterer. Use higher wattage bulbs for more heat if needed, but make sure the reflector is rated for higher wattage.

Lighting can keep water from freezing and the extra light can increase egg

Heated Pet Bowl on a Stand

A heated pet bowl can be used as a micro-flock winter waterer. It costs about a third of the price of a fount heater. It is thermostatically controlled to keep the water temperature at 40 degrees and has a waterproof wiring box and chew-proof cord.

production. But don't expose hens to too much or inconsistent light, as this affects their hormones.

Heated pet bowls can be used as chicken waterers. They range from 1 quart to 1.5 gallons. Some have automatic thermostats so the heater turns on only when necessary. The anti-chew spring is a nice touch and protects the supply cord, although I've never seen a chicken chewing on an electric cord.

Other waterer heaters sometimes mentioned in blogs include crock pots, electric woks, and electric skillets. Personally, I would be cautious about using items like these. They don't have temperature control and could burn your hens, and they are a potential fire hazard.

Hanging Metal Feeder

This hanging feeder holds about 12 pounds of feed. To discourage birds from sitting on it, put some kind of barrier on top, like this plastic top from a 5 gallon bucket. Hang the feeder as high as the chest of the smallest bird in your flock.

Wall Hung Baby Pig Feeders

Baby pig creep feeders make great wall feeders.

• Feed in the trough is regulated by a flow adjustment pane.

• The feed-saver lips help prevent birds from raking feed out.

• Trough dividers allow more than one bird to eat at a time.

The plastic feeder on the left holds about 20 pounds of pellets. The metal creep feeder on the right holds oyster shell.

An insulated lunch box (a small, personal-size box filled with warm water) will keep water thawed longer than a non-insulated container. The problem I had with this system is that they tip over easily and should be put in a holder.

My plastic cooler didn't crack when the water froze inside it, but smashing the ice in the cooler would crack it. Having 2 coolers and rotating them so that one is thawing while the other is in use safely solves the ice removal problem. Thrift stores are a good source for insulated boxes.

Thick rubber tubs are flexible even when frozen. When the water freezes you can kick or smash the ice with a hammer. You can actually "kick the bucket" and live to tell it.

Snow is a water source during the winter. Chickens eat snow. If they have access to plenty of clean ground snow, hens will forage for their water. Even so, make sure they get a good drink of fresh water every day to stay hydrated.

In summary, the design and management of your watering system(s) is the most important factor in hen housing. Hens can die quickly without water. Take time to set up and plan how your hens will have year-round access to water in such a way that watering chores are as easy as possible.

2. Feeders

Feeders come in various shapes and sizes and are easy to make. Some poultry keepers don't use feeders at all. The right type of feeder for your hens is determined by what kind of food they eat. There are 4 basic types of feeders: hanging, wall mounted, trough, and field feeders.

Hanging Feeders

Feeders that hang are convenient to use in chicken tractors and arks because they can be hung from a support beam. They are sanitary because they are off the ground, which keeps dirt and bedding out. Hanging feeders can hold 12 to 40 pounds of feed.

Hang the feeders hen-chest high so they can reach the food but not spill it on the floor.

One problem with hanging feeders is that birds tend to perch on them, dirtying feed in the process. You can hang an item (like a plastic lid) above the feeder to discourage perching.

Hanging feeders work best with pelleted and meal feed. Moist mash (which will sour if not removed) sticks like glue to these feeders, making them hard to clean.

Wall Feeders

Wall feeders are great space-savers. They are like hoppers that load from the top. My favorite is a baby pig creep feeder that has a feed-saver lip and a contraption to adjust the feed flow. Pig creep feeders hold about 15 pounds and are large enough to feed my flock of twenty chickens for 3 to 4 days, depending on the season and how much food they get from other sources. Pig creep feeders are available in plastic or metal.

Wall mounted feeders made for poultry are fine, but they tend to be smaller, less sturdy, and usually lack the adjustment contraption, which saves a lot of feed.

Wall feeders are also best for pelleted and dry meal feed. They have to be taken off the wall to be thoroughly cleaned, which is usually only necessary after moist mash is fed.

Trough Feeders

Trough feeders are long and narrow and have a barrier to keep the birds from roosting on them. They are a practical way to feed micro-flocks in a run or yard. Some trough feeders sit on the ground and others attach to a fence. They are portable and easy to clean. For that reason, trough feeders are the best way to feed wet mash, sprouted grains, and chopped vegetables. Moist mash quickly become sour and moldy, which can cause indigestion and diarrhea in chickens. Because troughs usually hold only a single day's supply of food, there shouldn't be any left over to spoil. Trough feeders should be deep enough to hold feed in and narrow enough so it isn't easily tossed or raked out by the searching beaks of picky poultry.

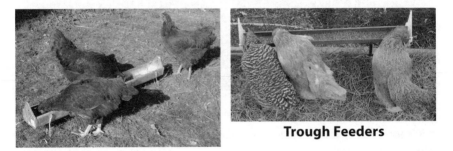

Trough Feeders

Trough feeders are the best choice when feeding moist mash because the feeders are portable and easy to clean. A ground trough feeder (left) allows birds to eat from both sides. A fence-hung trough feeder (right) lets you adjust the height and can be hung on an angle so that both short and tall birds can reach the feed.

It's better to have several small troughs than one big trough. Small ones are easier to carry and clean. My troughs are 3 feet long, and I bring them in to wash them before filling them up with mash. I put the rations in and add water directly to make a mixture in the trough.

Multiple small troughs help hens lower on the pecking order get their fair share of food as the flock shifts from one to another. Troughs are good to use when younger or weaker birds need to get away from the flock to eat in peace. Placing a field trough in an area of the garden helps get poultry workers to focus on where you want them to glean, clean, or fertilize.

Hang troughs on a nail or a hook when not in use. This keeps the long-bodied feeders out of the way, cleaner, and less subject to damage.

Creep Feeding

Creep feeding areas allow smaller and younger birds to eat without being bullied by larger and older members of the flock. This creep feeder is simply a gap in the fence that is too small for the larger birds to pass through. Fenced off areas like this one also let new hens get to know and be accepted by other flock members without getting hurt.

This creep feeding area has potatoes planted inside the run. Only a few birds can get in, so they can't damage the potato foliage, and they have a wonderful time eating the bugs.

Range Feeders

Range feeders are intended for outdoor use with large flocks. These are not the best choice for micro-flocks, because feed left outside overnight in urban areas will become a restaurant for raccoons, opossums, rodents, and stray cats and dogs. During the day, large range feeders become an expensive wild bird feeding station.

Creep Feeders

Creep feeders are feeding stations that allow young or smaller birds to eat separately from adults. This helps assure that they get their fair share and are not bullied by larger birds and forced to stay away from the feeders. Creep feeders are often sectioned off areas with entry ways that only smaller birds can pass through.

Scatter Feeding

Scatter (scratch) feeding can be used to supplement feeder feeding. Some poultry keepers don't use feeders at all; instead, they spread the pellet feed directly on top of deep mulch, forcing the chickens to scratch for their dinners. This encourages the chunkier hens to exercise — a poultry treadmill.

To get the best results from scratch feeding, scatter the pellets or grains on top of clean, dry bedding or mulch that is at least 4 inches thick. The top layers should be coarse enough to hide the feed, but not so heavy that

Floor Nest

Eggs are often laid in floor nests. It's common for hens to share the same nest. There is one wooden egg in this clutch.

Pet Carrier Nest

Hens will use almost any box-like cubby as a nest. In this coop, a pet carrier is a favorite spot.

the birds cannot move the bedding around with their feet to find the feed. If you use hide-and-seek feeding as the only source of feed, make sure the birds are eating enough by checking that they have full crops after they have gone to roost.

3. Nest Boxes

Chickens have been domesticated for thousands of years, yet much of their behavior remains instinctual. The more you know about their original habitat, the better you can understand how and why your hens do certain things. This especially applies to egg laying and nest boxes.

Nest boxes come in an amazing variety of styles, materials, and shapes. Hens don't seem to care if nest boxes hang on the wall or rest on the ground.

My mini-flock has an elegant arrangement of ten nest boxes with plastic nest liners — very upscale, and everything a hen could want in a perfect nest site. They are truly free range and explore their surroundings all day, including a storage barn and a large yard, yet they still choose to lay in the nest boxes. When a liberated hen decides where to lay an egg, she cocks her head, gazes around, utters a 3 syllable staccato sound, and wanders into a semi-dark, dry, protected, quasi-secret space of about 1.5 cubic feet of space.

I've found clutches of eggs behind bushes, in buckets, in corners of the barn, under lumber, and on top of shelves. I have found eggs in places where there should not be eggs. A free range hen will lay in one special nest box for a few days or weeks, then change her preference, randomly selecting a new spot every so often. This is the main reason I don't let them out until mid-morning, after most of the eggs have been laid.

Keep track of the number of eggs laid, and when the count goes down significantly, go looking for the hidden clutch. It's possible to have a non-Easter egg hunt any time of the year. Hens have a distinctive squawk when they are laying eggs, so one way to find wayward nests is to follow the distinctive hen-making-egg sound. It can lead you directly to the secret treasure trove and you will find her settling down. Often there is another hen close by, waiting to lay in the same spot. Several eggs might be already there.

While mowing or cleaning up brush, you might find a really old clutch of eggs. Danger! Old eggs can be explosive. If you pick up an old egg and it has even a hint of a sulfur smell, I suggest you treat it as gently as a live grenade. If possible, toss it to a place where it can explode on impact, but not offend anyone. I've had one explode in my hand, and several showers later I still smelled like rotten eggs. Rotten eggs are similar to skunks on the stink scale.

Hens like to share nests. Wooden eggs serve as an advertising sign to encourage a hen to "lay here". The faux eggs make her think that this nest site has the "hen seal of approval" from her sisters as a desirable spot. All this is an introduction to how hens think about nesting sites.

A hen's criteria for a nesting site include:

• Private and secluded. Hens like to hide when they are laying eggs. They want to be perceived as invisible, and yet still be able to keep an eye on what's going on around them.

• Dry with clean bedding and pest free.

• Large enough to turn around and settle down comfortably, 12"x14" for large breeds, 10"x 12" for small breeds, but not so large that 2 or more hens will crowd in at once and break eggs.

• Darkened so the hen feels protected (this also discourages egg-eating; a hen won't eat what she can't see).

• Accepted by and shared with other hens.

• Easy to access, with a perch in front of the nest for graceful entries and exits.

• Hanging the nest boxes high enough for hens to get under them offers refuge and safety to chickens at the bottom of the pecking order.

Egg management efficiency is greatly enhanced, or hindered, by the way nest boxes are constructed and located. Here are some considerations for the design and placement of nest boxes:

• Eggs should be deposited in the same place to eliminate egg hunt and waste.

• Nests should be easy to access for quick egg collection.

• Install one nest box for every 3 or 4 hens so they won't have to crowd in and break or soil the eggs in the process.

• They should be bright enough for you to see the eggs.

• The mats or bedding should be clean so you won't have to wash eggs.

• Nest boxes should be easy to remove for cleaning. Plastic and metal nest boxes discourage mites and lice from breeding, and are easier to clean and sanitize than wood.

After they lay their eggs, hens don't stay in the nest box long. They have better things to do, unless they are in the mood to brood, in which case they act like boomerangs and return to the nest box like homing pigeons.

External nest boxes are popular with owners of small coops because it's easy to raise the lid to collect eggs. Many micro-flock shelters are too small for humans to enter, so having a raised, easy-on-the-keeper's back, external nest box is convenient.

Large plastic lidded bins can serve as nest boxes. Cut a hen-sized hole in the large side, starting about 3 inches from the bottom. Cut a hole in the chicken pen to match the hole in the bin. The bin can sit 12 to 18" on a platform to get it off the ground. For an external nest box, attach the bin to the outside of the coop. Lift the lid to gather the eggs. Secure the top against the wind and predators, although a strong raccoon could still rip the plastic and enter the coop.

Wooden nest eggs or faux eggs show hens where to nest. They think that if another hen has already chosen a particular nesting spot it must be okay. This ploy usually works. You will often find real eggs clustered around the phoney ones. Sealing the

Nest Box Liner & Wooden Eggs

Plastic nest box liners have finger-like cushions that protect eggs from breaking. The liner is perforated so air can circulate and filter out dirt to keep eggs clean.

Wooden eggs encourage hens to lay in a particular spot. They have flat bottoms, which is useful because you can feel the real eggs from the faux ones at collection time.

wood with polyurethane or varnish helps to keep them from staining and makes cleaning easier.

You can use stone eggs, but they don't seem to be as believable, probably because stone is heavier than real eggs and colder, especially in the winter. Faux eggs can also help break the egg eating habit.

4. Nest Box Bedding & Liners

Hens like to have some sort of bedding where they lay eggs. Bedding keeps eggs from breaking and keeps them clean. When they make their own nests, hens naturally collect feathers, grass, or hay, or other soft things to tuck around themselves.

Plastic nest box liners, sometimes called plastic nest pads, are my favorite bedding because they are not expensive and last for years. They have flagella-like fingers to buffer eggs against cracking and a slotted, perforated base to circulate air around droppings and help them dry. Dried droppings fall through the liner; eggs stay clean and bacteria can't grow on wet nest material. The plastic is not attractive to mites, but if mites are a problem, the liners can be soaked for thorough cleaning.

To clean nests, pull the plastic liner out and whack it against a roost to dislodge dirt or dried manure. Then put it back.

Wood shavings tend to get raked out of nests. Unless a thick layer is applied, shavings don't provide a soft landing for eggs. Shavings are absorbent, but will clump and are messy to clean out. Mites can breed in wood shavings. Even though cedar shavings are insect repellent, the off-gassing is toxic to poultry.

Hay or straw will stay in nest boxes, but can be dusty and hard to clean. Straw has to be changed often, as fecal matter gets matted in it and doesn't dry quickly. Hens usually arrange all of the straw around themselves – not under – and lay eggs in a bare-bottom nest without egg impact protection. Mites and other insects can be a problem in hay or straw.

I've heard of other bedding material, like flax screenings, newspaper, and cardboard liners, but for easy, quick maintenance and egg protection, plastic nest box liners win the nest bedding contest.

5. Floor Bedding

Some type of absorbent material should be on the floor of all hen houses to mix with droppings. This keeps odors and the messiness of the chicken manure to a minimum. What's nifty is that the resulting high-carbon bedding is great for composting. Bedding makes the hen house easier to clean using a shovel or pitchfork.

Compostable bedding material includes wood chips, straw, dried grass clippings, and dried leaves. Shredded paper (mixed with wood shavings or straw) can also be used, especially under the droppings platform. This is a good way to dispose of old documents you have shredded.

Sawdust is too fine for bedding, and chickens will eat it. It has no nutritional value and can cause compacted crops. Don't use sawdust, except, perhaps, in the droppings pit, where hens don't have access. Never use sawdust from pressure treated lumber.

Pine and cedar shavings are widely used as poultry bedding because they are inexpensive, abundant, and convenient to handle as they come in compressed bags. Most chicken keepers are unaware of toxins in pine and cedar bedding. The aromatic compounds that give the shavings a pleasing aroma to humans can be irritating and caustic to lungs and elevate liver enzymes when inhaled over long periods. The odors irritate nasal passages, throat, and lungs, making way for respiratory tract infections. This is significant, because the most common poultry diseases are respiratory infections. Small animal laboratories don't use pine and cedar shavings because of the toxic effects .

Many poultry people say they have used pine and cedar shavings for years without any problems. Studies comparing different litters used for broiler production found that chickens kept on softwood litter had a higher incidence of respiratory infection.

I prefer taking the middle road by mixing a small amount of wood shavings, usually aspen, mixed with straw, leaves, grass clippings, or shredded paper, as bedding on my hen house floor. The wood shavings give a fluffy texture to the other organic matter.

Since I've switched to aspen bedding, very few of my baby chicks or chickens have had respiratory problems.

Cedar bedding is dangerous to chickens of any age. The phytochemicals in cedar that are poisonous to

Don't Use Cedar Shavings

The same aromatic compounds that give cedar and pine wood shavings their pleasant aroma and insect repellent qualities can cause respiratory problems in chicks and chickens. Aspen wood shavings used for bedding are considered safe wood-based bedding for poultry.

insects can be harmful to chickens. It's not recommended to use cedar bedding in brooders, nest boxes, or coops.

Cocoa-based products made from cacao bean shells are to be avoided completely in areas where chickens might go, both in your garden and in hen houses. It has a strong chocolate smell, like the factories it comes from. Cocoa mulch and related products contain high levels of theobromine and caffeine. Both are highly toxic and even lethal to dogs, cats, and probably chickens.

Ammonia odor indicates that it's past time to clean, move, or add a lot

Training Chickens Where to Roost

Chickens naturally return home to roost at night; they don't have to be herded in, as some people believe. To teach them where to roost, keep them restricted to their coop for a few days to let them bond with their shelter. Give them access outside just before dusk. They won't wander far and will put themselves to bed in the perches. Each day let them out a little earlier. After about a week they can be outside most of the day and still return to the coop to roost. You might have to teach them how to navigate the pop hole door.

more bedding to the coop. A buildup of ammonia is toxic and you don't want your hens breathing it. Neither do you want your neighbors complaining about the smell or the flying insects that often accompany it.

Wet areas on floors, over long periods of time, promote rot and might need the application of a drying agent. Commercial factory farms sometimes add lime to poultry litter for this purpose. Lime is very basic, and ammonia producing-organisms thrive in neutral or highly basic conditions, so adding lime increases the generation of ammonia from litter and floor soil. Lime is caustic and comes as a fine powder. When scratched up by the hens, it can cause respiratory problems, irritate their eyes, and burn their feet. The use of lime in small chicken coops is not recommended.

The same is true of using lime in garden soil to raise the pH. Mix it in with a rake or hoe instead of recruiting your chickens for the task.

Absorbent clays are an alternative to lime. They hold 5 to 10 times more water than lime. Clay-based kitty litter (often with ammonia-reducing additives) is one type of absorbent clay. Some litters are made by mixing clay with a cellulose material like sawdust, rice hulls, or wheat dust, and then pelletizing and packaging the mixture.

If excessive moisture is a chronic problem in your coop, increase the ventilation by opening or installing windows, roof vents, or adding a fan.

When mixed with bedding, sand keeps smells down and keeps the bedding dry. It is also great for compost.

6. Roosts and Perches

Hens prefer to roost off the ground – the higher the better. The roosting perch is the hens' version of a bed. Some hens prefer to sleep apart from others, especially when it's hot. Others cuddle and snuggle, often putting their heads under the wing of another. Depending on the weather, a flock's volume expands or contracts to self-regulate body temperature.

Roosts can be made from lumber or tree branches. Perches should not be too wide or too narrow; in either case, birds can't get a good grip and will be off-balance and uncomfortable. An ideal perch is about 2" wide for larger birds and 1" wide for smaller breeds. A 2" x 4" cut in half makes solid roosts and ladders. If you don't have a saw, a lumber yard will cut it for you, usually for free.

Rounded roost edges are preferable to square corners, because they are more comfortable and easier to grip. A wood router will round the corners on a 2" x 2" piece of wood. Hens can get splinters from rough wood, so if you notice a hen is gimpy, check her feet for splinter or thorns.

Don't use metal or plastic for roosts. Both are too slick for hens to grip, and in cold temperatures hens' feet can freeze to metal and cause frostbite. Bamboo perches are round, but they are slick and difficult to hang on to. When my hens had both bamboo and wooden roosts, they always chose the wooden ones.

Perch Height and Arrangement

Perch height and arrangement are important to chickens, both socially and physically. If you have more than one perch, make them all the same height if you can. Chickens prefer the highest available perch, and there will be some jostling for the prime positions. I've seen hens peck and push others off the perch for the best seats. Having a variety of perch heights can cause unnecessary and potentially dangerous feuds. Even in micro-flocks, the competition can be intense.

If your space is limited, a stair-step roost arrangement can work, but beware of squabbles. Separate perches vertically and horizontally by 12".

Don't put perches directly under each other because the hen above will poop on the hen below.

Some poultry keepers let their hens roost in rafters, with ladders to help the birds get up and down. In most coops, installing perches 2' off

Pop Hole with Ramp

A pop hole is a chicken version of the front door. A ramp with cleats helps the birds come and go gracefully.

There is an automatic door keeper on the inside of the coop, so there isn't an exterior door to close.

There is a photo of the interior automatic door keeper earlier in this chapter, in the section on coop arrangement.

the ground is sufficient. Heavy breeds and older birds might have trouble getting up and down gracefully. If roosts are too high, the landing can be hard on their legs and feet. Some of my hens are not athletically inclined and plop from roosts with the grace and velocity of a winged concrete block. A ladder gives them a welcome leg up (and down).

Putting perches higher than nest boxes – preferably on different walls – discourages nest sleeping. Otherwise, hens tend to snooze in nests and get the bedding dirty. About half of a hen's manure is released after dark.

If perches are too close to a wall, the hens can bend and break their tail feathers while they roost. Installing the perches about 6" to 12" away from the wall offers sufficient rear-end space.

Some roosts are staggered, with rungs arranged like ladders, about 8" to 10" apart. They should be close enough so hens won't fall through while roost hopping.

7. Pop Holes

Pop holes are small, chicken-size doors that let hens in and out of the coop. They range in size from 9" to 13" wide and 13" to 20" high. Pop holes should close tightly every night shortly after sunset, when all the ladies have retired to their roosts.

If the pop hole is more than a few inches off the ground, add a ramp the width of the door with cleats every 4" to 6" to help the birds get up to it.

Acting as a door keeper is part of a keeper's daily routine; it must be opened every morning and closed shortly after dusk. Hens are night blind and totally defenseless against nighttime killers.

The novelty of being doorkeeper will wear off quickly, especially during the winter months when days are short and it's cold, dark, and icy out.

Automatic Door Keepers. An automatic pop hole operator with a timer is a wonderful daily management item. It can be powered by batteries or hooked up to a solar panel. You can schedule the door to open and close anytime. When day length changes, it's easy to reprogram the timer accordingly.

The automatic door opener might seem expensive, but the convenience and peace of mind it gives more than offset the expense. There were times I'd get home late and find a predator had killed one, or all, of my hens.

One night, around 10 o'clock, a long-bodied critter that looked like a ferret got through a 2" gap in the coop wall. The slinky, masked murderer massacred all but one of my hens. I found her cowering in a corner of the coop. She had a severe case of post traumatic predator syndrome (PTPS). It was over a week before she would go back to the coop to roost.

Losing hens is heartbreaking, and finding replacements isn't easy. I lost my feathered friends and my free-flowing egg supply. Installing the automatic door keeper stopped the nighttime chicken killers, so the investment was more than worth it to me.

The door keeper motor unit is a small box that can be installed directly over the pop hole, either inside or outside the coop. The motor has a sensor that knows the door is down when the weight on the line slacks. Pulleys can be installed if your coop doesn't have enough room directly above the pop door for the motor. The door moves once inch per second, so there isn't any danger of decapitating or squashing a hen. My hens know the motor sound and when they hear it, they position themselves to race outside, many of them sliding under the door, fluttering and clucking.

The door keeper has a built-in photocell light sensor that activates the motor to open or close the door according to the ambient light.

The electronic door keeper lets me control when the hens go outside. I prefer to keep them inside and close to their nest boxes until most of them have laid their eggs, usually 2 to 4 hours after sunrise.

The electronic door keeper I bought is distributed in North America by Foy's Pigeon Supply; www.foyspigeonsupplies.com.

8. Ramps to Pop Holes & Ladders to Roosts

To make your coop accessible to hens of all ages, install a ramp with cleats leading up to the pop hole door. Slope the ramp at a 30 degree angle (or less) and make the ramp at least as wide as the pop hole.

Ramps are great to help hens reach pop holes, but ladders are better for access to the roosts. The open space between rungs allows manure to drop through, and the ladder can be used as a roost for hens not able to get higher.

9. Dust Box (Chicken Spa)

A dust bath is as important to hen health as a water bath is to humans. Like humans, hens enjoy bathing and seem to make it a ritual. They often bathe in groups of 2 or more and make low, satisfied groaning and chortling sounds.

Hens free ranging will make their own dust boxes in the best dirt in the garden, or in any dirt available. If that isn't possible, dust boxes are easy to make. They are like chicken-sized sand boxes and should be large and deep enough to hold sand in while the chickens roll around.

By wallowing in dusting material, a hen removes old scales and dirt from her skin. The feathers benefit from the cleaning as well. If a hen has lice, dust baths are her way of self-medicating and ridding herself of the pesky insects.

Dust bath material should be light, fine, and dry. The chickens prefer tiny matter that gets next to the skin and cleans with sandpaper abrasion. Sandy loam mixed with diatomaceous earth (DE) is a good combination. They also seem to seek out potting soil which, combined with equal parts of sand and DE or wood ashes, makes a high quality hen soap.

A dust box can be used to treat lice infestations. Add one or more of the following miticides to the dust box matter.

- Diatomaceous earth
- Rotenone powder
- Ash or coal cinders
- Dried leaves of bamboo, white cedar or neem.
- Powdered herbs of pennyroyal, pyrethum, and worm wood.

Place the dust box in the sunlight. Keep it clean and free of litter and other foreign material. Your hens will have a 4-star, luxury bathing experience, the likes of which factory farm hens can only dream of.

Chickens often combine dust bathing with sun bathing and stretching. There are several distinct chicken stretching movements that are remarkably yoga-like. The most prevalent is a poultry version of lying-down triangle pose (Utthita Gallusasana) where the hen stretches her chunky thighs backward and points her toes. She raises her top wing and stretches backward toward her feet, elongating her spine and extending her neck, head and gaze are upward.

The chicken warrior pose (Gallus Virabhadrasana II) is often done immediately after a hardy shake from the dust bath. The hen steps boldly forward with one foot, while stretching the hind foot and one wing backward (almost horizontal to the ground). She simultaneously extends her neck and thrusts her head high and proud, her intense, laser-focus gaze is at the wing-tips. At the peak of this pose, hens usually

Chicken Warrior Pose
Gallus Virabhadrasana II

utter a single-syllable grunt, ("urrh") and roosters are often unable to hold back a forceful challenger crow.

10. Flat Surfaces in a Coop

Shelves in a coop are handy. I have 2 in my chicken house to store bedding, seasonal waterers, feed supplements, and brooding items. But hens view flat, horizontal surfaces as places to roost. This translates to poop being where poop shouldn't be. If you have flat places in your coop use lattice, chicken wire, or other objects to prevent hens from flying up and plopping themselves down to poop.

11. Lighting

Commercial poultry production uses artificial lighting year round to stimulate egg production. Using artificial lighting is an old and common practice among poultry owners. It changes the hens' circulating hormones. Different hormones increase and decrease depending on light exposure, which corresponds to day length.

During the shorter days of fall and winter I use full-spectrum 40 or 60 watt bulb in the coop in the evenings and early mornings. This gives the hens about 12 hours of combined natural and artificial light per day and helps with egg production. The light also seems to keep them healthier. Research is increasingly showing that sunlight is as important to animal health as vitamins; vitamin S=sunlight. Hens kept without artificial sunlight keep laying all winter, although production is about half to one third what it is in the summer.

In the summer, the light timer is set to come on just before twilight and only stay on for about 1/2 hour after dark. This entices the birds into the coop keeping it light enough for the birds to be able to see and get their positions on the roosts. My coop is dark with both doors closed and there have been times when 1 or 2 birds get locked out as the automatic door closed before they got in.

In August or early fall some birds might go into neck molt, losing some feathers around the neck and back as the days get shorter. Adding light eases the transition. If you decide to provide extra light, here are a few suggestions:

• Don't use candles or an open flame. The dust and bedding are very fire-prone.

• For maximum egg production, most producers provide photoperiods (light timing) such that lights are on 14 to 16 hours a day. A light period of 12 hours is more natural.

• Using artificial light in the mornings, rather than the evenings, gets hens in the coop and bedding down with the natural evening tide.

• Lighting can be a combination of daylight and artificial light and should be continuous (one block of lights on, one block of lights off).

• Decreasing or changing the photoperiod can decrease egg production and induce molting (loss of feathers).

• Increasing the photoperiod for growing pullets can force the young developing hens to lay earlier than they are ready and result in pullet eggs (undersized eggs).

• A 40 watt, full-spectrum bulb with a reflector positioned about 7' above the floor will provide enough light for about 200 square feet.

• Multiple lights distribute light evenly throughout the hen house.

• Make sure your lights are properly installed and maintained to avoid fire hazard. Mice can chew on electrical cords, so periodically check them for damage.

12. *Heating*

Except in the severest of cold temperatures (in the single digits and below zero), healthy, fully feathered hens won't need extra heating. Hens cuddle together and share body heat.

In extremely cold weather, positioning roosts out of the wind and supplying a heat source to maintain the temperature between 0 and 30 degrees should be adequate. Routinely check for frostbite and remedy the bird, and the coop, accordingly. Select winter-hardy breeds with pea combs and feathered wattles.

13. Runs, Fencing & Backyard Day Ranging

A run is a fenced area attached to the hen's living quarters where chickens can walk around, stretch wings, and enjoy the sunlight in safety. The run space is usually very small, but it keeps birds happier and healthier than being inside their entire lives.

Backyard day ranging involves letting hens access fenced areas beyond the run. Permanent, temporary, and portable fencing allows you to control hens so they can experience life "outside the box" – the box being their coop, chicken tractor, or arc. Day range fencing allows hens to "express their chicken-ness" as Joel Salatin would say.

Every city code states that chickens must be fenced in and not be able wander freely. "Free-to-leave" is not what backyard day ranging is; hens are still restricted to an area within a yard or garden.

It doesn't take much to contain a small flock of hens. A 3' to 4' high fence will keep the birds in, especially if there are no perching opportunities on the fence posts. Clipping wings can help prevent hens from flying over a fence. A perimeter fence around the whole yard, as one would have for a dog, doubles the security. Backyard day range fencing systems are discussed in detail in the chapter on gardening.

14. Tools and Supplies for Chicken Care.

There are a few simple items that are handy to keep in, or around hen housing. These items include:
- Pitchfork and shovel to move bedding and mulch
- Trowel to scrape perches and get bedding out of tight spaces
- Arm extender to collect wayward eggs
- Chickens' Tool Box
  Scissors to trim wings, cut twine, and open things, like feed bags
  Pliers (2 pair to tighten hoses)
  Screw driver
  Hammer
  Hose washers (extra for field waterers and garden hoses)
  Plumber's Teflon tape to prevent leaky hose connections
  Wire cutters for fencing
  Flashlight for after-hours checking

Coop Building Materials and Green Design

Most urban coops and pens are small and don't require much lumber to construct. This makes them ideal candidates for reclaimed materials. If recycling painted wood, don't use wood with lead paint. Chickens peck at anything, including old paint flakes that drop onto the ground around the coop. Use durable materials and construction methods, especially if you plan to move the pen. Screws make construction tighter than nails and allow the coop to be taken apart easily if modifying, relocating, or rebuilding is necessary.

Daily Hen Care

When your system is set up, the daily routine of caring for chickens only takes a few minutes. Below are some tips about how chickens think and behave, and how to interface your day with your chickens' day to create an easy and fulfilling lifestyle for both parties.

1. Make sure the hens have access to plenty of clean water. Automatic waterers cut down on chore time and give peace of mind.

2. In freezing temperatures, hand carry fresh, warm water to the chickens each day. In severely cold climates, protect your hands with waterproof gloves because wet hands can freeze to frozen metal handles. Check the hens for frostbite on combs and wattles and treat accordingly.

3. Provide enough of the right kinds of feed and supplements for the season.

4. Move the pen to fresh ground or manage manure so it is covered with some type of organic matter to keep flies off. Clean the coop when necessary to provide proper sanitation.

5. Collect eggs. Most eggs are laid by late morning, so if you let your hens free range, letting them out mid-morning or later encourages them to lay eggs in the nesting boxes. Collecting eggs daily keeps them cleaner and minimizes breakage. In the winter, collect eggs before they freeze.

6. Keep enough feed and supplements on hand so you don't run out.

7. Keep a minimum of 3 hens. Chickens are social animals and need companionship. A hen alone will be an unhappy hen.

That's it. Daily care is simple and fast once your systems are in place.

Vacation Care

We all have occasion to get away for a few days or longer. In our neighborhood, we trade critter sitting. My neighbors, and their kids, are delighted to look after my flock while I'm away. Here are a few tips to help hen sitters:

- Have your stand-in keepers come over before you leave. Show them where things are stored and walk through chores with them.

- Write down the chore checklist and notes on the daily routine. Post in the hen house.

- Leave a list of flock members. For example:

> 5 All black (Australorps)
> 2 Black and white (1 Barred Rock & 1 Silver Laced Wyandotte)
> 9 Reddish (Rhode Island Reds and Buckeyes)
> 4 Blond (2 Buff Orphingtons and 2 Araucanas)
> 3 White (Leghorns)
> 23 birds total

- Label the feed containers: layer feed, scratch, grit, etc.

- If you use timers for lights or automatic pop doors, show your hen sitters how they're set and what to do if they don't work or have to be reset because of a power outage.

- Leave instructions about what to do in an emergency.

The most important vacation tasks are filling waterers and feeders, collecting eggs, and making sure the flock is safely locked in for the night.

6 Feeds and Feeding Micro-flocks

What your hens eat directly affects their health, quality of life, and the quality of their eggs. Since you will be eating their eggs, your hen's diet is important because what is in the hen's feed will pass one step up the food chain to you.

Whether you mix your own or buy commercially bagged feed, the best advice is not to buy the cheapest feed available. Instead, find and buy the freshest, cleanest, and most nutritious feed. You will pay more initially, but you will save in the long term by having healthier, more productive hens and more nutritious eggs for yourself.

Chickens don't have an omnivore's dilemma when it comes to eating. Although they have carnivorous preferences, they eat almost anything: grubs, grass clippings, worms, pizza, leftovers, you name it and they will probably eat it. They can be as competitive as Olympic athletes and dive head first or jump 5 times their height to get their beaks around that delicious something before any of the others beat them to it. They will snatch food from each other's beaks and run away chirping – the chicken equivalent of interception to touchdown. And this is when they are not even hungry.

Animals are fed fodder or animal feed. This consists of plants cut and carried to animals, as opposed to that which they forage on their own. In agricultural vernacular, fodder includes kitchen scraps, pelleted feeds, oils and mixed rations, sprouted grains and legumes.

How Chickens Eat

Chickens have a distinctive way of eating. They don't eat like dogs or cats, putting their heads down and snarfing food. Chickens move their heads back and up so they can focus on potential food bits with both eyes. Once the position of the grain or grub registers in the hen's brain, she will stab at it quickly and repeatedly, seemingly smashing her face on the ground. After wiping her feet (usually one swipe for each foot), the chicken will raise her head, get a visual reference, and strike again. If the food bits are clustered together and the bird is sure of her target, she will keep her head close to the source, raising it only an inch or so before taking the next stab. About every 3rd to 5th bite, she will scratch the ground again, even when eating at a feeder. Occasionally she will daintily tidy herself by wiping her beak, using the ground as a napkin.

The way chickens eat and their preferences for particular food sizes affect what they eat and how you feed them. Given a choice, they would rather eat large grains than fine crumble. Chickens have a pouch called a "crop" that holds and softens food before passing it to the stomach. Perhaps larger pieces of food fill their crops faster and create a feeling of satiety, or maybe the larger pieces are just easier to see.

A chicken's preference for a certain food size depends on its size and age, and usually varies among breeds. Perhaps the most desirable food size is determined by the chicken's past experience, sense of hunger, and crop size.

The preference for larger grains and food bits might explain why chickens tend to sift through feed and push it out of their feeders. They pick through the feed as if in search of something special. It's harder for them to push pelleted feed aside, but they still slop a significant amount out of the feeder. There are two problems with chickens pushing feed out of feeders:

1. It is wasteful. The feed becomes mixed in with the bedding and is on its way to compost before passing through the chicken's digestive tract.

2. If feeding meal or crumble, many of the nutrients are in the finer, almost dust-like portion of the ration. The hens won't be getting the value of this nutrition if they don't eat it.

The are four ways to decrease the amount of feed hens scrape out of feeders.

1. Use a feeder that has a ledge or lip in front. This helps catch the

grain, keeping bits in the feeder instead of on the floor (Chapter 5 on Housing, Furniture & Interior Design).

2. Raise the feeder so the birds can't get as much leverage to use their beaks like a shovel. Position the feeder so it is just a smidgen above chest high. If you have younger birds mixed in with adults, put in a block (the chick version of a step stool) so they can reach the feeder. To assure that smaller birds get their ration, create a separate feeding area that only they can squeeze or duck into. (See creep feeders).

3. Put a litter-free area under the feeder. Try a 12" x 12" patio block on top of the litter. This helps keep feed out of the bedding and available for the chickens to see and clean up. Clean the block often.

4. Put a catch pan below the feeder. Putting the pushed-out feed back in the feeder won't increase its chances of being eaten; the chickens will simply push it out again. The finer feed contains nutrients such as probiotics and vitamins, so it's valuable and you want the hens to consume it. Add water or milk to the rejected food in the catch pan to make a dough-like mash and the chickens will gobble it down greedily.

5. Scatter feed. Some poultry keepers don't use a feeder at all; instead, they toss feed on the ground. This method of feeding is unsanitary if the ground is muddy or contaminated with manure.

How Much Does a Chicken Eat?

It's hard to know exactly how much a chicken will eat. Many factors affect a bird's appetite: its size, health status, laying rate, environment, and ambient temperature are a few examples. A general rule of thumb for consumption rates are:

• Day old chick (bantam): 0.1 ounce (3 grams) per day; quantity increases rapidly as the chicken grows

• Day old chick (standard): .17 ounce (5 grams) per day

• Bantam (adult): 1.7 to 2.4 ounces (50 to 70 grams) per day; 3 to 4.5 pounds per month

• Layer (hybrid or heritage) 4 to 4.8 ounces or 120 to 140 grams per day; 7.5 to 9 pounds per month

Generally, it takes 100 pounds of feed to raise a day-old chick to her first laying cycle, about 50 to 60 weeks from birth. She will begin laying eggs when she is 5 to 6 months old and will keep laying until she is about 1 year old and goes into molt to grow new feathers. This is considered the first laying cycle.

This formula gives a guideline for how much a point-of-lay hen is worth. One method is to double the cost of feed to account for it, labor, and overhead, plus extra costs like shipping. The price of hens has not kept up with inflation and producers tend to undervalue their chickens.

One full-size production hen will eat about 110 pounds of feed per year, including foraging. This is about 9 pounds of feed per month/hen. If you have 3 hens, you can budget for about 50 pounds of feed every two months. Feed consumption will vary depending on the chicken's size, age, access to kitchen and yard biorecycling, amount of exercise, and free range access.

The Healthy Hen Diet

The nutritional requirements of hens include the categories of carbohydrates, fats and oils, proteins, supplements (including vitamins, minerals and other co-factors), and roughage.

There is a wide spectrum of nutritional quality, and biochemical differences, in these categories that impact the chicken's health and functionality in the short- and long-term. For example, all "protein" is not created equal. Neither are all "fats" and "oils" equal in value or health benefits. It is essential that the nutrients you feed your flock are of uncontaminated, high quality.

Just like there is "junk food" for people, there is even worse "junk feed" for animals. Feed and nutrient quality are discussed more later in this chapter.

It is very important not to feed a grain-only diet. Grains and cereals only contain about 10% protein which is not an adequate amount for chicken growth or maintenance. A grain-only diet doesn't offer the right balance of vitamins, minerals, oils, salt, probiotics, and other nutrients necessary for optimal chicken health and egg laying.

Growing birds and layers require 15 to 18 percent quality protein in their diet. Baby chicks require about 20 percent protein to support their rapid growth.

Protein is based on 11 essential amino acids and a score of other non-essential amino acids. Remember that all protein is not created equal? Here is an example of why: chicken feathers are considered "protein" in cow feed. Yes, you read that right. Chicken feathers are used as protein in commercial cattle feed. I've never seen a cow chasing a chicken to munch on its feathers.

Feed bags usually list the percentage of "crude protein" in the feed. Crude protein is an estimation of the feed's protein content. Protein contains 16% nitrogen, so the crude protein percentage is calculated by determining the amount of nitrogen in the feed and multiplying that number by 6.5. The crude protein percentage includes both true protein and non-protein nitrogen, so it doesn't give any information about the quality or bioavailability of the protein content. Even more worrisome is that there is no way to know where the "crude protein" comes from. Revolting ingredients like chicken feathers put the "crude" back into "crude protein".

Sources of Commercial Animal Protein

Protein used in animal feed comes from a variety of sources. It ranges from meats approved for human consumption to the slaughterhouse "by-products" and rendered protein. Meat is about 18 to 34% protein, depending on the species and animal part.

- Meat and bone meal can contain added antioxidants but they can also have contaminants, depending on how the animal was raised and fed.

- Fish meal is used as a protein source, although it may contain heavy metals.

- Milk powder is another common source.

Plant Based Protein Sources

If you are concerned about contamination of animal-source meats, the alternative is to feed protein from plant sources, although plant protein has its own set of problems and controversy. Plant protein sources include:

- Soy beans: 38% protein. There is growing controversy about feeding soy beans. The Weston A. Price Foundation (www.westonaprice.org) has more information about the problems with soy.

• Comfrey: 15 to 30% protein. Comfrey is a high-protein plant that can supplement feed either in dried form or as forage in comfrey cages. See chapter 3, Growing Food for Chickens for more information about this remarkable, easy-to-grow, perennial crop.

Feed Presentation: Pellets, Crumble, Meal, and Mash

It's important to match the age of a bird with the right ration, food bit size, and type of feed. Younger, growing birds require about 20% quality protein and smaller bit sizes. Older birds, including layers, need about 16% protein and usually take larger mouthfuls. Some forms of food will serve you and the birds better than others. The four types are described below.

Crumble is finely milled and tiny pieces that are easily eaten by even the smallest of baby chicks. Sometimes crumbles are pellets broken into smaller pieces. Usually chicks are fed crumble for the first 6 weeks or more.

Pellets are small kernels of compressed mash. The advantage to pellets is that the birds eat the whole blend and can't pick and choose. Wayward pellets pushed out of feeders are easier for a chicken to see and eat than are smaller bits or powders.

Meal has a variety of textures, from cracked grains to dust. It usually has the nutritional supplements, such as probiotics and vitamins, added in powder form. Because of its fine texture, meal can go rancid quickly, and should be mixed and ground in small batches that can be consumed quickly.

The problem with dry meal is that the chickens tend to waste a lot by using their beaks to sort through the mixture and find the best bit. They end up spilling the finer feed on the floor. Besides expensive wastage, the birds might not get balanced nutrition from the feed.

Meal is dry and dusty and takes longer for a hen to eat because she must stop and drink. The frequent drinking with meal-coated beaks quickly contaminates the water. The food settles on the bottom of the waterer and ferments. Especially when feeding meal, waterers must be cleaned on a daily basis.

Mash is a mixture of ground grains and nutrients. Liquid is added to it just before feeding so it becomes pulpy and has a dough-like consistency; it is basically moist meal. Hens can eat mash faster than meal. They seem to relish it more and waste less, although their beaks get covered with it and the water gets contaminated when they drink.

Sometimes the chickens' calorie consumption needs are elevated, like during the winter, when they need to eat more to stay warm. To increase calories, add grain (wheat, oats, and coarse ground corn are the most common) to the feed; you can feed up to 20% of the total ration in grain. Other grains (if available) can also be used, such as rye, barley, rice, sorghum, quinoa, and amaranth. Chickens will eat almost any grain or seed.

Commercial Feed

Commercial feeds have been formulated as complete rations and are sold in 50 pound bags. Commercial rations contain synthetic vitamins, minerals, and preservatives that oxidize with age. If commercial feeds are not fresh, the birds may not get the nutrients they need.

Commercial feed bags do not usually show the date of milling. If there is a date, it might be coded and hard to interpret. None of the bags I've seen have "use before" or expiration dates printed on them. The only way to estimate feed freshness is to taste it and use your best judgment. If you don't want to do that, sniff for rancidity, look for mold, or see if the hens will eat it, especially if they are given a choice between the feed in question and something they always like. I have had commercial feed that the hens wouldn't eat unless (in their estimation) they were about to drop dead of starvation. They made it clear the feed was bad and fussed at me to give them something else.

Organic Commercial Feeds

Organic feed is more expensive than non-organic feed by about 50%, but organic grains supposedly have greater nutritional value and fewer toxic tag-alongs. The producers of organic grains certify that their plants have not been exposed to toxic elements such as herbicides, insecticides, or caustic fertilizers that could leach into the grain. There is more microbial life in the soil where grains are organically grown. The synergistic relationship between soil life and soil chemistry enables plants to uptake and process soil nutrients more efficiently so they are more available up the food chain. Organic grains are not grown from genetically modified (GM) seeds. All these conditions make organic grains more valuable. As the old adage says, you get what you pay for.

Besides containing organically grown grain, certified organic rations often include quality vitamins and minerals such as kelp, probiotics, aragonite for calcium, alfalfa for greens, and flaxseed.

No matter what kind of chicken feed you choose, it is essential to provide enough and let the hens have easy access to it. Not enough feed will cause a drop in egg production, and feed that doesn't have the right vitamins, minerals, and oils can cause disease and, ultimately, death.

Antibiotics in Feed

Many commercial feeds include sub-therapeutic doses of antibiotics. Don't buy these medicated feeds for chickens. It's possible that antibiotic metabolites can pass through the hen and into the eggs she lays.

Antibiotics have been added to animal feeds for over 50 years. Most of the antibiotics fed to livestock are identical or related to antibiotics given to humans. These broad-spectrum "shotgun approach" antibiotics include penicillins, tetracyclines, cephalosporins, fluoroquinolones, avoparcin (a vancomycin 1st cousin) and many main–stream others.

The sub-therapeutic use of antibiotics is a cover–up for the poor hygiene and lousy management practices of most factory farms. Animals in these farms are often stressed and crowded in environmentally toxic housing. Antimicrobial agents are put in the food or water in sub-therapeutic doses to prevent infectious diseases. Some claim that antibiotics act as a growth enhancer by increasing feed efficiency and weight gain. I find this hard to believe. Antibiotics are actually keeping some of the birds from becoming classified as 4-D – dead, dying, diseased, or down – until they make it to slaughter so the total weight gain of the entire flock is increased. Thus, the U.S. industry standard is to treat the entire flock, rather than individual sick birds, so that environmentally stressed individuals don't become "inedible".

Here's the problem: any stressed animal is likely to get sick because stress suppresses immune systems. In the long run, this approach results in entire flocks of stressed birds with compromised immune systems. Sub-therapeutic doses of antibiotics in feed greatly contribute to our contaminated food supply system and have enormous unintended consequences for public health.

The Union of Concerned Scientists is a leading science-based nonprofit group working for a healthy environment and a safer world. They estimate that, each year, about 24.6 million pounds of antibiotics are fed to animals for non-therapeutic reasons. That's 14.3 million tons. Humans are prescribed about 3 million pounds (1.5 million tons) for therapy. So, 70 percent of antibiotics and related drugs used in the United States are used in animals. That's a staggering percentage. Antibiotic-resistant illness causes tens of thousands of early deaths each year and drives up medical costs for everyone. Do you hear the drug companies' cash registers ringing?

Antibiotics are not innocent; they can cause more long-term harm than good. Not putting antibiotics in animal feed is the best way to limit the development of antibiotic resistance. This keeps the drugs potent for human use.

The poultry industry is not the only guilty party, although antibiotics are put in the drinking water of entire flocks of chickens. Antibiotics are also routinely used in swine and cattle production. They are included in pig feed to abate illnesses caused by forced early weaning. They are given to beef cattle to "ease their transition from pasture grass-based diets to the feed-lot corn-based diet".

Other countries are using antibiotics more responsibly than the United States. In 2006, the European Union banned the feeding of all antibiotics and related drugs to livestock for growth promotion purposes in an effort to preserve the effectiveness of antibiotics for human use. There are other associated health risks of antibiotics that include:

- Decreased resistance to infection
- Increased autoimmunity (allergies, asthma, arthritis)
- Toxin-associated cancers
- Neurological damage and neuro-degenerative diseases (Parkinson's)
- Hearing loss (otoxic) aminoglycosides antibiotics
    such as streptomycin, tobramycin and gentamycin
- Hormonal disrupters and thyroid disorders

Tainted Commercial Feed

Grains are easy to identify. Wheat, oats, and corn have distinctive kernels. But with crumble and pellets, you have to take the manufacturer's word for what is truly in the mixture. The quality of animal feed is a profound factor in our food supply system and has an impact at each level.

Animal protein sources usually come from the rendering of slaughterhouse waste. These "by-products" consist of feet, heads, brains, blood, guts, hooves, grease, feathers, and bones. About half of every cow and a third of every pig is considered "unfit for human consumption" and sent to the rendering plant to be processed into pet and animal feed.

Additionally, rendering processes all the "down, diseased, dying, and dead livestock" into protein meal. This meal can also include euthanized animals from veterinarians and laboratories. Yes, your chickens might be eating dog and cat remains. You have no way of knowing what they are eating and the manufacturers won't tell. It is hard to know the true sources of "by-product meal," "crude protein," and "crude fat".

As mentioned earlier, protein quality is important and all protein is not created equal (remember the chicken feathers?). The protein used in animal feed can be from questionable sources. For example, there was the issue of melanine, a protein test-alike that killed so many pets.

In March, 2007, the Food and Drug Administration (FDA) announced that many pet foods were causing sickness and death among cats and dogs. I believe my beloved dog died because of melanine. She died of kidney failure at only 7 years old, which is young for a hybrid-vigor, mixed-terrier dog.

The FDA found melanine contaminates in vegetable proteins exported worldwide from China. A portion of the tainted proteins was used to produce animal and fish feed. Melanine is a by-product of coal; it's a cheap additive that passes as protein in laboratory tests. It does not provide any nutritional benefits and is toxic. It was the reason for the recall of 60 million packages of pet food and has been found in wheat gluten and rice protein concentrate used to produce pet and animal feed.

Concerns about food safety are not limited to pet food. The Associated Press on MSNBC, Saturday, September 20, 2008 wrote:

> *"Beijing, China: Milk and dairy products from 22 com-panies have been recalled after batches tainted with the industrial chemical melamine sickened more than 54,000 children and left 4 infants dead from kidney failure so far...this is one of the worst product safety scandals in years."*   THE ASSOCIATED PRESS

In July, 2008, I read that a class action lawsuit was filed against pet food companies for misleading consumers regarding the contents of pet foods. The lawsuit was filed in the U.S. District Court for the Southern District of Florida by attorney Catherine MacIvor of the firm Maltzman Foreman. She claims that major brand pet food companies spent millions promoting false and misleading marketing statements about the ingredients in their pet food.

The Defendant's advertising materials state that their pet food contains "choice cuts of prime beef, chunks of chicken, fish, fresh wholesome vegetables and whole grains". The Plaintiffs claim the food is actually made from "inedible" slaughterhouse waste products such as spines, heads, tails, hooves, hair, blood, and diseased animal parts. They claim that the protein rendering companies processing the waste also added other inedible waste such as euthanized cats and dogs from veterinarian offices and animal shelters, road kill, zoo animals, rancid restaurant grease, and other toxic chemicals and additives.

The goal of the lawsuit is to "force the Defendants (pet food manufacturers) to more accurately describe the ingredients in their pet foods and to offer more healthful pet food options that provide pets with food quality similar to that provided in human food products".

The bagged feed we often buy from the farm supply is "All Grain Poultry Layer/Breeder Pellets: a Complete Vegetarian Formulation Chicken Layer and Breeder Feed". Even though it's labeled as vegetarian you still cannot be sure what's in it; ingredients include things like "Feed Grade Fat Product". Fat usually comes from rendering the 4-D animals (dead, dying, diseased ,or down) along with road kill and other "animal protein" sources.

All this is a prelude to home-made chicken feed.

Home-made Chicken Feed

Mixing your own feed takes more time, but the rations can be designed for specific objectives and sometimes (though not always) it is cheaper than bagged feed. Questions to ask yourself about home-mixed feed include:

> • Are the commercially produced, and scientifically formulated, feeds you are using "better" or "worse?"

> • Do they have similar or higher quality ingredients than feed you can mix at home?

> • Do you have the time and energy to mix feed?

Your feed ration can vary greatly with different types and quantities of ingredients, but to be complete, feed has to include nutrition similar to that of commercial feeds: (1) carbohydrates, (2) proteins, (3) fats and oils (preferably non-hydrogenated), and (4) supplements to provide a range of vitamins, minerals, and nutritional factors.

To make a basic ration that incorporates all of these is not hard, but it takes planning to have the ingredients on hand. Like cooking, recipes can be creative and vary with the seasons and availability of ingredients. For example, in cold weather, use a higher portion of grains (carbohydrates) so the hens can stay warm. Only use good, wholesome materials and avoid like the plague any rancid, moldy, or musty feed.

The sequence for mixing feed is usually:

1. Make a pre-mix of the dry, finer-texture items like the probiotics, vitamins, kelp, etc. To save time, mix enough for several batches.

2. Coarsely grind the chunkier materials such as whole corn and peas.

3. Add the premix and oils to one of the grain ingredients and mix thoroughly. The object is to disperse the finer materials evenly throughout the entire batch.

4. Add all the ingredients together and mix thoroughly.

5. Mix small batches, enough for a few days only. Feed immediately or store in a covered container.

Each of the 4 groups (carbohydrates, proteins, oils, and supplements) are described below.

1. Carbohydrates in Home Made Chicken Feed

Cereal grains are the chief sources of energy. In the U.S., these are primarily corn, oats, wheat, barley, and rice. Root crops like potatoes and yams can also be carbohydrate sources. Grains can be given whole to poultry but are usually cracked, broken up, or sprouted to enhance digestion.

Grain by-products, such as bran and middlings, are valuable for supplying bulk and loose-textured feed. They also contain excellent nutrition. Bran is the outer layer of the grain and is particularly rich in dietary fiber and omega oils. It contains significant amounts of starch, protein, vitamins, and dietary minerals. The high oil content in bran makes it subject to rancidity, which is one of the reasons the bran is often separated from the grain before storage or further processing. The bran itself can be heat-treated to increase its longevity. Bran can be a flourmill by-product from any grass grain including wheat and rice. Grains are available in bulk or 50 pound bags.

2. Protein Sources for Home Made Chicken Feed

Protein is necessary for growth, tissue maintenance, and egg production. Sources of protein can be from animal or vegetable origin. The protein required by laying hens ranges from about 14% to 18%. The super-charged layer hybrids lay more eggs (290 to 300/year) and require more protein than heritage breeds that lay from 150 to 200 eggs per year.

Animal Source Protein

The most common ingredients in animal-source protein are fish meal, meat scraps, meat and bone meal, and dairy products, including dried skim and butter milk. For free-ranging hens, insects, worms, and grubs are also protein sources.

Fish meal is made from whole fish, including the bones and offal (guts and organs). It is a brown powder or cake obtained by rendering, pressing the whole fish or fish trimmings. Fish meal can contain high levels of heavy metals.

Meat and Bone Meal is a product of the rendering industry. Rendering is a process that converts animal waste products into stable materials. Most of the MBM comes from slaughterhouses but can also include restaurant grease and butcher shop trimmings, expired meat from grocery stores, the carcasses of euthanized and dead animals from animal shelters, zoos, and veterinarians. The most common animal protein sources are beef, pork, sheep, and poultry. Meat and bone meal has about 50% protein, 35% ash, 8 – 12% fat, and 4 – 7% moisture. It is used in animal feeds to boost the amino acid profile of the feed.

The problem with MBM is the potential toxic tag-alongs that can accompany the source material. This includes residues and metabolites of antibiotics, hormones, or whatever else was in the animal's food.

Dairy Products are often used as poultry feed. Agriculture texts written before the 1950s routinely discuss using dairy products to raise and maintain poultry. Chickens will eat dairy products in all forms. In fact, they like dairy products so much there is a danger they will over-eat. Too much dairy in a chickens' diet is bad because of its high fat content. Chicken livers are not designed to process high fat diets, and whole dairy products are high in fat. Only a portion of a hen's daily fat intake should come from dairy products. Like humans, chickens require different kinds of fine oils that contain a balance of the omega 3, 6, and 9 fats for complete nutrition.

The protein levels in whole milk range from 3% to 4% for bovine and slightly higher for goat milk. Fat in whole milk is about 3.25 % and is high in saturated fat. Dried whole milk has about 26% protein

Dairy products make excellent supplements for chickens; just don't feed too much. Milk alone can't provide enough protein for a hen, so make sure their rations have protein from other sources.

Micro-livestock as Protein

It's worthwhile to consider the protein content of worms and insects when they serve as poultry feed. Earthworms have 60% to 70% protein with a higher content of amino acids (lysine and methionine) than meat or fish meal.

There is an entire field of agriculture called "micro-livestock". This is the study and practice of growing worms, grubs, and insects as human and animal food. Micro-livestock provides an inexpensive and quality substitute for meat protein. Here's another term for you: entomophagy is the eating of insects. Chickens are dedicated entomorphagists.

As squeamish as this might make you, The Food and Agriculture Organization (FAO) of the United Nations stated in their November 8, 2004 Newsroom:

> *"For every 100 grams of dried caterpillars, there are about 53 grams (54%) of protein, about 15 % of fat and about 17 % of carbohydrates. Their energy value amounts to around 430 kilocalories per 100 grams. The insects also have a higher proportion of protein and fat than beef and fish with a high energy value."*

THE FOOD AND AGRICULTURE ORGANIZATION (FAO)

Naturally living in yards and gardens, insects are cheap, tasty for chickens, and are a natural protein source that require little, if any, land to produce. Common insects and their larvae include: ants, termites, beetle grubs, caterpillars, grasshoppers, moths, wasps, butterflies, dragonflies, beetles, cicadas, stick insects, moth weevils, house flies, spiders, and crickets.

What do insects eat? They eat green plants (hopefully organic ones).

Compare this with commonly eaten human food such as crabs, lobsters, and catfish. These scavengers are considered "bottom feeders" because they eat any kind of decomposing material. How appetizing is that? Many cultures have entomophage-based recipes that they think are delicious.

Protein from Plant Sources

The most frequently used vegetable proteins are in the form of meals from soy beans, peas, peanuts, cottonseed, linseed, and corn-gluten. Soy beans and field peas are most commonly available in bulk.

Soy beans can have up to 38% protein. Although soy is as abundant as corn, soybeans are not necessarily the best protein source for hens. Some organic formulations go so far as to label their products as "soy free".

Soy has been shown to reduce the assimilation of B12, calcium, magnesium, copper, iron, and zinc, and thus can stunt growth. High levels of soy have been linked to thyroid and autoimmune diseases. As if that weren't enough, soy foods contain high levels of aluminum, which can be toxic to the nervous system and kidneys. Only corn is mono-cropped more than soy, and there are many genetically modified soy bean crops grown for animal feed. If you want to research soy further, a good place to start is the web site of the Weston A. Price Foundation for Wise Traditions in Food, Farming and the Healing Arts. Their web site is: www.westonaprice.org.

Field peas can have up to 24% protein and are of the legume grain. These peas are also called cow peas (Vigna group) or field peas (pisum arvense). The energy content of peas is higher than barley and is close to that of corn and wheat. Peas can be bought and stored whole, then ground or sprouted before feeding. Standard size hens can eat peas whole, but for bantams, whole peas can be too big and you might need to grind them.

Comfrey is a protein source you might want to grow. It can have 15% to 30% protein. Comfrey can be harvested, dried, and stored for winter use when green feeds are rare. Additional information on comfrey is in the Chapter 3, Growing Food for Chickens.

3. Oils and Fats in Home Made Chicken Feed

Every single cell in an animal's body, including those in the brain, requires lipids to function properly. Most lipids are derived from oils or fats. Lipids have many key biological functions; they store energy, act as structural components of cell membranes, and function as intermediates in signaling pathways. The brain is composed primarily of lipids. "Lipid" is sometimes used as a synonym for fats, but fats are chemically a subgroup of lipids.

Oils are generally derived from plant sources and are liquid at room temperature. Fats are from animal sources and tend to be solid at room temperature.

Chickens require quality oils to make quality eggs and yolks. A variety of oils can be used including flax (linseed), vegetable, olive, peanut, almond, walnut, grape seed, avocado, and apricot. You will only need a tablespoon of oil per 5 pounds of feed.

It's good to rotate the oils so the hens get a variety and balance of the omega 3s, 6s, and 9s, along with other nutrients in the better oils. Mix or add oils in feed just before feeding so they don't get rancid. With pellets, mix the oil in as you feed.

Rancidity can be decreased, but not completely eliminated, by storing fats and oils in a cool, dark place with little exposure to oxygen or free-radicals. Heat and light hasten the rate of reaction between fats and oxygen that causes rancidity.

Cod liver oil is in a class all it's own. It has Vitamins A and D that are especially needed in winter months. Mix in 1 tablespoonful for every 5 to 10 pounds just before feeding. Periodically clean the feeder to remove the film of oil that will become rancid.

4. Feed Supplements in Home Made Chicken Feed

Extensive research has been done on the vitamin and mineral requirements of chickens at various stages of their lives. Feed companies use this data to formulate feed rations that give birds a balanced diet. In an ideal world, this feed would work perfectly. But even with the best feed, nutritional problems can come from many sources. A few of these are:

- The feed is not fed in sufficient quantities.

- Hens don't eat the normal amounts.

- Feed is old and its nutritional contents have deteriorated from age, improper storage, moisture, or mold.

- Feed is damaged by improper storage or exposure to moisture, heat, and rodents.

• Feed is not palatable to the hens; they won't eat it unless it's their only choice and they are really hungry.

• Hens have digestive problems, worm infestations, or other gut afflictions that affect proper digestion and/or nutrient absorption.

• Certain hens in a flock might be blocked by other hens and subsequently don't have adequate access to feed.

• The incorrect feed is given for the age of the bird.

Health problems are a red flag, telling you to look at the birds' diet and make sure they have sufficient access to optimal levels of nutrients. An optimal diet and supplementation program is the best way to prevent diseases and ailments. Each vitamin and mineral is responsible for multiple functions in the body and, with the exception of vitamin C, poultry get vitamins from their food. The fat soluble vitamins A, D, E, and K are stored in fat and the hens can use reserves for a couple of months before deficiency shows. A deficiency in the water soluble vitamins – especially the B complex – shows up much faster.

5. Supplements for Poultry Feed

Grains alone are not a complete ration. There are many supplements that both enhance general health and target specific nutritional needs. Popular ones are listed below:

• Vitamin & mineral supplements that come in a packet and/or as a powder

• Salt, preferably natural, without fillers

• Kelp or seaweed (trace minerals)

• Dried comfrey (protein, vitamins, and minerals)

• Bran (protein, zinc, manganese, and phosphorous)

• Brewer's yeast (B-complex) to help prevent leg problems

• Probiotics to aid digestion and boost the immune system

• Oils, especially flax and cod liver oil

Of all the supplements, probiotics are the least well-known in the poultry world.

Probiotics

Probiotics are live, naturally occuring, non-disease-causing bacteria and yeast that contribute to the balance of the intestinal tracts and health of the poultry. In effect, all warm-blooded animals are walking vats of fermentation necessary for proper nutrient breakdown and absorption. Think of yourself and your hens as living ecosystems that host 400 microbial species.

Probiotics are especially useful for baby chicks, because they reduce coccidia incidence and the associated diarrhea, as well as other digestive problems. Probiotics aid weight gain and overall health by creating enzymes and fermentation vitamins and breaking down nutrients so they can be absorbed by the body.

It has been shown that a lack of beneficial probiotic bacteria can cause inflammation. In Functional Medicine, probiotics are sometimes referred to as the "undiscovered organ" because they have crucial functions in the body. The mass of microorganisms must be property balanced for normal functioning of the gastrointestinal tract and immune system. The GI tract interplays with, supports, and affects the nervous system and endocrine systems. This is why diseases seemingly unrelated to the GI tract, such as allergies and arthritis, can be mitigated — and even cured by repairing leaky guts and establishing good bacteria. The disruption of probiotic balance is another reason why sub-therapeutic use of antibiotics in animal feed is so harmful to animal's long-term health (if they live that long).

There is evidence that probiotics in humans affect the immune system function, mood and anxiety disorders, and decrease inflammatory bowel disease and celiac disease. Probiotics can be given orally, added to water, or mixed with feed.

Prebiotics

Prebiotics are fodder for the probiotics. Prebiotics (pre = before and biotics = life) are also important for poultry health. The prebiotics feed the probiotics and help keep them established. Prebiotics modify the composition of intestinal microflora so probiotics can dominate through competitive exclusion. The good bacteria can overload and kill off the bad bacteria. Prebiotics are an insoluble fiber present in whole grains and plant matter.

Oils & Fats

Fine oils offer different nutritional benefits, depending on their source. I rotate between quality olive oil and other non-hydrogenated nut and seed oils for my birds. Oil helps the powdered vitamins and minerals stick to the grains, rather than ending up in a dusty heap at the bottom of the feed bin.

Flax seeds and flax oil help boost Omega 3 fatty acids. Omega 3 fatty acids, found in flax seeds and flax oil (linseed oil), are increasingly found to benefit poultry, their eggs, and human health. Other sources of Omega 3 fatty acids are cod liver oil, fish oils, and leafy green vegetables.

Flax seeds and their oil are especially rich in one of the most important Omega 3 fatty acids: alfalfa linolenic acid (ALA). Flax oil has 57% ALA and a high content of polyunsaturated fats. This makes it especially valuable in poultry (and people) nutrition.

Adding 2% to 10% flax seed to the poultry feed changes the fat composition of the egg yolk. Ordinary store eggs have about 0.4% ALA content. Flax-fed hens produce eggs with 4.6% ALA along with other valuable Omega 3 fatty acids.

The improvements Omega 3s make in hen health is evidenced by beautifully feathered biddies. Omega 3s increase vigor and disease resistance. In humans, omega 3s supplementation has been shown to help with dyspraxia (motor skills), dyslexia, attention deficit hyperactivity disorder (ADHD), dementia, schizophrenia, bipolar disorder, and depression.

Only use food-grade linseed oil. Food-grade oil is cold-pressed and obtained without solvent extraction. Non-food-grade oil is processed using solvents and metallic high-heat dryers to extract the oil. This is how linseed oil used as paint thinner and wood finish is made. It contains polymers and has been oxidized to make it thicker. You don't want to feed this oil to your hens.

In dry meal rations, chickens avoid whole flax seeds and eat them only when everything else is gone. Grinding the flax seeds and mixing them with the ration in a moist mash is the best way to get the seeds into the hens. Grinding flax seed drastically shortens its shelf life, and the oils can turn rancid faster. If you grind the seeds, do it in small batches to use over a few weeks and keep refrigerated.

Whole flax seeds store well and can keep for years when stored whole in a cool dark place. They seem to be resistant to weevils.

Apple Cider Vinegar (ACV)

ACV is an antique and proven supplement. I'm a long-term fan of apple cider vinegar and put it in water to drink with meals. Apple cider vinegar is one of those old folk remedies. Its documented use by Egyptians and Romans dates back to over 4,000 BC. We give it to all our livestock and pets as a tonic to supplement their health.

Apple cider vinegar is produced from apples and has lots of vitamins and minerals including potassium, sulphur, and phosphorous. Its acidity lowers the bird's body acidity, making it harder for certain bacteria, parasites, and yeasts to survive. ACV also helps with digestion by breaking foods down for easier assimilation.

Because ACV helps regulate calcium in the body, there is some evidence that it helps with soft-shelled eggs. Put 1 teaspoonful per quart of ACV in the water of baby chicks, beginning when they are 2 weeks old, so they can benefit from the vitamins and digestive aid. The biggest risk with ACV is using too much, which makes the water acidic and unpalatable. The hens won't drink yucky-tasting water and can become dehydrated.

Don't use ACV with metal waterers. The acidity will cause corrosion, which will damage the equipment and put toxins (especially aluminum) into the drinking water. ACV is wonderful. It is old fashioned, inexpensive, non-toxic, and when used correctly, beneficial to chickens.

6. Garden Grown Supplements

Chickens like to eat root vegetables (carrots, potatoes, sweet potatoes, Jerusalem artichokes, turnips, etc.). They will eat them raw, but it's easier for them to eat if the vegetables are in small bits and cooked. Put a few roots in the food blender and feed straight, or mix in with the flock's feed ration.

Make Scratch from Scratch

Scratch is a blend of whole and cracked grains. It's called scratch because that's what the chickens do to find the seeds. Scratch is usually broadcast on the ground where chickens can clean it up, such as around fruit trees or in bedding. When scratch is tossed on top of bedding the hens churn it, mixing manure (nitrogen) with the bedding (carbon) in the process. The bedding absorbs moisture from droppings and the scratching mixes the manure into the bedding and keeps it aerated. It also helps keep fly populations down because the hens eat the insect larva as they uncover it while looking for grains.

If there is a heavy layer of manure, it's better to remove it to the compost pile and add fresh bedding. Whenever the coop smells of ammonia, it's past time to clean out all the litter and send it to the compost pile.

In the garden, toss scratch in places you want your hens to work over, and they will spend hours turning the upper layers for you.

The formula for scratch varies, but it is usually a combination of the following grains and seeds, although some scratch contains only cracked corn and whole wheat. Scratch ingredients can include, but are not limited to:

- Cracked corn: 40% to 50%
- Whole wheat: 30% to 50%
- Whole oats: 20% to 50%
- Rice: 30%
- Sunflower seeds: 10%

Just as making your own granola can increase the value of the grains included in it, you can sometimes save money by mixing your own scratch, depending on the current price of grains and the time of year you are buying.

Scratch is not a complete feed in itself and needs supplementation to provide a balanced diet. It doesn't contain many nutrients or enough protein, so always have a complete ration available for continual feeding. The nutritional analysis of scratch is about:

Crude protein: 8.5%
Crude fat: 2.75%
Crude fiber: 3.0%

Grit and Calcium

Grit is composed of hard, abrasive granules. Gastrolith is the scientific term for grit, which is also called gizzard stones or stomach stones. These abrasive rock pieces are used by birds, crocodiles, alligators, seals, and sea lions to aid digestion by grinding food. The size of the gastrolith depends on the size of the animal and ranges from fine sand to rocks the size of cantaloupes; chickens prefer grit sized between sand and pebbles.

Grit is picked up naturally by free ranging hens. The stones enable the otherwise toothless (and lipless) chickens to grind tough plant fiber in their gizzard. Especially for birds not free ranging, a ration with calcium and/or a calcium source like grit should always be available.

There are two kinds of grit: insoluble and soluble. Insoluble grit does not dissolve. It is held in the gizzard where it helps grind food into smaller bits for digestion. Ground granite and flint are common types of insoluble grit.

Soluble grit is dissolved during digestion and its calcium is used by the bird for bone development and egg shell production. If your hens are producing thin-shelled eggs, they are not getting enough grit. Limestone and oyster shells are common forms of grit. Grit is divided by size:

- Chick grit (1 day to 3 weeks)
- Grower grit (3 to 10 weeks)
- Layer grit (11 weeks and older)

Grit that is too small for the bird's size can quickly pass through the gizzard and does not serve as a digestive aid. Always make sure your hens get enough grit. Free range birds with access to a gravel drive will get enough insoluble grit, but keep soluble grit available as a calcium source.

Feed Formulations

Feed formulations, like recipes, can vary greatly between batches, according to the seasons and what's available. Below is a basic formula containing at least 16% protein for laying hens:

- Cracked corn for carbohydrate: 30% to 50% (max)
- Field Peas (contains about 20% protein): 20%
- Fish meal (contains 60% to 70% protein): 5% of ration
- Whole oats/spelt or triticale (for carbohydrates): 10% to 20%
- Poultry Nutri-Balancer for vitamins and enzymes (per directions)
- Kelp meal for trace minerals/chlorophyll: 1%
- Salt for minerals: .01% or a dash
- Aragonite for calcium: .15% (or free-feed oyster shell)

Feed formulations are not set in stone. Vitamin supplements, such as poultry Nutri-Balancer, help assure that basic nutritional requirements are available to your poultry.

Seasonal Adjustments of Poultry Rations

Winter Feed Adjustments

During the winter the hen's calorie needs increase as temperatures drop; they need more carbohydrates to keep warm and can crave greens and vegetables. Include more grains – like corn, wheat, or oats – in the ration. This can be accomplished by scattering extra scratch. Greens and vegetables can be enhanced with beet pulp and alfalfa pellets.

Vitamin D Supplement in Winter

Vitamin D supplementation is warranted when days are shorter and hens get less sunlight. Pour about 1 tablespoonful of cod liver oil over each bucket of feed.

Beet Pulp

Beet pulp is what's left over after beets have had simple sugars extracted to make table sugar. The sugar extraction is so complete that the beet pulp has hardly any sweet taste at all. Some feed manufacturers add back molasses to increase the palatability. Most beet pulp contains only about 5% of molasses which is only 8.6 grams of simple sugars. This is about the same amount of sugar contained in several apples.

For livestock, feed is categorized as a forage, an energy feed, or a protein supplement. Feeds containing 18% or higher crude fiber are forages. Forages include hay (including alfalfa and pellets made from alfalfa), soybean hulls, almond hulls, and ground corn cobs.

Feeds with less than 18% crude fiber and less than 20% crude protein are energy feeds and include all cereal grains, wheat and rice bran, fats, and molasses.

Feeds with less than 18% crude fiber and more that 20% crude protein are protein supplements. Protein feed includes meals derived from soybean, linseed or cottonseed, brewers yeast, fish meal, sunflower seeds, and dehydrated milk.

Understanding the definitions of

Alfalfa Pellets and Beet Pulp

Alfalfa pellets and beet pulp supply extra greens and nutrition, especially in winter months.

feeds that are forage, energy, or protein is helpful when comparing commercial feed mixes. Beet pulp has 10% crude protein and 18% crude fiber. As such beet pulp is in the gray zone of being a low protein, and high fiber and high energy feed. Most livestock nutritionists consider beet pulp as a forage. But here's something interesting. Beet pulp contains fewer calories than cereal grains. Does this mean that beet pulp is less valuable as a source of calories than grains? No, because beet pulp has a lower glycemic index than any of the cereal grains, which is good.

The glycemic index is a relative value given to carbohydrate rich foods that shows the average increase in blood glucose levels that occur after a food is eaten. Feeds with a high glycemic index, such as corn, convert to glucose very rapidly and quickly elevate the blood glucose levels.

More importantly, a diet with too much high glycemic index feeds affects the functioning of the entire body, including the pancreas, kidneys, and gut. High glycemic index foods foods over the long term affect the mix of populations of good bacteria (probiotics) versus bad bacteria which can cause diarrhea, acidosis, and enterotoxicity.

Beet pulp contains both soluble and insoluble fiber so that energy is released more slowly after microbial fermentation in the gut. Because of this, beet pulp can be safely fed in relatively large amounts, especially as a winter calorie/vegetable supplement.

Beet pulp is better soaked prior to feeding to make it more palatable. The dry pulp is remarkably hard and dusty, causing the chickens to ignore it. Beet pulp soaks up quite a lot of water, expanding its volume. My chickens prefer their beet pulp mixed in with moist mash with a dash of flax seeds.

When introducing beet pulp into a poultry diet, the chickens might be leery at first. You can start by initially soaking a small amount of pulp until it becomes juicy (and more palatable), and then mixing it with grain or other already accepted feed.

Alfalfa Pellets

Alfalfa pellets can supplement a flock's "greens" for the winter or if the flocks doesn't have access to fresh graze. Alfalfa pellets come in 50 pound bags and are the same as those fed to horses and guinea pigs. Feed sparingly as alfalfa pellets don't seem to be a favorite food of chickens. Nutritionally, alfalfa pellets contain:

- Crude Protein   14%
- Crude Fat   1%

- Crude Fiber   32%

Summer Ration Adjustments

During the summertime chickens need less energy to stay warm. Decrease the proportions of corn and barley in rations. A typical summer formula for laying hens is:

- Corn (cracked or meal): 10%
- Wheat: 25% (chickens tend not to like whole wheat)
- Bran: 10%
- Oats or barley: 10% (not greater than 15%)
- Sunflower seeds: 5%
- Dried greens of some sort (such as clovers or alfalfa) 10% or more
- Dried or fresh comfrey leaves up to 5%, or free choice in comfrey cages
- Add vitamin and mineral supplements (readily available in packets) to the grains and seeds.

Sprouting Grains and Peas for Feed

There are enough advantages to sprouting grains and peas that you might want to consider it, especially in winter. Chapter 3, Growing Food for Chickens, describes how you can sprout grains easily and inexpensively for your flock.

Sprouted grain is food you can grow for yourself, and your hens, no matter if you live in an inner city, on top of a mountain, or in a boat. There are no other crops that can be grown without soil, indoors, and harvested in 2 days to a week. Sprouts are simple to cultivate, harvest and can be eaten raw, without cooking. Sprouts were eaten by ancient cultures for over 5,000 years. Sprouting is written about in the Bible, in the Book of Daniel. Daniel and his friends refused to eat of the king's luxurious diet. They preferred seeds, and probably sprouted seeds which contain up to 600% more nutrition than cooked food.

In the 1700's, many sailors suffered and died from scurvy during their 2 to 3 year ocean travels. The famous Captain James Cook, from 1772-1775, made his sailors eat limes, lemons, and varieties of sprouts. All these foods are high in vitamin C. Citrus fruits and a continuous program of growing and eating sprouts were credited with preventing, and curing, the scurvy problem.

During World War II, a nutrition professor at Cornell University referred to sprouts as:

> "A vegetable that will grow in any climate, will rival
> meat in nutritive value, will mature in 3 to 5 days, may be
> planted any day of the year, will require neither soil nor
> sunshine, will rival tomatoes in Vitamin C, will be free
> of waste in preparation and can be cooked with little fuel
> and as quickly as a . . . chop."
>
> — DR. CLIVE M. MCKAY

Dr. McKay, and a team of nutritionists, studied the properties of sprouted soybeans. They found sprouts retain the B-complex vitamins present in the original seed, and showed an average 300 percent increase in vitamin A and a 500 to 600 percent increase in vitamin C. In addition, in the sprouting process, starches are converted to simple sugars, thus making sprouts more easily digested than un-sprouted seeds.

Sprouts also contain more amino acids, phytochemicals, and enzymes than un-germinated seeds. The elevated water content in sprouted seeds aids digestibility, and the nutrients help support immune function.

Sprouting, also called germination, is the process of growth emerging

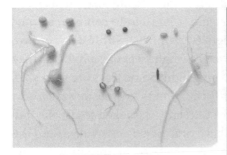

Sprouts & Sprouting

Sprouting grains and field peas only takes a few days and dramatically raises the nutritional value.

For poultry feed, sprouting can be done easily using a series of rotating buckets, one for each day. Not all sprouts are good for eating. Kidney, lima, and buckwheat sprouts can cause health problems.

after a period of dormancy. In English usage, germination can imply something expanding into a greater being from a smaller existence. The rate of germination depends on water availability, temperature, oxygen, and light or darkness.

The seeds most commonly sprouted for chicken feed are field peas, oats, wheat, and corn. Other seeds that can be sprouted for feed and human food include soybeans, sunflower, barley, triticale (a hybrid of wheat and rye), mung, and just about any kind of bean.

But be careful. Before you delve into a sprouting frenzy, check which seeds and sprouts are edible. Some are toxic, like kidney and lima bean sprouts, which contain undesirable growth-inhibiting chemicals. Buckwheat greens contain fagopyrin which, when eaten in large amounts, can cause the skin of animals (and people) to become hypersensitive to sunlight (phototoxic).

You need to know if the seeds are intended for sprouting or planting. Seeds for planting are usually treated with chemicals you don't want to eat. Seeds are often heated, making them non-viable for sprouting.

How to Sprout Grains

Sprouting starts with an initial soaking of about 8 hours or overnight. Larger seeds, like peas or corn, might need to soak slightly longer. Drain and rinse seeds at regular intervals until they germinate (sprout). Usually rinsing is done twice a day. Sprouting can be done in any size batch and in many kinds of containers, from quart jars to buckets, to commercial production trays. The seeds swell and begin to germinate within a day or two.

Sprouting in trays is the best way to grow light-seeking sprouts, as they provide a large surface area for absorbing light. Some trays are stackable, which saves space. Stacking also allows the sprouts to grow upwards.

Cover the bottom of the tray with a thin layer of soaked seeds. The tray should be at least 2 inches high and have drainage holes smaller than the seeds. When the roots have something to attach to, the sprouts will grow better. Mesh in the bottom of a tray provides a foothold for the sprouts. Size the mesh to match the seed.

Sprouting in a mini-greenhouse raises and regulates temperature and slows water loss. A mini-greenhouse can be as simple as a clear plastic bag; just leave enough room inside for air.

Water is the trigger for germination. It causes the seed coat to swell and break, so viable seeds will sprout when there is enough water to moisten them. All seeds contain starch, proteins, and oils to nourish the growing sprout. Once the seeds start to sprout, the food reserves are used quickly and the sprouts need a continuous supply of water for nutrients and light for photosynthesis. Photosynthesis provides the energy needed to sustain growth.

Temperature affects growth rates. The grain-sprouting seeds used for chicken feed generally do best at room temperature or warmer.

Regardless of the container you use, you must allow water to drain. Sprouts that sit in water will quickly spoil and rot.

Rinse sprouts at least once a day. Some advise rinsing 3 to 4 times a day, depending on the temperature and amount of seeds in the batch. Warmer temperatures cause sprouts to grow faster, so they need the growth toxins to be flushed away more often.

In about 3 to 5 days, the sprouts will be 2 or 3 inches long, suitable for consumption. Or you can let them grow longer and they will develop leaves. Now they are baby greens. The growth of any sprout can be slowed or halted by refrigeration.

Grazing for Feed

Really good grazing conditions that include lots of greens, seeds, and bugs can only provide 5% to 20% of a hen's daily intake. It is unrealistic to expect chickens to get all, or even most, of their feed from pasture. The quantity of forage from pasture varies widely. Common sense dictates that there is more forage in summer and less in winter. Spring and fall are somewhere in between. Access to lots of different plant varieties affords a hen many choices, whereas bare, compacted ground offers no food, save the occasional wayward insect. The following is a list of factors that affect the availability of foraged food:

- Age of the chicken (chick, pullet, laying hen, mentor hen)
- Time of year and growing season
- Rainfall and soil moisture
- Seasonal temperatures
- Insect populations and type of insects
- The depth and fertility of the soil supporting the earthworm and soil dweller populations
- Amount of area to forage
- Stage of plant growth
- Whether or not the chickens like the plants

- Quality and quantity of commercial feed
- Hen's preference for different foods
- Availability of food residuals from the kitchen or food establishments

Even if your hens get hardly any of their food from free range, rotational ranging is probably the most important portion of their diet. The exercise improves muscle tone. They are exposed to sunlight which supports overall health, bone integrity, and egg development. They also get contact with the earth which offers small grit for digestion and the many nutrients and co-factors found in healthy soil. Chlorophyll and fiber, contained in green plants, are detoxifiers that keep the bird's digestive tracks moving regularly and act as prebiotics to feed good gut bacteria. All these factors combine synergistically to make happier birds and healthier, better tasting eggs.

Access to many varieties of plants, grasses, legumes, and weeds allow a hen to choose different proteins, lipids, and carbohydrates. Given a choice, a hen will eat plants that best suit her needs.

Grasses and weeds are moister, more tender, and more succulent in the morning than any other time of day. If you are using bottomless pens, this is the best time to move the pens to let the chickens graze on something fresh.

My hens semi-free range. Fences and poultry portals direct them to different areas of my property, depending on where I want them to be. As soon as I open the pop door in the morning, my hens stampede down the gangplank, scurry through the poultry portals, and fan out to hunt for early worms and the choicest greens.

I've added different clovers and grasses to my lawn and small orchard to enrich the flock's grazing. This is what Joel Salatin calls a "salad bar" yard, rich with varied plants. No mono-culture in my yard. I let the grasses and clovers in my yard go to seed before mowing certain sections, and let the chickens have access to the newly mowed sections; they mosey about gleaning the seeds from the grass. Gleaning means that they pick up grains (or vegetables) that were missed or unwanted during a harvest.

With proper management, a yard can act as a mini-field and grow grain (seeds) for a small flock of hens. This only works if the yard's farmer doesn't insist on golf-course-type grass on his or her lawn. Most people don't realize that as the height of the mowed grass decreases, the root depth decreases, and maintenance increases. Grass that is not mowed to the crown is healthier for lawns, roots, soil dwellers and grazers.

Storing Feed

Storing feed correctly is essential to maintaining its quality and keeping rodents out. Here are a few tips:

• Store feed in sealed bins. Plastic trash cans or plastic storage containers make good feed bins and seal tightly. Plastic buckets, such as the ones kitty litter comes in, are a handy size. A kitty litter bucket will store about 25 pounds of feed and is much easier to handle than 50 pound bags. Metal trash cans are initially more expensive, but they last longer and are more pest-proof. It's hard to chew through metal, whereas sharp rodent teeth can sawzall through plastic fairly easily. Plastic also tends to crack with extensive outdoor use and is especially brittle in cold weather.

• Secure the storage bins with bungee cords. This helps keep the nighttime marauders from plundering feed. Plus, if the bin tips over, the lid stays on.

• Keep feed bags off the floor in a cool, dry place. Don't store feed in open bags or let them sit directly on dirt or concrete floors; the bag will wick moisture out of the ground, and moisture causes mold and grain deterioration. Moist paper-bottom bags get weaker and can break when you hoist them up, spilling feed all over the floor.

• Keep bins and bags sealed and as airtight as possible. Feed will get stale and spoil faster when exposed to air. Tightly sealed bins help keep out grain weevils (*Sitophilus granarius*). Weevils are discussed below.

• Make small batches. Mix only as much feed as your hens will eat in about 2 weeks. Feed can get stale and rancid quickly, particularly in warm weather.

• Feed loses its nutritional value as it gets older. Mark the purchase date on each feed bag and use the older bags first. Use all the feed from one bag before you open another one.

• Add supplements just before feeding, especially the oils.

• Never, ever use rancid oils. They do more harm than good. If you are not sure what rancid oil smells like, put some peanuts in a closed jar and leave it in the sun for a week or so. The nuts will develop a distinctive, unpleasant, rancid odor.

Grain Weevils

Weevils are common and cause grain loss and spoilage, so they are worth knowing about. Weevils are small grub-like larvae that grow inside seed grains. You can buy a weevil-infested bag of feed and see no evidence of their presence until the adult weevils emerge from the grains. Tiny holes in feed bags are a sign of weevils.

Weevils can infest all types of grains and flours including oats, corn, wheat, barley, flax seed, rice, flour, popcorn, saved garden seeds, and breakfast cereals. Web-like lumps of white material on feed and feed bags are evidence of weevil life. Cocoons form in the folds of the bags and adult weevils will flitter around. If you see small moth-like bugs flying out of a cupboard or around the feed room, you have weevils.

3 Ways to Store Feed

1 A metal trash can keeps rodents and moisture out.

2. A pallet stops feed bags from absorbing moisture from the floor.

3. Plastic buckets with lids are easier to handle than 50 pound bags and offer more rodent and moisture protection.

You can still use weevil-infested feed for your hens. The weevils are high in protein and your hens will enjoy eating them. Use the infested feed first. Weevils are harmless to everything but grain products. They don't carry diseases or bite. They won't infect people, clothing, furniture, or pets. They will not eat wood or house structures.

You can kill all stages of weevils by freezing them for about 3 days. Heating weevil-infested grain to 140 degrees for 15 minutes also kills weevils. Adding diatomaceous earth (DE) to feed helps keep the weevil population down and is non-toxic and safe for hens to eat.

14 Ways to Cut Feed Costs

"Chicken feed" is a term used to imply that something was cheap. Not anymore. Corn is becoming less available due to ethanol production, human population rates are rising, and weather patterns are changing; these factors (and more) will cause feed prices to keep rising. Feed costs can be a limiting factor in keeping hens. Most micro-flocks are considered pets so (like the cat and dog) their food costs are part of the family budget. At least a hen's feed costs are offset by the eggs she produces.

Start thinking outside the feed bag. Until now, most folks only thought of feed as something they bought in a bag from the farm store (they also thought that milk comes from cartons). However, the "eat local" movement and local food self-sufficiency efforts are changing how people view their (and their chickens') food sources.

Below are some alternative ways to lower feed costs.

1. Food Residuals (Waste) as Feed

If people will eat it, hens will eat it. Now that you know this handy fact, "food waste" is a meaningless term for you. Last night's dinner leftovers are this morning's hen breakfast. Make arrangements with your neighbors to collect their yard waste and table scraps for your chickens. Give them eggs as a thank you.

When Karl Hammer sees the waste management trucks bringing in tons of food residuals, he doesn't think of it as stuff to get rid of. He thinks of the residuals as stored energy to be transformed into cash flow. Karl is proud of transforming all that food (most of which is imported from outside the state) for local agriculture use. Similarly, you can decrease your feed bills by seeking out ways to bring in "residuals". Find potential chicken food sources that would otherwise be sent to the landfill and truly wasted.

2. Adopt a Restaurant

We've been collaborating with two restaurants: McCoy's Deli and Masa-moto's Sushi. The logistics of this arrangement are simple: food residuals are collected in 25 pound buckets which are easy to transport (remember the kitty litter bucket suggestion? Most have handles and are easy to carry). Almost any food-serving institution produces chicken-palatable waste, although I wouldn't feed my flock fast food.

It's essential to educate restaurant workers about the necessity of sepa-rating food residuals from ALL other trash such as paper products (nap-kins, plates and cups) and, even more importantly, plastic. This includes the small labels stuck on fruit. Some plastics and paper are compostable, but these products require a special composting process that is out of the scope of backyard composting. Even after these paper and plastics break down, nobody knows what residues they leave in the soil. You don't want the plastic by-products of long-chain polymers and other chemicals in your garden beds. Another problem with plastics is the hens eat bits that get caught in their crops. This can cause crop compaction and can kill a hen.

3. Collect Yard Residuals

When you mow your lawn, collect the clippings in the mower bag and give it to your hens. This is fresh green fodder and hens love it. You can dry grass clippings and store them for winter use. If a yard has overgrown and the grass has set seed, clippings make an abundant green and grain.

I've made a deal with friends that if they bag their yard waste (leaves and grass clippings), I'll pick it up and return the bags for refilling. I give them a couple dozen eggs as a thank you. If kids do the bagging, I pay them a small fee. I only take yard waste from households that don't use pesticides and herbicides. Since I began this project, I know that several local hom-eowners have quit using toxic yard treatments after they learned how the toxic tag-alongs can transfer to the soil and up the food chain.

If you are an exceptionally enthusiastic biomass collector, keep a very large and heavy-duty contractor's trash bag in the car. Whenever you see a pile of leaves, stuff them in your bag for use as bedding or compost.

4. Feeding Chickens Weeds

Most "weeds" have chicken feed power. The most common ones are dan-delion, lamb's quarter, burdock, and nettle, just to name a few. The mineral content varies in different plants, so feeding a variety will help balance their nutrition. In a corner of my yard (or in the chicken run) I pile up weeds and clippings from the yard and garden. This isn't a compost pile; it's a biomass pile so greens are available to the hens. They chortle the way hens do and scratch through the stack, eating as they go. Every mouthful of weeds and yard residuals is one less mouthful of purchased grain.

5. Wild Bird Seed as Chicken Feed

Wild birds can be wasteful with seed in feeders. Some birds shovel out seed from the feeding station while they look for better bits. To benefit from wasteful wild birds, place bird feeders in places where the chickens can clean up leftover seeds. My feeders are hung from the railing on my deck. When I let the chickens range the yard they head for the "wasted" seeds below the feeders and snatch them up. Not only do they eat the seed and any inhabitant bugs, they also mix the seed shells into the grass to enrich the soil. Two of my bird feeders hang over garden beds and, in the spring, some of the rejected seeds turn into next year's sunflower crop.

6. Yards as Mini-pasture

If you were a chicken, only about 16 inches head high, the square footage of an urban yard seems as big as acres of pasture and a huge place to explore. If your local laws and hen safety allow, let your flock free range in your yard. This free range area would preferably have a perimeter fence to keep critters out, even more then keeping hens in; they rarely stray very far from home.

Chicken manure on your lawn won't be a problem as long as your micro-flock is not there a long time. Most of their droppings will be absorbed quickly, usually overnight or in a few days by earthworms and soil dwellers. You'll be amazed how quickly the droppings disappear.

Chickens will probably head for the mulched areas, so be prepared to have your mulch spread even further. The hen's carnivorous appetites will clean the yard of many pests. Chickens can keep flea and tick populations down to almost zero; I'd love to see a study quantifying their effectiveness. It would also be fascinating if the incidence of Lyme disease decreased because of chickens eating ticks. I've never seen a tick embedded in a chicken. I suspect chickens are able to get them off with their beaks and sharp toenails, or perhaps the ticks hide under their feathers.

If your yard isn't fenced, you can use temporary fencing to restrict the hens to an area, but hens will not range far from their coop. Twenty to fifty yards is about the farthest they will wander to search for food. At dusk they will return to their roosts and you won't have to look for them, unless they become disorientated or decide to spend the night in a tree.

Be alert to neighborhood dogs that don't have good chicken manners. A dog's instinct is to chase and kill fowl. Most city laws require that chickens be restricted and dogs be on leashes.

7. Deep Litter Chicken Runs

To keep chicken runs from becoming toxic from too much manure, use a deep litter system. Put boards around the run, like a raised bed, as a threshold to keep the litter in. Add 6 to 12 inches of biomass, which can include wood chips, leaves, old hay, or straw. The litter will break down and the hens will help by scratching and making increasingly smaller bits. They will also enrich it with their manure. Keep adding bedding to build up the base. The deep litter becomes increasingly bioactive with soil-dwellers that the hens will seek out and devour. After a while, you will have a wholesome, nitrogen-enriched soil amendment that you can harvest (move) and use in your yard or garden.

8. Growing Small Grains in Small Places

Many grains are easy to grow but hard to harvest. Grow grains as a cover crop and allow the seeds to form. Then let the birds in to glean the grains. Wheat, oats, barley, rye, and other grasses can be used. Other homegrown fodder for rotational grazing can include kale, mustard, salad greens, and rape. Cleaning cold frame beds is an option in the winter.

9. Gardening for Chickens

There is an entire chapter on growing food for chickens, which can cut your commercial feed bill. Plantings such as sunflowers, amaranth, and comfrey can make niche and edge plantings. Hens will eat root crops, although they prefer them cooked or in tiny pieces. Squash, pumpkin, melon, and other seeds are high in protein and can be collected, dried, and stored for winter feed. Save the pumpkin seeds from those Halloween jack-o'-lanterns for your hens.

10. Milk and Dairy Products

Anita Mason raises replacement hens for small-flock owners. She has 2 Guernsey cows producing milk that she can't legally sell for human consumption. Having more pasture than cash flow, she has her dairy cows eat grass and convert it into milk, which she then feeds to the chicks as a protein supplement. The chicks drink every drop and Anita doesn't have to buy much protein to feed her younglings. Getting excess dairy products from local dairies is a time-honored tradition among hen keepers. It shows up in quality eggs.

11. Worm Bins

Worms are considered micro-livestock and can be raised in small bins as chicken feed, while producing compost at the same time.

12. Not Feeding Rodents

Rodents are persistent, pervasive, clever finders of food, and they can grow very fat on chicken feed. Chapter 15 on Primary Poultry Health Care talks about rodents as a health problem and describes ways to deter rodents from freeloading on your feed tab. An old Norwegian proverb advises: "It's better to feed one cat than many mice". Perhaps even if it's your neighbor's cat.

13. Culling

In micro-flocks, hens are usually regarded as pets. Has her prior service merited social security benefits? It could be that she carries her weight as a mentor for younger flock members and as a worker in the garden. It's hard to dispose of a pet before her time, but be aware of how much it costs to keep that older hen.

14. Cutting Back on Feed Available

There are many opinions about the practice of cutting back on feed to encourage foraging. It is tricky to make sure a hen has enough feed for egg production when you reduce the amount of commercial feed available so she will get off her feathered behind and forage. How much a hen can eat by foraging off the landscape and gardens depends on the many factors discussed earlier in this chapter.

When my flock goes out in the morning they immediately start searching for their breakfast. Their preference is usually not to eat out of the feeder. If there is a particular place I want them to focus on, a little scratch in that area will direct their attention. Scratch can act as chicken caffeine to get them jump started and out foraging.

In the late afternoons, about an hour before roost time, I make sure my birds got enough to eat during the day. I want them to go to bed with full crops. I want the growing pullets to eat enough so their growth isn't stunted. My flock always has free choice feed available.

How to Tell if a Chicken is Getting Enough Food

The easiest way to assess if a hen has eaten enough is to feel her crop when she is roosting at dusk or after dark. If the bird's crops are full in the evenings, they are getting enough to eat.

However, if many of the birds' crops are not full, then there isn't enough food available. If only one or a few of the crops aren't full, then something else might be happening. There might be too much competition between the birds. The birds with empty crops might not be getting access to the feeder. Sometimes a bird will be kept from eating because she is at the bottom of the pecking order; this often happens in small areas. Separate the bullied birds from the others so they can eat without fear of interference. Put them in a pet carrier, creep feeder, or fenced-off area with food for at least twenty minutes so they can eat their fill in both the morning and late afternoon. Twenty minutes is about the amount of time a hen needs to dine. They will soon be healthy enough to compete with the others for their fair share of the fare.

Empty crops and/or a chicken off its feed might indicate an illness. This merits further investigation and treatment.

A decrease in egg production is another way to tell if adequate feed and proper nutrition are available. With micro-flocks it's easy to know which birds are laying how many eggs. If you reduce feed to encourage foraging and have a drop in egg production, then more feed is necessary. Egg production naturally declines as days get shorter, and as the weather gets colder, your birds will need more to eat to keep warm.

Decreasing the amount of feed available to encourage foraging is a delicate balance. It requires a keen eye, thoughtful observation, and tracking the nutritional status of individual birds. It requires experienced poultry management. It's best to keep feed available all the time.

The quality and content of commercial feeds can be questionable. If you are concerned about feed quality, then buying cheap feed is not the answer. There are many ways to enhance your flock's diet with fresh, natural feeds from sources other than the feed bag. Some of the alternative feeding methods described in this chapter can help you save on feed costs, produce higher quality eggs, and have healthier birds.

7 Chicken Quest: How to Find Your Flock

Congratulations! You've made the decision to keep a flock of City Chicks. All the necessary arrangements have been made. Your uptown hen housing is ready and all of the hen furniture (water, feeder, nest boxes, ladders, pop holes) is in place. You are ready to add hens. The next step is to decide where to get the hens, which breed best suits your goals, and which breed will thrive in your climate. The breed (or breeds) you choose has a huge impact on your poultry experience. Some breeds are better able to handle cold weather, others are better foragers, and some are more docile than others.

Before getting just any hens, ask yourself: why am I keeping chickens? The answer to this question will help you decide which breed to buy. Some reasons to keep chickens are:

- Have fresh, healthy eggs

- Employ chickens as garden helpers

- Utilize chickens as solid waste management workers

- Exhibit at poultry shows

- Preserve heritage breed genetic pools

• Create rich compost and bio-recycle

I've chosen which types of chicken breeds to buy based on the objectives I had for my micro-flock. They are listed below, in order of importance:

1. Egg production. Eggs are my primary source of protein. I want to know that the eggs I eat are from happy, day range hens that eat nutritionally superior food.

2. Helping in my garden and enriching the topsoil. I'm an organic gardener and it's important to me to grow healthy foods in healthy soil.

3. Entertainment: I genuinely enjoy having chickens in my yard.

There are poultry keepers who prefer to own purebred hens because they are proud of the breed. Others have eclectic flocks composed of several breeds, a few hybrids, and a few of unknown origin. I like to keep heritage breeds with utility strains bred specifically for egg laying, such as the Rhode Island Red.

If eggs are your only goal, purebred chickens are not necessarily the best choice. Some purebred hens lay only about 90 eggs per year; not much compared to the commercial hybrid chickens, which are genetically selected for high-quantity egg production. A young commercial hybrid hen at the height of her career can lay up to 300 eggs per year – almost one egg a day. Production decreases with age.

What is a hybrid? When two distinct breeds with a desirable characteristic mate, they produce offspring that are superior in the desired trait. The offspring are called hybrids. Chickens have been hybridized to enhance many characteristics, including:

• High-quantity egg production
• Variety of eggshell colors
• Size of the bird
• Ease of weight gain and good feed conversion
• Feather patterns and colors
• "Hybrid vigor"

I keep a few hybrid hens along with the heritage birds. The hybrids are great layers, active foragers, and brave hunters. Some have strong wills, charming personalities, and a poultry sense of loyalty.

Remember that you will probably have your hens for years, which is not the goal of industrial chicken operations. In factory farms, hens are considered production "units" and are disposable, like widgets. In commercial production, a pullet (a hen under one year of age) begins to lay eggs at 16 to 18 weeks, or 4 to 4.5 months old, before she is fully grown. The hen has been stressed by a high-protein, antibiotic-laced diet and overcrowding. Chicks and young pullets are sometimes caged with other birds in such small spaces that they lose the ability to walk. Hens live inside around the clock, often with continuous light. They burn out quickly; a hen's productive life is usually over by the time she is just a little over a year old (60 weeks). Heritage breeds develop slower than hybrid chickens and begin laying eggs at 5 to 7 months. They lay fewer eggs than hybrids, but continue to lay for years.

The hens you get will hopefully be raised in a gentle environment, and not at a fast-track commercial pace. One of my friends has a hen that is 12 years old and still lays about one egg a month. At a commercial farm, this hen would have been culled years ago, but she earns her keep in her backyard flock by working in the garden and mentoring younger flock members. She's Headmistress of the Poultry Finishing School.

Julie Scherbarth of Oregon wrote in *Backyard Poultry Magazine* that she has a healthy, active, 14-year-old hen that lays one egg a year in the spring. Whether you choose hybrid layers or purebred show stock, every old, young, purebred, or hybrid chicken can be productive by eating food scraps, providing manure for compost, and/or fertilizing soil directly.

There has been a lot of discussion about the foraging ability of different breeds; some people think certain breeds are naturally better than others. Personally, I think that when any chicken gets acclimated to free ranging, they will innately and vigorously express their chicken-hood by joyful foraging. It might take a while for cage-raised birds to adapt to having more space to exercise, flap weak wings, and graze, but they will eventually.

Cold Climate Considerations

Some chickens cope with cold weather better than others. Healthy chickens have heavy coats of feathers to insulate them from the cold. The combs are the most cold-sensitive part of a chicken and are most likely to suffer frostbite. Wattles, feet, and toes are also at risk.

Large combs are more susceptible to frostbite than small ones because of the surface area exposed. Breeds developed in very cold climates, like Canadian Chanteclers, have tiny "pea" combs that lay close to the chicken's head. Some breeds with pea combs are Wyandottes (from New York State), Dominiques (a Heritage breed from Colonial times), and Buckeyes (developed in Ohio).

Feathered Legs and Feet

Some breeds have long, fine feathers on their legs and feet. These birds are often very pretty and distinctive-looking. If you decide to include these fancy feathered friends in your flock, keep in mind that they require slightly more maintenance and care than bare-legged chickens. You need to make sure that mud or manure does not get caked on the feathers, as this can cause painful toe balls and make it hard for the hens to walk. Trimming the leg feathers is also an option.

How Long to Keep a Hen?

Your urban micro-flock will probably have hens of various ages. You have to ask yourself if it's practical to keep older hens whose egg production

Combs and Wattles as a Consideration for Cold Tolerance

Breeds with combs that fit closely to the head (pea combs), feathered wattles and ear lobs like the Araucana hen on the right are more resistant to frostbite than breeds with larger combs, wattles, and exposed ear lobes.

has waned. Do they justify the cost of their feed? Should you enjoy them as loyal pets or as a chicken dinner? These are hard questions that only you can answer.

I take a middle-of-the-road approach. I buy pullets and occasionally adopt orphans. I think that older hens are valuable flock members because they mentor younger chickens, forage for garden pests, and provide manure for fertilizer. I don't know exactly how much feed the older hens eat, but I suspect it's less than the 100 pounds/year average that a production hen consumes. If egg production were my only objective, keeping an older hen wouldn't be practical or cost effective.

Where to Buy Chickens

There are many sources, both local and national, where you can buy chickens. Many of these are explored below.

Local Poultry and Egg Producers

Natural and organic food stores, food co-ops, and farmers' markets are good sources of contact information for local poultry owners. Farmers' markets have a wealth of information about who has what. Ask the vendors selling eggs if they have any hens to sell; they usually have a few to spare. Sometimes they would like to sell – or even give away – older hens whose egg production has dropped. These hens still have plenty of productive life (hens usually lay until they are 8 years old), and will lay larger eggs than young hens, but not as many.

Local regulations often don't allow live animals to be sold at farmers' markets. If you find a vendor who has hens, you will probably have to pick them up at their farm.

4-H Fairs and Poultry Shows

Each year, 4-H Poultry Club members raise a batch of 50 to 100 chicks. When the chickens are grown, they take the best ones to the county fair, which is a wonderful place to meet poultry people and learn about different breeds. At the end of the fair, they host an auction and sell the 4-H chickens. The money from the auction goes into a pool kept by the USDA Extension Service to purchase chicks for next year's club project. This perpetual funding allows children to participate in 4-H whether they can afford chicks or not. 4-H Poultry Club members are required to sell the best 10 of their brood and keep the rest. This means there are usually extra pullets ready to begin laying. I often buy young hens from 4-H members as replacement hens for my micro-flock.

Poultry Breeders

In almost every county or region, there is at least one breeder raising his or her favorite poultry breeds. For these poultriers, breeding is usually a passion and hobby, combined with a small backyard business. They maintain breeding flocks and often have incubators to hatch eggs. You can support them by buying their pullets as replacement hens. You might have to network to find these poultry people.

Livestock Auctions

Use the Internet to find a livestock auction in your state or county. In season there will often be an auction every week, or at least once a month. In the fall many breeders thin their flocks so they don't have to carry them over the winter. This is an especially good time to find deals at the poultry auctions.

Heritage Breeds Associations

American Livestock Breeds Conservancy (ALBC) and the Society for the Preservation of Poultry Antiquities (SPPA): Both groups maintain a list of poultry breeders and will be glad to connect you with breeders in your area.

Swap Meets

In many places, poultry swap meets are regular events. These informal and often word-of-mouth swaps are usually tailgate affairs at someone's private farm.

Network Sources of Poultry People

•*Farm Supply Stores.* Check with the local farm supply store; they will know who buys chicken feed. In the spring, farm supply vendors will often order chicks and have full grown hens available for sale.

• *USDA Extension Office.* Call the U.S. Department of Agriculture and talk with the agricultural agent. He or she can let you know who is involved with the 4-H poultry club, follow the list they give you.

• *Farmers' Markets.* Sleuth who is selling eggs at the market and ask if they have any hens to sell. Also, find out where they get replacement hens. As the City Chicks Movement expands, selling hens could become a stream of income for small farms. There is a farmer in our area who loves brooding baby chicks and raises custom batches; customers tell her what breed they want, and she purchases the chicks and broods them until they are ready to live in a coop.

• *Local Poultry Clubs.* If there is a local poultry club, its members will be pleased to help you find hens.

• *Breed-Specific Clubs.* There are clubs and societies dedicated to promoting specific chicken breeds. Some of these are listed in the Resource Section. Search the Internet for the breed you are interested in.

• *American Livestock Breeds Conservancy* provides members with a directory listing contact information for heritage breeds of all livestock.

Finding the chicken breeds that best suit your needs might involve some trial and error. Be realistic about what kind of chickens are best for you and be patient while trying to find them.

Having an appropriate number of the best breeds of hens at the right age for your goals will make your City Chicks flock easy to care for and fun to own. Don't be afraid to ask for help during your chicken quest. Chicken lovers are generally a helpful, friendly bunch, and even the grumps will be happy to answer your questions. You will probably get more information than you bargained for.

The following article by the ALBC discusses the importance and advantages of keeping heritage chickens. It also explores how essential it is to preserve the continually dwindling genetic pools of of these valuable and rare historical breeds.

The ALBC Chicken Breed Comparison Table on the following pages summarizes many of the particular features of heritage breeds. Much of this information was gleaned from the catalog published by the Sand Hill Preservation Center, Calamus, Iowa. Some of the information is subjective, especially about temperament and foraging skills. Both these factors are dependent on owner handling. and having forage access.

This table contains the breeds tracked by the ALBC as part of their breed preservation efforts. Many of the breeds are on the verge of extinction.

All of these breeds have fallen out of favor to the high-power production hybrids. Production hybrids are bred specifically for quantum weight gain, or early laying, and high quantity of eggs laid in the first two years of life.

The heritage poultry can't compare with the speed of early weight gain and high egg production of the factory hybrids. But like the story of the race between the turtle and the hare, the heritage breeds might be slow to start, but they will be of service for a long time. Many of the heritage breeds are kept for both meat (table birds) and egg production which is not in the production hybrid genes.

Antique breed preservation is kept alive by small, independent, flock keepers. Heritage breeds cannot be saved, or even preserved, in an enclosed, artificial environment of the factory farms. That is not what they are bred for, and there isn't incentive for a commercial factory farmer to have interest in heritage breeds. The bottom-line economics that demand the fast-to-market flock don't allow for slow growth of the old time pioneer breeds.

To preserve a breed, environmental conditions and selection need to be similar to the original environment, purpose and development of a breed.

There are many organizations dedicated to preserving the heritage poultry genetic pools. An Internet search will help you get in contact with other heritage breed preservation minded folks in your county. The American Livestock Breeds Conservancy has a dedicated staff tracking endangered breeds of all livestock. I encourage you to become a member to support their valuable efforts.

ALBC has members all across the country and internationally. Members have a wide variety of backgrounds, skills, experiences, and perspectives. Through your membership you can network with other ALBC members who share common interests.

As a micro-flock owner, you are in a unique position to help support the heritage poultry breeders and keep the past alive for future generations.

> *"...when the last individual of a race of living things breathes no more, another Heaven and another Earth must pass before such a one can be again."*
>
> — WILLIAM BEEBE

The American Livestock   Breeds Conservancy

8 Raising Heritage Chicken Breeds in the City

BY JENNIFER M. KENDALL AND
THE ALBC STAFF

Do you have an interest in preserving rare breeds of livestock to protect the genetic diversity of the species and to contribute to the greater conservation effort? If so, the American Livestock Breeds Conservancy (ALBC) suggests raising a heritage breed of chicken if at all possible.

Founded in 1977, ALBC is a nonprofit, membership organization working to protect over 150 breeds of livestock and poultry from extinction.

Traditional breeds of livestock and poultry are threatened because agriculture has changed. Modern food production favors the use of a few highly specialized breeds selected for maximum output in intensively controlled environments. Many "heritage" breeds do not excel under these conditions, so have lost popularity and are threatened with extinction.

These older breeds are an essential part of the American agricultural inheritance. They are remnants of our cultural heritage, but more importantly, they possess genetic material critical for the continued existence and future adaptation of on-going changes in agriculture. Like all biological systems, agriculture depends on genetic diversity to adapt and respond to an ever-changing environment.

Heritage breeds retain essential attributes for survival and self-sufficiency making them good options for those interested in raising chickens in the city.

Heritage breeds possess foraging ability, longevity, resistance to diseases and parasites, maternal instincts, fertility, and the ability to mate naturally. As modern agriculture adapts and evolves, we will need these genetic resources to draw on for a broad range of uses and future opportunities. Once lost, genetic diversity is gone forever. It can never be recovered. You can play a part in the survival of these breeds. Buying a few from a hatchery or local breeder gives them a reason to keep the breeding flock.

If you are interested in heritage breeds we encourage you to take a close look at the American Class of chickens. All of these have been developed in the North America, and, as U.S. citizens, we have a special responsibility for them. Some examples of American Class breeds include the Dominique, Plymouth Rock, Wyandotte, Rhode Island Red, Jersey Giant, and New Hampshire. For more information on ALBC and its work, visit www. albc-usa.org.

What is a Heritage Chicken Breed?

The American Livestock Breeds Conservancy has recently defined a heritage chicken as the following:

Purpose of Breed

Chickens have been a part of the American diet since the arrival of the Spanish explorers. Since that time, different breeds have been developed to provide meat, eggs, and pleasure. The American Poultry Association began defining breeds in 1873 and publishing the definitions in the Standard of Perfection. These standard breeds were well adapted to outdoor production in various climatic regions. They were hearty, long-lived, and reproductively vital birds that provided an important source of protein to the growing population of the country until the mid-20th century. With the industrialization of chickens many breeds were sidelined in preference for a few rapidly growing hybrids. The American Livestock Breeds Conservancy now lists over three-dozen breeds of chickens in danger of extinction. Extinction of a breed would mean the irrevocable loss of the genetic resources and options it embodies.

Therefore, to draw attention to these endangered breeds, to support their long-term conservation, to support efforts to recover these breeds to historic levels of productivity, and to re-introduce these culinary and cultural treasures to the marketplace, the American Livestock Breeds Conservancy is defining "Heritage Chicken". Chickens must meet all of the following criteria to be marketed as Heritage.

Definition of a Heritage Breed

Heritage Chicken must adhere to all the following:

1. American Poultry Association (APA) Standard Breed

Heritage Chicken must be from parent and grandparent stock of breeds recognized by the American Poultry Association (APA) prior to the mid-20th century; whose genetic line can be traced back multiple generations; and with traits that meet the APA Standard of Perfection guidelines for the breed. Heritage Chicken must be produced and sired by an APA Standard breed. Heritage eggs must be laid by an APA Standard breed.

2. Naturally Mating

Heritage Chicken must be reproduced and genetically maintained through natural mating. Chickens marketed as "heritage" must be the result of naturally mating pairs of both grandparent and parent stock.

3. Long Productive Outdoor Lifespan

Heritage Chicken must have the genetic ability to live a long, vigorous life and thrive in the rigors of pasture-based, outdoor production systems. Breeding hens should be productive for 5-7 years and roosters for 3-5 years.

4. Slow Growth Rate.

Heritage Chicken must have a moderate to slow rate of growth, reaching appropriate market weight for the breed in no less than 16 weeks. This gives the chicken time to develop strong skeletal structure and healthy organs prior to building muscle mass.

Chickens marketed as "heritage" must include the variety and breed name on the label.

Terms like "heirloom," "antique," "old-fashioned," and "old timey" imply "heritage" and are understood to be synonymous with the definition provided here.

Definition of a Heritage Breed Egg

A Heritage Egg can only be produced by an American Poultry Association Standard breed. A Heritage Chicken is hatched from a heritage egg sired by an American Poultry Association Standard breed established prior to the mid-20th century, is slow growing, naturally mated with a long productive outdoor life.

The American Livestock Breeds Conservancy has over 30 years of experience, knowledge, and understanding of endangered breeds, genetic conservation, and breeder networks.

The ALBC and its guidelines is endorsed by the following individuals:

Frank Reese, Reese Turkeys, Good Shepherd Turkey Ranch, Standard Bred Poultry Institute, and American Poultry Association.

Marjorie Bender, Research & Technical Program Director, American Livestock Breeds Conservancy.

D. Phillip Sponenberg, DVM, PhD, Technical Advisor, American Livestock Breeds Conservancy, and Professor, Veterinary Pathology and Genetics, Virginia Tech.

Don Bixby, Independent Consultant, former Executive Director for the American Livestock Breeds Conservancy.

R. Scott Beyer, PhD, Associate Professor, Poultry Nutrition Management, Kansas State University.

Danny Williamson, Windmill Farm, Good Shepherd Turkey Ranch, and American Poultry Association.

Anne Fanatico, PhD, Poultry Program Specialist, National Center for Appropriate Technology.

Kenneth E. Anderson, Professor, Poultry Extension Specialist, North Carolina State University.

Which Heritage Chicken Breed is Right for You?

Below is a chart that compares some of the heritage chicken breeds. Some of the information provided on this chart is subjective. Production characteristics will vary by source as different lines will have been subjected to different selective decisions. To find information on ALL heritage chicken breeds on ALBC's conservation priority list, visit www.albc-usa.org/cpl.

When reviewing the chart, here are some things to consider:

• Flightiness: Heavier breeds that have been bred more for meat than eggs tend to be less flighty than the lighter weight egg producers. When keeping birds in the city, you may prefer a bird that is less flighty.

• Rate of Lay: All chickens lay eggs, but some will lay more – nearly one a day - and for a longer season. If you have a small family and don't eat eggs every day, a meat or dual purpose breed will serve you well.

• Temperament: When choosing a breed, consider temperament. Some breeds are calmer than others. If raising chickens in a backyard in the city, you may prefer a calmer breed. Some birds are more people friendly than others. This varies both by breed and by breeder. Some breeders won't tolerate aggressive birds in their flocks, so those lines will be safer for families with small children.

• Space: If you have a small space in which to raise the birds, choose breeds with a calmer temperament and avoid birds that are listed as "active" or "likes to range". These breeds will not be happy in close confinement.

• Geography: Consider geography when selecting a breed. In colder areas of the country, consider raising heavier breeds. In hotter areas, consider lighter weight birds. Some birds on the heritage breed list have been specially bred for cold climates. Look for those with rose- or pea-combs that won't be as susceptible to frostbite.

• Special Markets: Two lists are developing to promote both culinary and cultural attributes of endangered chicken breeds. Slow Food USA's Ark of Taste identifies breeds their expert panel believes to have superior flavor. Renewing America's Food Traditions (RAFT) is a collaborative effort of Gary Paul Nabhan, ALBC, Chefs Collaborative, and Slow Food USA to identify and promote endangered seeds and breeds that have provided this nation with its food and fiber.

• Check your ordinances and neighborhood covenants before buying. Most cities will only allow you to keep a limited number of hens based on city ordinances.

Heritage Chicken Fun Facts to Know and Tell

1. Buckeye – Ohio 1890s. Pea comb. Brown eggs. Cold tolerant. Most active American breed. Only American breed developed by a woman – Miss Nettie Metcalf of Warren, OH.

2. Chantecler – Quebec 1908. Cushion comb no wattles. Brown eggs. Super-cold tolerant. Only breed developed in Canada.

3. Delaware – Delaware 1940. Single comb. Brown eggs. Developed as a commercial breed. Delaware and Delaware X New Hampshire most popular commercial broilers on the Delmarva Peninsula for approximately 20 years. Fast rate of growth and good layer. Noted for good personalities.

4. Holland – New Jersey 1934 by scientists at Rutgers University. Single comb. White eggs. Developed as a dual-purpose farm chicken that could produce white eggs. One of the two rarest American breeds.

5. Java – East coast and Missouri 1835 +/-. Single comb. Brown eggs. America's second oldest breed of chicken. Noted for slow rate of growth and the favorite for capon production in the 1800s. Ancestor to the Jersey Giant and Plymouth Rock. Long rectangular back – longest back in the American Class.

6. Cubalaya – Cuba, first in US in 1939 but much older. Pea comb. Tinted eggs. Lobster shaped tail. Often lacks spurs (historically). Noted as being extremely tame even at hatching. Excellent forager. Slow grower up to 3 years to reach final size.

7. Dominique – America's first chicken breed. Rose comb. Brown eggs. Ancestor to the Plymouth Rock. Barred coloration good for hawk avoidance (Hawk colored). Active forager with very good "chicken sense". Ancestor to the Barred Plymouth Rock.

8. Jersey Giant – New Jersey developed between 1870-90. Largest chicken breed, roosters 13 pounds, hens 10 pounds. Single comb. Brown eggs. Fine flesh favored once for capons (replaced the Java for this niche).

9. New Hampshire – Massachusetts and New Hampshire 1935. Single comb. Brown eggs. Developed from the Rhode Island Red by selecting for fast feathering and rapid growth. Once popular as a broiler or in crosses to produce broilers.

10. Rhode Island White – Rhode Island 1888. Developed by Mr. J. Alonzo Jocoy. Rose comb. Brown eggs. Excellent layer. Used in crosses to produce sex-linked layers.

11. Plymouth Rock – America mid 1800. Single comb. Brown eggs. Very hardy. Excellent producer of meat and eggs. Barred variety once the most popular and extensively bred chicken breed (in US). Excellent farm chicken. White variety used very heavily in the production of modern broilers (Corn Rock cross found in every supermarket).

12. Rhode Island Red – Rhode Island and Massachusetts late 1800s. Single or Rose comb. Brown eggs. Once second most popular American breed. Noted for fine flavor of meat. Body profile should resemble a brick. Excellent layer that adapts to a variety of climates.

13. Wyandotte – New York late 1800s. Rose comb. Brown eggs. Compact body. Good producer of meat and eggs. Often able to survive under rugged conditions.

14. Lamona – Maryland 1921. Single comb. White eggs. Developed at the Beltsville, Maryland, experiment station (of the USDA) by Harry Lamon, senior poultryman. Excellent layer. Old birds noted as having fine flesh. Good forager.

15. Nankin – England prior to 1800. Rose and Single comb. Tinted eggs. One of the oldest of bantam breeds. Ancestor to many others including Sebrights. Friendly personality. Excellent broody hens. Very rare for 150 years yet still exists.

ALBC Chicken Breed Comparison Table

Breed	APA Class	Use	Egg Shell Color	Egg Size	Rate of Lay	Brood-iness	Skin Color	
Ameraucana	All Other Standard Breed	Meat, Eggs	Green, Blue	Large			White to Yellow	
Araucana	All Other Standard Breed	Eggs	Blue	Medium to Large			White to Yellow	
Buckeye	American	Meat, Eggs	Brown	Large			Yellow	
Chantecler	American	Eggs, Meat	Brown	Large			Yellow	
Delaware	American	Eggs, Meat	Brown	Large to Jumbo	Very Good		Yellow	
Dominique	American	Eggs, Meat	Brown	Medium to Large			Yellow	
Holland	American	Eggs, Meat	White	Large	Good		Yellow	
Java	American	Meat	Brown	Large			Yellow	
Jersey Giant	American	Meat, Eggs	Brown	Extra Large			Yellow	
New Hamp-shire	American	Eggs, Meat	Brown	Extra Large			Yellow	
Plymouth Rock	American	Eggs, Meat	Brown	Large			Yellow	
Rhode Island Red	American	Eggs, Meat	Brown	Large			Yellow	
Rhode Island White	American	Eggs, Meat	Brown	Large to Extra Large			Yellow	
Wyandotte	American	Eggs, Meat	Brown	Large			Yellow	
Cornish	English	Meat	Tinted	Medium to Large			Yellow	
Dorking	English	Eggs, Meat	White	Medium to Large			White	
Orpington	English	Meat, Eggs	Brown	Large to Extra Large			White	
Sussex	English	Meat, Eggs	Tan to Brown	Large			White	
Australorp	English	Eggs, Meat	Brown	Large	Excellent		White	

Market Wt in Pounds	Rate of Growth in Weeks	Early Maturity	Tempera-ment	Special Characteristics	Special Markets
4.5-5.5	16-20 Weeks		Calm to Gentle	Egg Color	
3.5-4				Egg Color	
5.5-8	16-20 Weeks		Active, yet gentle	Winter Hardy, Extra Meaty	RAFT, Slow Food Ark of Taste
5.5-7.5	16-20 Weeks		Gentle	Winter Hardy, Can be Eaten at Any Age, Winter Layer	RAFT
5.5-7.5	12-16 Weeks	Yes	Gentle	Fast Growth, Can be Eaten at Any Age	RAFT, Slow Food Ark of Taste
4-6	16-20 Weeks		Likes to Range	Good Forager	RAFT, Slow Food Ark of Taste
5.5-7.5	16-20 Weeks		Likes to Range	Good Forager	RAFT
6.5-8	20-24 Weeks		Likes to Range	Excellent Forager	RAFT, in 1800s Considered Finest Table Fowl
8-11	16-20 Weeks		Calm	Large Size	RAFT, Slow Food Ark of Taste
5.5-7.5	12-16 Weeks	Yes	Calm	Rate of Growth, Early Feathering	RAFT, Slow Food Ark of Taste
6-8	12-16 Weeks		Calm	Meat Qualities, Early Feathering	RAFT, Slow Food Ark of Taste
5.5-7.5	16-20 Weeks		Calm, but Likes to Range	Flavor - Once Considered Finest Flavored	RAFT, Slow Food Ark of Taste
5.5-7.5	16-20 Weeks		Calm, but Likes to Range	Used in Producing Hybrids	RAFT
5.5-7.5	16-20 Weeks		Calm	Good Cross with Plymouth Rock	RAFT, Slow Food Ark of Taste
6.5-8.5	24+ Weeks		Calm	Cornish Hen, Superb Table Qualities	Famous for Flavor
6-8	20-24 Weeks		Likes to Range	Winter Layers, Good Forager	Famous for Flavor
7-8.5	20-24 Weeks		Calm	Excellent Rate of Growth in Some Lines	
6-7.5	16-20 Weeks		Calm	Good All Around Table Bird	Famous for Flavor
5.5-7.5	16-20 Weeks	Yes	Active, yet gentle	Productive and Fast Growing	

Breed	APA Class	Use	Egg Shell Color	Egg Size	Rate of Lay	Brood-iness	Skin Color
Ancona	English	Eggs	White	Medium to Large	Excellent	Non-Setters	Yellow
Buttercup	English	Eggs	White	Small to Medium		Non-Setters	Yellow
Blue Andalusian	Mediterranean	Eggs	Chalk-White	Large	Very Good	Non-Setters	White
Catalana	English	Eggs, Meat	White to Tinted	Medium			White
Leghorn	Mediterranean	Eggs	White	Medium to Large	300 +/-	Non-Setters	Yellow
Minorca	Mediterranean	Eggs	Chalk-White	Very Large	Excellent		White
Penedesencas		Eggs, Meat	Deep Dark Brown	Medium	160+/-		
Spanish	Mediterranean	Eggs	Chalk-White	Large		Non-Setters	White
Barnvelder	Continental	Eggs, Meat	Deep Terra Cotta	Medium to Large	150 to 200		Yellow
Campine	Continental	Eggs	White	Medium to Large	Excellent	Non-Setters	White
Crevecour	Continental	Meat, Eggs	White	Medium to Large	Good		White
Faverolle	Continental	Meat, Eggs	Tinted to Light Brown	Medium to Large	Excellent		White
Hamburg	Continental	Eggs	White	Medium	Excellent	Non-Setters	White
Houdans	Continental	Meat, Eggs	White	Large		Non-Setters	White
La Fleche	Continental	Meat, Eggs	White	Medium to Large			White
Lakenvelder	Continental	Eggs	White to Tinted	Medium	Good	Non-Setters	White
Maran		Eggs, Meat	Deep Terra Cotta	Large	150 to 200		
Polish	Continental	Eggs	White	Medium to Large	200+	Non-Setters	White
Welsumer	Continental	Eggs, Meat	Deep Terra Cotta	Medium to Large	up to 200	Non-Setters	Yellow
Naked Necks	All Other Standard Breed	Eggs, Meat	Brown	Large	Good		Yellow
Langshan	Asiatic	Eggs, Meat	Very Dark Brown	Large		Setters	Yellow

Market Wt in Pounds	Rate of Growth in Weeks	Early Maturity	Temperament	Special Characteristics	Special Markets
4-5		Yes	Very Active	Noted for Hardiness and Vigor	
4-5.5	Good		Active		
4.5-6	Good		Active, yet gentle	Hardy, Good Forager, Good Layer, Excellent Layer	
5-6.5	Good		Active		
4-5		Yes	Very Active	Noted for Hardiness and Vigor, Excellent Layer	
6.5-7.5	Slow: 20-24 Weeks		Very Active, Likes to range	Great Foragers, Large Frame	
	Slow: 20-24 Weeks			Egg Color, Fair Meat Qualities	
5.5-6.5	Good		Active	Good Layer	
5-6	20-24 Weeks			Egg Color	
3.5-5		Yes	Very Active	Vigorous Forager	
5.5-7	20-24 Weeks		Calm		A French Favorite
5.5-7	16-20 Weeks		Active, yet gentle	Very Productive when in contact with the ground	Dish - Petite Poussin (Small Breast)
3.5-4			Very Active	Flighty, Roosts in Trees	
5.5-7	20-24 Weeks		Calm		French Favorite
5.5-7	24+ Weeks		Active	Good Forager, Large Breast for Size	French Favorite
3.5-4			Active		
	16-20 Weeks			Egg Color, Good Meat Qualities in Some Lines	
4 - 5 lbs		Yes	Calm, but Flighty when Disturbed		
5 - 6 lbs	20-24 Weeks		Calm, Likes to Range	Egg Color, Good Foragers	
5.5 - 7.5 lbs	20-24 Weeks		Calm	Super Disease Resistance, Heat Tolerance	
6.5 - 8 lbs	Slow: 24+ Weeks		Very Calm		

Copyright © ALBC 2009

9 **Brooding Peepers to Layers**

The rapid development of a chicken — from peep to chick to pullet — is nothing less than a miracle. Watching the chicks grow at warp speed into adulthood is like watching life in fast forward. It's as awesome as sitting quietly in a corn field and listening to corn grow. Yes, I admit that I've heard corn grow. I am an agri-nerd and proud of it.

The daily growth of the chicks is fascinating and humbling. During the brooding process you are their keeper and guardian angel. It's a life and death responsibility; the first few weeks of a chick's life set the stage for their short-term survival and long-term development.

Brooding – the process of raising chicks from peeps to pullets – can be scary for the first-time chick owner. You are responsible for growing those fragile, one-ounce peeps into 7 to 10 pound hen. Although the entire process is exciting, the first month is the most dramatic, as peeps grow amazingly fast, shed their fluff, and sprout feathers.

The growth rate for chickens can vary depending on the bred. A guideline for time to "market" is:

Fast = 12 to 16 weeks, or 3 to 4 months

Medium = 16 to 20 weeks or 4 to 5 months

Slow = 20- to 24+ weeks or 5 to 6 months

This chapter is an overview of my experiences, observations, and unconventional methods of raising chicks. As with parenting children, there are a multitude of opinions about chick rearing and as many books on the subject. There is no one right way to brood chicks.

Many of the methods and feeds I use with my micro-flocks are too expensive and impractical for brooding flocks of hundreds or millions. Commercial growers don't have the time to pamper their peeps. In fact, commercial brooders might pshaw my system as being silly. With micro-flocks, individual birds become more like pets. Personalities and preferences emerge when they get individual attention. And something about a baby (of any species) brings out my maternal instinct. Watching my fluffy nurfs cheeping, chirping, huddling, stretching, and flapping awkwardly gets me in a protective, mother-hen-nurturing mode.

Raising babies is not the fastest – or most economical – way to get chickens, especially if you want eggs produced immediately. But if you can wait for eggs, mail ordering chicks gives you a wide variety of breeds to choose from and is a good way to get replacement hens for your flock. You might even have a few extras to sell and potentially cover your costs.

The dark side of brooding is that it is heartbreaking when something goes wrong. In my early days of chicken care I felt like a murderer, guilty of involuntary manslaughter when I found a baby dead, smothered, or drooping, almost lifeless. Worse yet if they are just missing, having fallen prey as a predator's dinner. I'd feel true sorrow and mourn my missing or lifeless peeps.

There are a lot of things that can go wrong; chicks are tiny and extremely vulnerable during the first few weeks of their lives. Chicks have suffered because I didn't know about brooding pitfalls or because I neglected to check on them routinely. Hopefully the lessons I learned from my misfortunes can help prevent losses in your brooding experiences.

Brooder Systems

A brooder system is an environment for baby chicks to thrive in. You are creating a chick biosphere that requires several systems to moderate crucial environmental components.

The most important factor in a brooder is temperature control. Chicks must not get chilled, too hot, or exposed to drafty breezes. What exactly does this mean? While preparing to raise my first brood, here's a summary of the instructions I read. I've since found they are sorely lacking.

> *"Baby chicks can be raised in a box with a 60 watt bulb*
> *for heat. Keep chicks at 95 degrees for the first week. Then*
> *lower the temperature a maximum of 5 degrees each con-*
> *secutive week until the temperature is the same temp as*
> *outdoors temperature. Then you can move them outside.*
>
> *If the chicks are huddled together for warmth and peeping*
> *plaintively then lower the heat lamp. If they are panting*
> *and have their little wings spread and are as far from the*
> *heat lamp as possible, then they are too hot. Raise the*
> *lamp until normal comfort body behavior and chirping*
> *contentedly resumes."*

That was it! Cut and dry, black and white, and sorely inadequate. What I have learned is that for optimal results and low – to zero – mortalities, you have to design a comprehensive, anticipatory peep nursery system that involves more than a bulb and a box. A successful brooding system is dynamic and interrelated. Equipment and space requirements change as chicks grow, so the system must be easy to monitor, adjust, and maintain.

The brooder system should make the keeper's chores easy and efficient and afford high quality of life for the chicks. There are multiple interdependent factors to consider, including:

1. Brooder Location

2. Dynamic Brooder Set-up

3. Environmental Control Systems

4. Watering System

5. Feeding System

Each of these systems is described below, followed by my daily journal of rearing a micro-flock of 30 mixed, heritage breed chicks.

1. Brooder Location

Brooders are easy to set up, and can be placed in almost any enclosed area. We have set up various types of brooders made from many different materials. One was a commercial brooder (in a pole barn) built from cardboard skirting and several propane heat lamps. When the free-range poultry farm was in high gear we used a commercial battery brooder. This brooder was the chick equivalent of a five-story building; each floor housed 100 chicks. Once we brooded poultry in an old trailer house, bought at an auction for a song. We have used a plywood brooder set up in our garage and a bathtub in our guest room.

So, brooders can come in all shapes, sizes and configurations. While deciding where to put your brooder, keep in mind that chicks stir up a thick layer of ultra-fine dust. It will land everywhere: on tools, towels, cars, windows. The bathtub brooder worked well — although guests would spend curiously long times in the bathroom — but dust settled everywhere in the house and made housecleaning more time-consuming and difficult. Plus, the intermittent cheeping was noisy, especially at night.

2. Dynamic Brooder Set-up

Brooder Size and Shape

Even a few chicks, anywhere from 3 to 30 will, over the course of 4 to 6 weeks, require about 15 to 30 square feet (3' by 5' or 3' by 10') or more. This is enough space for the birds to get away from the heat source so they can self-regulate their temperature, and have room to play, which develops wings and muscles.

Have at least ½ square foot per chick for the first 2 to 3 weeks. Then increase the square footage to a minimum of 1 square foot per bird. You might need even more space if your birds have to be kept inside longer than planned because of cold weather. A good reason to wait until late spring or early summer to buy chicks is the weather; warm temperatures make it easier for chicks to transition from the brooder to outside hen housing.

A big advantage to cardboard skirting is that the oval walls can be adjusted as the body mass of the brood increases.

Brooder Sub-bedding

Using several layers of newspaper as a sub-bedding on the bottom of the brooder makes the final cleaning much easier and gives the babies something to read. Chicks will spend enormous amounts of time looking at and pecking the letters and punctuation marks, seeming to research every word to find out if it has something to do with food.

All the bedding can be composted and made into future garden soil. Only use newsprint and not the shiny, glossy flyers. Shiny paper stuff is slick and a peep can slip – causing straddle leg – and glossy paper isn't good to put in compost.

Brooder Bedding

Many kinds of bedding can be used in brooders. In the wild, a mama hen collects grass, feathers, and other soft stuff for her nest. You can use wood shavings (not cedar or pine), grain by-products, chopped hay or straw, and bedding made from paper products. Don't use pellet bedding, as chicks will eat it and it's not good for them. Sawdust is too fine for bedding; chicks will eat it and not get any nutritional value.

Most animal owners don't know about the toxins in pine and cedar shavings. Pine and cedar bedding is inexpensive, abundant, and convenient. But most laboratories have stopped using these wood shavings in animal cages because of the toxic effects, and you don't want to use them in a brooder.

There is strong scientific evidence that both pine and cedar shavings are harmful to small animals, including baby chicks. The same aromatic compounds that give cedar and pine their pleasant aroma can be toxic. These compounds (phenols) are caustic and acidic. Living in the shavings – like chicks do in brooders – can cause irritation to the nasal passages, throat, and lungs, making way for bacterial infections and pneumonia. This has

Aspen Shavings for Brooder Bedding

The type of bedding used in brooders can have long-term affects on chick health. Pine and cedar shavings are toxic to baby chicks and can cause respiratory problems. Aspen shavings don't contain aromatic oils and are scent free, which is healthier for tiny baby lungs and immune systems. One bag expands to 4 cubic feet.

significant implications for chicks, as respiratory infections are the most common chicken diseases. In a confined space like a brooder, using pine or cedar bedding will give your young babies phenol exposure that can result in respiratory problems later.

Pine shavings are widely used for larger animals, such as goats and horses, and probably safely so. Pine shavings are probably safe to use with adult hens. I use a thin layer of pine shavings on my hen house floor under the droppings pit to collect the manure. Very few of my layers have had respiratory problems. However, I do think that pine bedding has caused problems with previous batches of chicks we've raised, and we no longer use pine in brooders.

Cedar bedding is dangerous for chickens of any age. The phytochemicals in cedar that repel and are poisonous to insects are also poisonous to chickens. Don't use it as bedding.

Aspen bedding is the best choice for chick bedding. It should be available at your local pet or farm supply store. They can order a bag if it's not in stock. One compressed bag of shavings expands to 4 cubic feet, which is more than enough for brooding 25 chicks.

When putting bedding in the brooder, put about ½ inch layer of aspen on top of the newspaper. Put down some fine sand and crumble feed on top of the aspen so that when the chicks arrive they can scratch and find food right away to fill their tiny craws. Chicks will start scratching as soon as you put them in the brooder. As the weeks go by and odors go up, remove old bedding and add new layers when your nose tells you it's time.

A lot of books say to put in a deep bedding down of 2 to 4", but with micro-flocks, I prefer a small amount of bedding for the first few days of brooding, and then add more as needed to keep the surface clean. I think this gives them a better opportunity to find food because the crumble doesn't get as lost as it does in deeper bedding.

Brooder Screen

When they are about three weeks old, the chicks will want mightily to give their new feathers a test flight. Their take offs usually result in crash landings; earnest, but out of control. A brooder screen on top of the brooder keeps the chicks from flapping out. Any kind of screen will do, including 2x4 wire fence, window screen(s), chicken wire, etc.

Training Roosts

Chicks have a natural instinct to roost. If they practice on a training roost in the brooder, they acclimate faster to roosting in the hen house. Training roosts are not commercially available but they are easy to make.

3. Environmental Control Systems

Throughout brooding, don't let your chicks get chilled. You want them to use their energy for growth and development, not to stay warm. Your system has to control temperature, ventilation, and humidity, the requirements of which will be adjusted as the chicks grow.

Heat Source

Heat lamps are the most practical source of heating. There are several heat lamp wattages available, but the 75 watt bulbs are my preference. I think the 150 and 250 watt bulbs are too hot, bright, and harsh for the small area used in micro-brooding of only 25, or fewer, chicks.

I've used the higher wattage for larger broods and in larger areas. They scare me because I've burned my fingers while adjusting the lamp height. The lower wattage bulbs also seem safer to use if children are involved in the brooding experience.

Heat shields are made to handle different wattages. Be extremely careful about matching the right wattage of lamps to shields. It's easy to put a 250 watt bulb in a heat shield only rated for 60 or 100 watts. That creates a fire hazard that can burn the brooder, or house, down.

I use 2 heat lamp units so I have a backup if the first bulb blows or breaks. At the beginning of the brood, use both lamps to keep the temperature at the required 95 degrees. Eventually, one lamp will suffice as the chicks feather out and the temperature (at chick level) can be lowered gradually — a total of about 5 degrees over each week.

Hang the heat lamps from the ceiling for precise temperature control. Place them at least 16" above

Training Roost

It is easy to make chick-size roosts with perches for tiny feet. Using a roost at a young age seems to help the youngsters acclimate to adult sleeping quarters faster after brooding.

the brooder floor and put a thermometer on the floor, directly under the heat lamp, to get a chick-level reading, Raise or lower the lamp as needed to get the right temperature. A brooder blanket can also help with temperature control.

Red heat bulbs are less harsh on the babies than white lamps. The chicks seem to sleep better under a red lamp and don't fight as much. If a chick gets pecked and is bleeding, the others can't see the blood under a red lamp. Pecking each other and drawing blood is a sign that the chicks are not getting enough protein, and you need to raise the protein levels in their feed as described in the chick nutrition section.

Brooder Blanket

A brooder blanket helps control air temperature, humidity, and air circulation. It would be impractical to use with large numbers of chicks raised in larger areas, but in my stock-tank-turned-brooder, a piece of

Brooder Set-Up

There are many ways to set up brooders. This photo shows the basic features: rounded corners to prevent babies from piling up and suffocating, a heat lamp hung for easy adjustment, a screen to keep birds from flitting out, a Reflectix brooder blanket and towel to control air circulation and temperature, and boards to keep the screen in place.

Reflectix aluminum foil is just big enough to cover the top. It's also pliable, and I can easily fold back part of it to allow more ventilation. A piece of plywood over the Reflectix provides rigidity in case a cat jumps on it.

Moisture condensing on the underside of the brooder blanket indicates that humidity is too high. High humidity can cause fungus and mold to grow in the bedding, and that's bad. To decrease humidity, pull back the blanket to leave more open space across the top. This increases air circulation, which decreases condensation and humidity levels in the brooder.

The brooder blanket also helps you to control temperature. Chicks will use body language to tell you how to adjust the cover. If the chicks are huddling together for warmth, then cover a little more of the brooder top so that more warm air stays in and the temperature rises. If the chicks are scattered with their little wings spread out and beaks open and panting, then they are too hot. Pull the brooder blanket back to let hot air out. Never cover the brooder top completely with a blanket; this inhibits air circulation and the brooder will get too hot.

QUIK CHIK
Vitamins and Electrolytes for Poultry

Net Weight 4 Ounces

Ingredients per pound:

Vitamin A (as palmitate)....................5,000,000 IU
Vitamin D3 ....750,000 IU
Vitamin E..2,500 IU
Riboflavin..500 mg
d Pantothenic Acid..............................4,000 mg
Menedione Sodium Bisulfite complex......2,000 mg
Folic Acid..125 mg
Thiamine Mononitrate..........................250 mg
Potassium and Sodium as Chloride Salts
CAUTION: Keep out of reach of children.
          For oral use in poultry only.
DOSAGE: Put 1 teaspoon in each gallon of water.
     Can be used with sugar in the water.
     Use until gone. Mix fresh daily.
     This package can also be mixed in 250 pounds of feed.
     Keep package sealed
     Can still be used if the powder gets hard.

Manufactured For
MURRAY McMURRAY HATCHERY
Webster City, IA 50595
Phone 515-832-3280

Chick Starter Supplement

Adding electrolytes and vitamins to chicks' water supply gives them a healthy start, especially if they have traveled by mail and are stressed. The packet is usually tucked under the bedding mat of the mailing box.

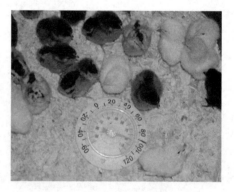

Brooder Thermometer

Brooder temperature must be controlled at chick level. This easy-to-read thermometer is on the bedding directly under the heat lamp so it will give an accurate brooder floor reading. Don't leave it there long, as it will get too dirty to read.

My dream brooder would have a radiant floor to warm the chicks from the ground up. My dream house would have one, too, to warm cold feet and paws.

Thermometers

Any easy-to-read thermometer will monitor the brooder temperature. The round ones with face dials are especially good because you can set them on the brooder floor, directly under the heat lamp, and they will give you an accurate temperature reading at chick-level. Don't leave the thermometer on the floor because it will get very dirty from chicks sitting on top of it.

Draft and Ventilation Control

The how-to poultry books say that you should have a brooder that is "draft-free with good ventilation". But what does this mean? Last year one of the 4-H poultry club members had their brooder set up according to standard guidelines, and chicks started dying. Only after about half the chicks died did they figure out the chicks were wind-chilled from a floor draft.

People don't realize that drafts can be different around their upper body than they are around their ankles. A breeze so slight that it

Brooder Temperature

These photos show how chicks self-regulate their body temperature. In the photo above, the chicks are nicely scattered. The temperature is just right.

The chicks below are cold and huddled together to share body heat. To increase the brooder temperature, lower the heat lamp, or cover some of the brooder with the brooder blanket.

Chicks that are too hot will be as far away from the heat source as possible to get cooler.

escapes a person's notice feels like a whistling, arctic wind to a chick on the floor.

Using a match or votive candle, check for drafts 1/2" to 2" off the ground. This is knee-high to a chick, where drafts can turn into wind-chilling chick killers.

4. Brooding Water System

Chick waterers are designed so chicks can't fall in and get wet or drown. As they grow from peeps to pullets, chicks will use three sizes of waterers.

1. A baby waterer is usually quart size for the first week or so.

2. A pullet waterer provides beak space, but not so much they can fall in.

3. Adult waterers that come in various shapes and sizes (see Chapter 5 on Housing and Furniture & Interior Designs).

First Water

The first few drinks a chick takes helps it rehydrate and stabilize its body temperatures. A chilled, dehydrated, 2-day old chick that just came

Brooder Waterers

Waterers should be small enough so the babies won't fall in, but big enough to provide at least a day's worth of water. The metal waterer on the left holds a quart (peep size), and the plastic waterer on the right holds a gallon (big chick to pullet size).

in the mail will need water to help her acclimate to the brooder temperature. If she is cold and needs warmth, give warm (not hot) water for the first drink. If a chick arrives in hot weather and looks overheated, give her cool water. After the first batch of water, refills can be at room temperature.

City or treated water with heavy chlorine can discourage the chicks from drinking and disrupt their digestive system. If you only have access to chlorinated water, use bottled or filtered water during the first week.

To give chicks an extra boost, add an oral rehydration powder with vitamins and electrolytes to their water. Quik Chik is one brand. This will help the chicks overcome transit stress and fill in any nutritional deficiencies. One pack lasts for at least several weeks and might be enough for a few micro-flock broodings.

Some poultry keepers put a quarter cup of sugar per quart of water to give chicks a boost when they arrive. After the chicks have had their fill of food and drink, refill the waterer with fresh, plain water. Just like with humans, too much sugar isn't good for chick systems; you don't want hypergly-

Brooder Feeders

Chick feeders come in different sizes and shapes. They can be made of metal, plastic, or wood. Some hang on walls and others sit on the floor. It is important that the feeder have a lip so the chicks don't scrape the feed out easily and waste it.

Chick's Sand Box

Baby chicks eat fine sand from the very beginning. Sand helps with their digestion. In about a week they will outgrow this box and a larger one will take its place.

cemic, hyperactive, attention-deficit chicks dashing around because their small systems are filled with sugar.

5. Brooding Feeding Systems

Feeding chicks has one basic rule: have enough of the right food available at all times. A baby chick's metabolism is remarkably high and they need correct nutrition for their explosive growth. They eat at all hours – even in the middle of the night – so make feed available around the clock.

Chick feeder(s) are designed with two things in mind:

> 1. Chicks must be able to reach food, but shouldn't be able to get on top of the feeder. There is usually a bar across the feeder to keep them from perching and pooping on the feed.

> 2. Decreasing the chances of chicks tossing feed out of the feeder and wasting it.

Chicks are attracted to bright and shiny objects. To help them find food quickly, put some crumble on a piece of aluminum foil. Use non-medicated (no antibiotics) feed with at least 20% protein in crumble or mash form. It's irrational to give healthy chicks antibiotics because antibiotics kill the both bad and good bacteria in the chick's gut: anti (against), biotic (life). This compromises its digestion and immune systems.

Fine Sand & Sand Box

Sand (grit) in the brooder helps the chicks digest food and, by doing so, helps prevent pasty butts. There is some controversy about using sand in brooders, but I've had good results with it and the chicks, from the very beginning, peck the sand with such gusto that sand-eating seems to be an innate chick tendency.

Spread about a half pound of fine sand with the initial ½ inch of bedding. You'll see your babies constantly pecking at the sand and consuming it at an amazing rate. Later, as fresh bedding is added, add a small cardboard sandbox with 2" tall sides; this will hold about 1 pound of sand and will help you monitor when the chicks need more. While living in the brooder, 25 chicks will consume (and scatter) about 15 to 20 pounds of sand.

Sand is also useful in composting. It adds trace minerals and structure to the finished product.

List of Top Nutritional Supplements for Brooding Chicks

There are many chick supplements available to optimize nutrition. Chick crumble is formulated to meet the general needs of growing babies, and they will do fine eating only the commercial ration. Having said that, I prefer to give my chicks as many nutritional advantages as possible. Below is a list of my favorite brood supplements:

1. Oral rehydration and vitamin mix added to the first batch of water. Supplements can be added to the chicks' waterers during the entire brooding process but be sure not to overdose them with too much.

2. Organic hard-boiled eggs (mashed up) for quality protein.

3. Flax & sesame seeds (first ground into a meal and later fed as whole seeds) for fiber, omega-3 oils, lignans, and prebiotics to support intestinal bacteria.

5. Liquid B-complex added to the water if there are any signs of leg problems; preparations like Quik Chik contain B-complex vitamins to help avoid leg problems.

6. Apple cider vinegar added to the water after the second week; about 1 tablespoon/gallon provides trace minerals and is slightly acidic to aid digestion. Do not use in metal waterers.

Brooder Checklist

[ ] Brooder structure and space (garage, cardboard box or stock tank, etc.)
[ ] Newspaper — enough for several layers under the bedding
[ ] Bedding (aspen, NOT pine or cedar)
[ ] Two heat lamps (one is a backup)
[ ] Red or whole spectrum 75 watt bulbs and proper heat shields
[ ] Thermometer that is easy to read at chick floor level
[ ] Chick waterer designed so chicks can't get wet, and yet reach the water
[ ] Chick feeder designed for easy chick access

A Complete Micro-flock Brooder

This photo shows an example of brooder design and furniture placement. A brooder with round corners, like this stock tank, helps prevent chicks from piling up in the corners and suffocating those on the bottom. This can happen if the temperature goes to either extreme, or the chicks become unusually frightened. This brooder is draft-free at chick level.

Starting at the top center is a gallon waterer, which sits on blocks of wood to keep the waterer chest-high to a chick. As the chicks grow, more blocks are added for extra height.

The red bulb heat lamp hangs from the ceiling and can be raised or lowered to control temperature. The heat lamp has a safety line so it won't squash a chick if the clamp fails or start a fire. The lamp is placed at one end of the brooder so the chicks can self-regulate their body heat by getting directly under the lamp for warmth, or moving away from it to cool off. A training roost is at the edge of the heat zone.

There are two feeders, a round metal base feeder and a trough feeder. The trough feeder is used for supplements and moist mash.

The ever-popular sand box is in the lower left corner.

Non-toxic aspen wood shavings are used for bedding. Underneath the aspen shavings are layers of newspaper for easy clean up. Not shown in this photo is a brooder screen and brooder blanket. These chicks are two weeks old.

[ ] Chick feed – non-medicated with at least 20% protein or higher
[ ] Fine sand and chick-size sandbox
[ ] Training roosts
[ ] Feed supplements
[ ] Brooder cover to control temperature, humidity, and air circulation
[ ] Brooder screen to keep chicks from getting out

At first this might all seem complicated and overwhelming. It's not. Just take one step at a time and it will all fit together. The chicks will tell you what's right or wrong if you listen to them and observe their body language. Let them show you how to become a brooder extraordinaire.

Micro-Flock Brooder Journal

We've been brooding, raising, and keeping poultry for about 20 years. We've raised small home flocks for personal egg use and flocks of hundreds for commercial production. We tracked data like growth rates, feed consumption, morbidity, and mortality statistics for those flocks, but I'd never kept a personal journal about brooding experiences.

Brooding is a personal journey, rife with emotions from joy to sorrow. It inspires respect, awe, and appreciation. It's a sacred responsibility to care for and protect the babies. It's mid-February and I can't stop thinking about ordering chickens. The seed catalogues are arriving and I'm tired of cold weather. Thoughts of spring and green landscapes invoke images of my perfect garden, filled with a variety of happy hens, chortling as they graze and glean.

The Murray McMurray Hatchery poultry catalogue is dog-eared and rumpled from rifling through the pages as I look at the photos and read breed descriptions. The catalogue is well written and has fairy-tale descriptions like:

> *"These birds are especially endowed with "spizzerinktum," and are unusually handsome and vigorous."*

> *"The history of these can be traced back to Japanese paintings over 300 years ago."*

> *"These little birds are quite a treat for the eyes. White plumage and mulberry colored comb, face, and wattles.*

The earlobes are a light blue turquoise and the skin is dark bluish/black."

There are many breeds to choose from. Some have fanciful names: Buttercups, Lakenvelder, Silkie, Sebright, Dorking, Mottled Houdans, and Phoenix. They even have a chocolate turkey breed. I wondered if they lay chocolate eggs like Easter bunnies. The hatcheries even have creative names like Cackle Hatchery, Ideal Poultry, Stromberg's, and Larry's Poultry.

Looking on the Internet, I find there are even more hatcheries that sell specialty breeds than I thought were available. This is reassuring for the preservation of lesser-known breeds; I hope that more hatcheries will sell them in the future. There are breeds of chickens that are at risk of becoming extinct. Their preservation fell by the wayside when commercial growers concentrate on breeding highly efficient production hybrids. The American Livestock Breeds Conservancy (ALBC) and the Society for the Preservation of Poultry Antiquities (SPPA) were formed to help preserve rare and endangered breeds of fowl. They publish lists of breeders and poultry fanciers all across the U.S., with some located close to where I live.

After mulling over the catalogs for hours, I decide what to order: 25 mixed-breed brown egg layers and 5 Araucanas (also called Easter egg chickens because they lay eggs in several colors, from turquoise to olive). 30 birds is more than I have room for in my hen house, so if they all survive, I'll sell 10 of the pullets to keep my flock around 25; my current flock has dwindled to five hens.

Until this batch, we had only raised high production hybrids because they grew quickly, had good feed conversion and were fast to market and profit. This is my first time raising heritage breed chicks. I'm expecting some behavioral and growth rate differences between them and the mixed-breed brown layers. I want more from this brood than just high egg production. I also want to raise garden helpers. I want to get experience raising heritage breeds and see how they can be integrated into gardening systems. They will be research hens: feathered guinea pigs. In addition to all those good reasons, the real reason for this batch is that I genuinely enjoy keeping chickens.

I put in my order and request delivery on March 24th, the first Monday after Easter. It's still cold in March and April, so I might have to keep chicks in the brooder until the nights are warmer. March chicks will be laying hens by August or September, so I have about 5 months to wait before any eggs touch down – just in time for football season.

I'm setting the brooder up in the lower level of my house, which is currently used as an office and work room. If it was later in the spring or summer and warmer, an unheated garage or garden shed would have worked just fine.

A 150 gallon rubber stock tank with 15 square feet of floor space will be the brooder this time. The stock tank has draft-free sides 2 feet tall; high enough that chicks won't flutter out, yet not so high that I can reach in to tend them without needing to see my chiropractor.

In about four weeks (some breeds grow faster than others), these chicks will outgrow the stock tank brooder and require more leg and wing room. I'll move the larger chicks to the hen house and give the slower-developing chicks space and time to feather out. Before moving to the coop, babies should be in a suit of full-body feathers. Their heads can still have a little baby fluff, as these are the last feathers to grow in.

A few days before peep arrival, I get the brooder cleaned and set up for a test run. I make sure the heat lamps are at the proper height to keep the temperature between 90 to 95 degrees. I get the chick furniture out of storage: waterers (one quart-size and one gallon-size), heat lamps, feeders, thermometer, and training roost. One of the heat lamp bulbs has burned out, so I'll have to buy a new one.

Week 1 of Brooding

Days One and Two of Life. At the hatchery little beaks with a single egg tooth are chipping the way to freedom. On Easter Sunday the chicks poke out of their shells in the incubator and dry off. Within hours of hatching each chick is packed into a shipping box with a minimum of 25 others; 25 chicks is enough to keep the flock warm and buffered from impact so they aren't hurt as the box is jostled around.

The specially designed chick transit box is 12" by 10" by 6" high. Air holes are evenly spaced on the

Chicks in Shipping Box

This is how chicks arrive in the mail. There are 30 packed in this mailer. The chicks keep each other warm. A packet of vitamins and electrolytes is under the shipping mat.

sides and on top of the box. LIVE BABY CHICKS, PLEASE HANDLE WITH CARE. RUSH is printed on all four sides of the box. There are clear instructions to "Call Upon Arrival" with my name and phone number in bold. I've already notified my local post office that I'm expecting chicks and to call me as soon as they arrive, no matter how early in the morning it might be.

The post office will be happy to call because the chirping is only cute for about a nanosecond. The chicks clearly announce their discomfort, displeasure, hunger, and terror by making cheeping noises that have the pitch of a fire alarm or a backing up bull dozer.

Day 3 of Life. March 24th, Monday, and day three of chick life. My post office calls at 6:30 AM to tell me the chicks have arrived. I can hear the chicks shrilling alarms in the background. "Thanks, on my way; be there in 20 minutes," I reply. Because I am an early riser, I've already been up for an hour and have turned the heat lamps on to get the brooder toasty warm. Just before dashing out the door, I triple-check the brooder floor temperature directly under the heat lamp. The bedding and ambient brooder temperature is a cozy 93 degrees.

I can hear the chicks from my car as I drive up to the post office delivery ramp. The postal worker greets me with a grin and hands me the box that was crying RUSH AND HANDLE US WITH CARE.

The box is light — only about 4 pounds — and vibrating with high, shrilly, siren cheeps from mega-hungry, thirsty, and impatient chicks who don't yet know that a sanctuary awaits them. I drive back gently so they don't lose their balance in the box. My mixed terrier dog Woody is peering through the air holes and wagging his tail.

Once home, I open the box to see if any need emergency care or — even worse — are dead. I'm thrilled that they all look healthy, just hungry and stressed. The packet of Quik Chik with electrolytes and vitamins is tucked under the mat in the box. I add a quarter teaspoonful to the quart waterer and take each chick out individually to dip their tiny beaks in the water. Each one quickly masters drinking; remember, this is the first drink of their life.

When their thirst is satiated, they start a frantic search for food. They whip their feet backwards, scratching and searching. I've already put some crumble on aluminum foil to help them find nourishment fast. I also put chick starter feed on the newspaper and thinly-spread bedding so they can scratch and look for food. The majority of chick deaths occur within the first three days, so getting them off to a good start minimizes mortality. Chicks that survive to 4 days have a good chance of living to pullet-hood.

After they are acclimated and twittering contentedly, I get a baseline weight of each chick by weighing them on an electronic scale that measures fractions of ounces. The average weight is 1.2 ounces.

I have company during the check-in. Three cats and two dogs are peering around the door, trying to see what's going on. Like kids at a parade, they are drawn to the noise and fascinated by the activity. Woody, a terrier-cross adopted from the Humane Society, licks his chops when I hold a chick close to his nose for a sniff. "No, these are not treats or toys," I remind him. He ducks his head, wags his tail, and seems to understand, but keeps licking his chops. This is the dog who couldn't keep his mouth off guinea hens. No matter how I scolded him, if he found an opportunity to fluster and flush the guineas, the chase was on, often ending in a pile of feathers and a dead bird.

Brooder First Day with Chicks

This is the initial brooder setup. Half the brooder is blocked off until the chicks get acclimated to their new home. The two heat lamps with red bulbs keep the temperature at 95 degrees; one is on and one serves as a backup. The bright foil attracts chicks to the crumble feed spread on top. Take each baby carefully from the shipping box and dip its beak in the water for its first drink. Two baby waterers assure that everyone has water access. Aspen bedding is thinly spread over newspaper.

Guineas are not hard to flush and the entire neighborhood can hear when they are flustered; they are in the same noise-making league as roosters.

The three cats are watching but not as intently as Woody. Perhaps the cats have been scolded so much for bothering birds they are conditioned to be disinterested, because birds = an unpleasant experience. Until the chicks get to pullet-size of about 2 pounds, they can be easy finger food for cats.

We had one cat who was always interested in anything chick-related. He was a Maine Coon named General and thought it was his responsibility to protect baby chickens. We made sure he was present whenever baby chicks arrived. I'd hold up a baby chick to his nose and say, "General, you have to protect these". He seemed to know what I meant, and always looked at the chicks with that wide-eyed Maine Coon look.

He would watch the peeps for hours, but never harmed a single one. On the contrary, he guarded them. One sunny afternoon some pullets were following their mama hen across the garden. General had been watching them, and he suddenly went into battle mode, puffing up to three times his already formidable size, hissing, growling, and glaring. I couldn't figure out why until I saw that a neighbor's cat had been stalking the young birds. The dogs heard General fussing and came to his assistance, chasing the predator cat off our property. General saw the danger and acted decisively to protect his flock.

By 9 AM the babies are rested, fed, content, and falling asleep standing up or wherever they stop. Narcoleptic fuzz balls. They tend to cycle together and are usually all awake or all asleep. I check the brooder several times during the day. This is the cutest the chicks will ever be: fluffy, clean, and delightful. All is well in the nursery and the peacefulness makes me sigh gratefully and smile; the pride of broody.

Day 4 of Life (day 2 in brooder), March 25th. The chicks have grown overnight.

Chick with Egg Tooth

Chicks have an "egg tooth" that penetrates the eggshell so they can hatch. This hard, white tooth drops off after a few days. Don't attempt to pull it off.

They are already showing wing feathers and the buds of tail feathers. Some have lost the egg beak – a calcification on the end of their beak – that they use to chip their shells and free themselves from the egg. There are three chicks with slightly pasty butts. A pasty butt is when manure sticks to the chick's rear and blocks the passage of more feces. It's reverse constipation. If it isn't taken care of, the chick can become sick and die.

The cure for pasty butt is simple; put probiotics in their water and crumble. It is also necessary to remove the pasty from the butt so that feces can pass. It's not as bad as changing a diaper. Get a paper towel wet with warm water and hold it against the little behind. This softens the caked mass. Gently – extremely gently – pull it off. If you pull too hard, some of the tissues around the anus can be torn. If the pasty still sticks, let it get moister and softer so it will come off easier. Sometimes scissors are necessary to cut off the chunks. The fluffy behind will be replaced by feathers soon, so don't worry about raising bare-butt chickens. Their feather growth will just be "a little behind" for awhile.

Adding bedding gives the chicks a clean surface to play on. They have survived the first four days of life and their chances of reaching layer-hood have improved dramatically. Today is a milestone in their brief lives.

Day 5 of Life, March 26th. Every chick is doing great. No more pasty butts. They are tearing around the oval tank like Roman race horses. Sometimes they run and stop abruptly, as in the Road Runner Cartoons. The body language of the babies staying far away from the heat lamp tells me they are too hot.

I put the thermometer at chick ground-zero. It reads about 100 degrees: too hot. Raising the bulb about 3" higher lowers the ground level temperature. Soon, the chicks gather directly under the heat lamp and indicate the temperature is just right. The thermometer shows 93 degrees; just the right temperature for the first week of life.

Day 6 of Life. March 27th. Today the babies are preening their rapidly growing wing feathers. They separate and nibble on each one, and I wonder if such rapid growth makes them itchy. Developing muscles need toning and coordination, so chick bombing has begun. This is where which one brash chick races up and smashes into a sleeping group of chicks. It is an incredibly obnoxious sport by human standards, but the molested sleeping chicks rustle, annoyed, and quickly fade back to sleep.

All their digestive tracks are stable and there isn't any sign of pasty butts. This is when I like to give the babies a boost of high quality protein in the form of organic hard boiled eggs from the hens in my home flock. The chick ration from the feed store says that it contains "crude protein," but doesn't elaborate about the kind of protein or its source. To guarantee that your chicks get good-quality, easily-digestible protein, dice up some hard boiled eggs. The chicks love it.

People have asked me if feeding poultry eggs to chicks is cannibalism. My response is no. The eggs are unfertilized and the white is the placenta. Many animal mothers will eat the placenta after giving birth because of the nutrients it contains. I could feed duck or turkey eggs if they were available, but I know that my hens receive excellent nutrition and their eggs provide the best protein for a chick's rapid growth.

Over the next few days, I'll add some ground flax seed meal and sesame seeds to the chopped eggs for good lignans, oil, and fiber.

Day 7 of Life, March 28th. A friend comes over and we sit around the brooder for an hour, chatting and watching the chicks' antics. Their tail feathers are coming in for the ugly duckling phase. Some of the chicks have fuzzy heads and pinfeathers, making them look like punk teenagers with spiked do's.

Week 2 of Brooding

Day 8 of Life, March 29th. It is the chick's one week birthday. A dear friend, Barbara Lane, is visiting from Vermont, so it's party time. We have about 25 people over for a potluck and long-overdue good time. Throughout the evening almost everyone, either alone or in groups of twos and threes, descends to the basement to see the chicks. I had chairs set up around the brooder — a good place to sit and talk about life and the interesting political and economic times we are in. Chick-snuggling puts a smile on all my guest's faces.

There is a heavy dust layer on the heat lamp; you could write "clean me" on it. The brooder is starting to smell musky and animal-like. It's time to dust, sweep, and tidy up the room and the brooder.

Adding fresh bedding brings the litter up to about 2 inches. Quite a bit of it is feed they have

Chick Supplement: Hard Boiled Eggs, Ground Sesame Seeds, and Flax Meal

shoved aside while looking for better bits. Frugality is not a chick priority. I wonder if they know about carbon footprints and starving chicks in India?

Day 9 of Life, March 30th. I start feeding fine bits of vegetables. Today, they get steamed broccoli stems left over from dinner. A food processor twirled the veggies to fine bits, which I mixed with the chopped eggs and crumble. They swarm over it like piranhas, clearing every bit in minutes. There is a particular allegro cheeping pitch that corresponds to happiness about food. They are going through about a pound of sand a day, some of it kicked away, but most of it seems to have been eaten. It's time to gradually lower the temperature a few degrees to the low 90's or high 80's. The chicks are starting to flap their wings, knocking themselves off balance as they try to coordinate flapping and walking. The pecking order is getting established, and soon it will be set in stone as power plays become serious — even deadly. The flock leader's motto is, "Walk softly and carry a big beak". For now, their mild aggression is amusing for those of us bigger than a tennis ball.

There was a heavy rain and I collected earthworms for the young carnivores. They stabbed at the wriggling mass with dead-on accuracy. Once securing a worm in their mouth, a chicks would rocket around, announcing loudly that she had something others would want.

Day 12 of Life, April 2. The chicks are going through sand like kids in a playground. I replace the little cardboard sandbox with a larger, deeper one. The flock has gone through about 3 pounds of sand so far and has eaten quite a bit of it.

Day 13 of Life, April 3. The first chick made it over the rim of the stock tank before landing on the floor with a thud. She quickly realized she was in foreign and possibly hostile territory. I heard her distress call from upstairs (an advantage to

Brooder TV

Watching chicks grow is multi-species entertainment. All the elements of a good show are there: comedy, tragedy, athletic stunts, power struggles, and tenderness.

Chick Growth and Feather Development

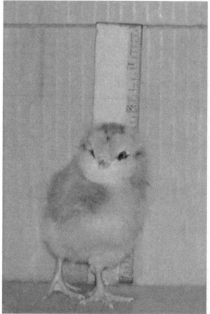

3 day old, Araucana/Americana chick that just arrived by mail. She weighs 1.2 ounces – just a puffball with legs. She is 4 inches high standing on her tippy-toes. This is the cutest a chick will ever be.

6 days old. Notice the wing feathers development. Chicks are learning to flap and spread their wings but they can't fly yet.

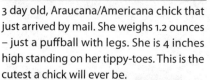

At left is our chick, now ten days old, and above at 13 days. She has full wing feathering but her tail feathers are still stubbles. At this age, chicks can flutter enough to get out of the brooder but don't have controlled flight or soft landings.

28 days old (4 weeks) and almost ready to move out of the brooder. The nighttime temperatures and daytime winds are too cold for this young one to be without supplemental heat and shelter. Some breeds, like this heritage chicken, mature more slowly than the production hybrids.

18 day old chick. Our gal has tail feathers that are somewhat ready for flight. She is able to flap up high enough to get out of the brooder.

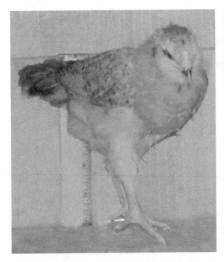

35 days old (5 weeks). The brooder is too crowded, our model pullet hen and others are ready to move to the coop as temperatures are warm enough not to stress the fledglings. They will still need supervision and close monitoring.

Our gal is 24 weeks (6 months) old, and fully grown. She weights 5.8 pounds and has started laying her fashionable, turquoise colored eggs. She's a tame and personable hen.

having the brooder within hearing range). I put her back in the brooder before a cat decides to use her as a toy. It's time to put a brooder screen on top to keep the chicks in. I'm using some 2"x4" wire fencing that is 3' tall – just wide enough to cover the brooder. A large window screen would have also worked.

Day 14 of Life, April 4. It's clear the fledglings can flitter up to perch on top of the waterer. To discourage this, I put a small cardboard box on top of it, so there is no space for a chick landing.

Week 3 of Brooding

Day 15 of Life, April 5. We had beet peels left over from dinner. I shredded them in the food processor and mixed them with hard boiled eggs and nutritional yeast; the chicks loved it and ate every last smidgen. The next morning, I pulled back the brooder blanket and was horrified to see red in the droppings. It looked like an epidemic of coddicidia, a disease in which bloody stools are the signature symptom. Then, chuckling, I remembered the beets they ate the night before. The red passed through to the stools, just as it does with people. Checking stools is a good animal husbandry habit to get into, as what comes out tells you a lot about the health and general condition of any animal.

Day 18 of Life, April 8. The temperature has been gradually lowered to the low 80's. The chicks need more air circulation because they are larger and giving off more body heat and humid breath. The underside of the Reflectix aluminum foil brooder blanket is damp, indicating too much moisture in the brooder environment, so I push the brooder blanket back so that it covers less of the top and air flow improves. If I were brooding during the summer, a small fan might be useful to help improve air circulation.

Week 4 of Brooding

Day 22 of Life, April 12th. Because the chicks are giving off so much body heat, the heat lamp can be turned off during the day. Switching off the light terrifies the chicks, so it's better to turn the lights down slowly. They are less afraid if the darkness fades in a little at a time, giving them a chance to select their sleeping places and settle down.

Day 25 of Life, April 15. The brooder has become too crowded. The smaller chicks are getting pushed around by the bigger clucks. Even though the nights are still cold outside, I've got to give all of the chicks more room. None of the younglings are ready to be totally un-brooded, so I retrofit a large size

(20" x 34" x 24") pet travel kennel with a feeder, waterer, and roost. My hen house is small and there isn't enough floor space for this size kennel, so I put it on top of the existing roosts.

Even though I have a contractor's license, I'm a lousy carpenter. Power tools intimidate me. I summon my inner chicken-carpenter and make another training roost. This one is bigger than the first, with 2"x4" joists and perches ½" by 18". It is just small enough to fit through the pet carrier door and sturdy enough to be put on top of the regular roosts later. The bar across the front and back will keep chicks from getting underneath.

Twelve of the largest chicks graduate to the kennel in the hen house. They will keep each other warm at night. The hen house is cozy even though there is a chilly spring wind blowing, but to make sure the chicks are warm enough, I use an old towel as a brooder blanket and hang it over the front of the kennel. It makes me feel better.

The 18 chicks in the brooder are clearly relieved to have more space and less domineering companions. Like wallflowers coming out of their shells, they begin to be more active and adventuresome. Their submissiveness to the old rank-and-file fades as the pecking order gets reorganized.

Week 5 of Brooding; Chicks are One Month Old

Day 28 of Life. April 18. The chicks in the dog kennel have been there for 2 days and are ready for more room; they are bored and restless. I open the door and get each of them to perch on my finger, then gently let them down to the floor. A whole new world awaits them. The five older hens aren't pleased with the youngsters and club them on the heads. The Araucana hen has turned broody and started sitting on eggs. Having chicks around must make her want some of her own.

The slower-developing chicks still inside are sleeping in the dark. They sleep together and use each other for pillows. They are not cramped, like cold chicks huddling together for warmth; instead, they are cuddling and cooing to each other as they drift off. Their feathering is almost complete, and they are wearing full downy suits. With the absence of the domineering chicks, the smaller ones have become more self-assured and express themselves with gusto. In a few more days they can join the others in the coop house. They will acclimate in the dog kennel for awhile before they are let out to mingle with the flock. Although the heritage chicks are growing more slowly than the commercial hybrids, they will eventually mature to about the same size and weight.

Day 34 of Life, April 24th and just under 5 weeks old. The last of the batch moves to the hen coop. It's a relief to get them out of the basement. The first wave of chicks to hit the hen house are bold enough to go outside for their first peek at sunlight. Unfortunately, they haven't figured out how to get back in. At dusk I had to put them back in the hen house; luckily, they are easy to catch. A deep cleaning of the basement is scheduled for tomorrow. The brooder bedding will go in the compost bin, and the chick furniture will be put back in storage.

Week 6 of Brooding

Day 36 of Life April 26th. Today a mid-afternoon storm blew up. As I was closing up the house against the gusts of rain-filled wind, I heard a mound of shrilly, screaming chicks outside the hen house, 3' from the pop hole. They were getting soaked in the storm. Out in the downpour and lightning, I gathered the wet blobs and put them in their dry abode. I'm soaked; they are safe.

The ambient air is warm and the coop is draft-free, so I'm not worried about them getting chilled. If they were chilled, I'd bring them in and dry them with a hair dryer, chickadee-do style. A lightening bolt snaps-crackles-and-pops just as the pop hole door begins its automatic descent. I check inside to make sure the chicks are not shivering. They are wet but contentedly settling down in the straw, murmuring to each other about their near-death experience.

I'm so grateful that I was home. Without human help, they would have stayed in that pitiful corner of the run, in the rain. With the high winds, some of them might have gotten chilled or suffocated from crowding. That has happened before. When writing *Chicken Tractor*, I was concerned about putting chicks out too early in the spring before they had acclimated to cold temperatures. Young chickens on wet ground is a recipe for chilling and catching colds, so while writing *Day Range Poultry* we switched our system from bottomless pens to mini-barns with floors. This kept the chickens off the ground and better sheltered from wet, cold, windy weather. The mini-barns paid for themselves with decreased mortality rates and faster weight gain, due to less stress.

Day 37 of Life April 27th. The chicks are almost 6 weeks old. Graduation! It's a beautiful spring day. Red buds, forsythia yellows, and gleaming greens cover the hillsides. When the pop hole door opens, the young chick-keteers proceed cautiously into the sunlight for a full day of playing chicken.

They've grown from an average of 1.2 ounce brooder-bound chicks to 1 pound pullets — a 13 fold increase in body weight. One of them would not fit in the shipping box that held thirty of them just 6 weeks ago.

I spread scratch on the ground to help them celebrate. They are thrilled and let me know it by going after the grains with the chortling and enthusiasm of a soccer team at the World Cup. All of them are optimally healthy and ready to begin apprenticeships as garden helpers and compost makers.

This officially ends the brooding phase. But even though the chicks have graduated from the brooder, they still don't know the flock culture, the rigid structure of the pecking order, the coop or the great outdoors. They don't know how to come in at night and will pathetically huddle together outside, a few feet from the ramp and pop hole. They don't know how to get up on the higher perches to roost, and still bed down in a corner on the floor. These things they would know if there had been a mama hen to tutor them.

For several weeks, I'll have to monitor them to make sure they find their way back to their coop at night and that they have access to enough food in the creep feeder. Some of them will need a lift to the roosts so they stop sleeping on the floor. A few of them get the routine quickly, and the rest follow their lead. Like the 100th monkey, a seemingly odd activity can become an ingrained group habit. Be patient. Older chickens can be hard on the newbies and even injure them. Keep a keen eye out for any wounds that need treatment.

Your chick-sters still need diligent oversight and protection and will for the rest of their lives. In return, they give you superior eggs, slapstick entertainment, garden help, fertilizer, and provide stories for many conversations and magic moments to come.

Congratulations! Finally, the flock is with you!

*"It may be hard for an egg to turn into a bird: it would be
a jolly sight harder for it to learn to fly while remaining
an egg. We are like eggs at present. And you cannot go on
indefinitely being just an ordinary, decent egg. We must
be hatched or go bad."* — C.S. LEWIS

10 Incubating and Hatching

Sometime during your flock-keeping career you might be tempted to hatch
eggs. You can buy fertile eggs of almost any breed from poultry breeders,
so hatching can be a good way to add various breeds to your flock. Small-
scale incubators are inexpensive, and you witness a miracle when chicks
start pecking out of their shells.

Hatching is not all cute and fluffy, though. Even if you monitor the incubator religiously and follow instructions to the letter, you can still wind up
with chicks dead in the shell or some that hatch but are malformed, dis-
figured, and must be put down. Sometimes mortality can be attributed to
the way eggs were handled before you got them; sometimes it is due to an
incorrect environment in the incubator. Be prepared to handle any nonvi-
able eggs or chicks before you commit to hatching a batch.

If you decide to hatch eggs, you should count on incorporating the
resulting chicks into your flock. Don't expect to sell the chicks for a profit;
you will probably be disappointed.

Setting up your incubator

Incubating eggs is not difficult, but there are several factors that must be
carefully monitored, particularly during the first couple of weeks: the type
of incubator you buy, where you set it up, temperature, relative humidity,
ventilation, and egg turning will all impact your hatching success.

The Incubator

Incubators can be purchased at local feed stores, farmers' cooperatives, or through online or mail-order poultry supply stores (see the Resource section for a list). They can be as simple as an $18, three-egg incubator, or as elaborate as a $1,300 incubator/hatchery combination with automatic temperature and humidity monitoring. Ask the salesperson which one fits your needs best. Some people move eggs from an incubator to a separate machine called a hatcher three days before eggs are scheduled to hatch. This usually isn't necessary when hatching eggs on a small, City Chicks scale. You will probably end up spending around $150 for an incubator that holds 30-40 eggs.

The Incubation Room

Put your incubator in a location with a constant temperature. Un-insulated buildings, like garden sheds or garages, are usually not good spots for incubators.

We put our incubator in our office. This room has an most even temperature. The hatching process was easy to monitor because the incubator was so close to our daily activities.

Incubator Temperature

Chicken eggs need a constant temperature of 99.5 to develop properly. Temperature is probably the most important factor in hatching success.

Incubator Humidity

Having the correct humidity is also essential to a good hatch. The incubator you buy will probably have a built-in humidity monitor, but if it doesn't, there are several inexpensive humidity monitors available in poultry supply catalogues. The Resource section lists several sources.

Incubator Ventilation

Air circulation is important because eggs give off carbon dioxide, which can accumulate to toxic levels without proper ventilation. Many commercial incubators have a fan to circulate air, but convection will cause air to move slowly if your incubator doesn't have a fan.

The main challenge with a fan-less incubator is that the temperature can differ by as much as 5 degrees between the top and bottom of the incubator.

Be sure to follow the manufacturer's guidelines when setting the temperature, and take temperature readings at both the top and bottom of the incubator.

Turning Eggs

Turning eggs keeps the yolk from floating to the top of the egg, which can kill the embryo. If chicken eggs are not turned at all, a small number of them will hatch anyway, but you won't get nearly as many chicks. Your hatch rate will be better if you turn the eggs at least twice a day. Stop turning them three days before they are scheduled to hatch.

Buying Eggs to Hatch

Think about the goals you set for your micro-flock, and choose chicken breeds that best fit your needs. How many eggs you get will be determined by the size of your incubator, the size of your hen house, and the time you have to care for eggs, chicks, and hens. Remember to factor in the expense of brooding and feeding chicks, too. The Resource section lists some egg sellers, or you can find others by searching the Internet, participating in online chicken chat rooms, or visiting with local chicken folks about where they get their eggs.

Egg Transport Options

In a perfect world, you would pick your eggs up at the hatchery and drive them straight home in your car. The second best choice is to buy eggs that can be delivered via the mail. Eggs can suffer during air shipment because of changes in air pressure, temperature fluctuations, long waits on loading docks, and lack of oxygen in the air as they fly. Federal Express and the United Parcel Service (UPS) won't deliver live poultry, so the U.S. Postal Service is the only carrier option. Each year fewer airlines agree to fly poultry for USPS; half of the commercial airlines won't handle live animals at all. The limited options for air transport increase the chances that your eggs languish on a loading dock somewhere while awaiting shipment.

We once shipped eggs to a customer with the utmost assurance from our local post office that they would be delivered no later than 3 P.M. the following day. 24 hours later, the eggs hadn't arrived. We called the USPS toll-free line, punched in our tracking number, and were told by a pleasant, recorded voice that "your package is on the way to its delivery destination".

The eggs were never delivered. We have heard nightmarish stories from other hatcheries about lost shipments and eggs shattered on arrival.

Having said that, most egg deliveries arrive safely at the correct destination. Just be prepared for the possibility of a few broken ones.

When Hatching Eggs, Timing is Everything

When you purchase eggs from a hatchery, you are counting on the hatchery to do their part to make sure the eggs are healthy when you get them.

Eggs awaiting incubation should be stored between 40 and 60 degrees Fahrenheit (55 is optimal). At room temperature (68 degrees), the embryo begins to develop slowly, which weakens the budding chick and make it less likely to hatch. Relative humidity should be between 70% and 90% to keep the egg from dissipating moisture into the air. Eggs should not be older than 3 or 4 days when they are mailed. Egg hatchability decreases sharply if they are not incubated by 7 days. Allowing 3 or 4 days for shipping is a good idea, even if the USPS promised 24 hour or next-day delivery.

Hatcheries should avoid shipping on Friday or Saturday to avoid a Sunday egg layover. Eggs should be shipped in boxes with plenty of packing material.

The Hatching Process

Eggs hatch 21 days after incubation begins. Begin counting the day you put the eggs in the incubator.

What To Do When Eggs Arrive

Have your incubator ready and running at incubating temperatures before the eggs are delivered to make sure you don't have mechanical problems when they arrive. Upon receiving your eggs, discard any that are small, misshapen, or have cracked or thin shells.

Never incubate cracked eggs. Bacteria invades even tiny hairline cracks, so overly dirty or cracked eggs are likely to absorb bacteria and get rotten. A rotten egg can explode in your incubator, which smells unbearable and contaminates other eggs. If an egg oozes or smells like sulfur, it is rotten. Carefully remove the rotten egg and bury it. Candling eggs at days 10 and 19 will identify eggs that are not viable. You can cull these to minimize rotten disasters. When the eggs arrive, let them set at room temperature for 6 to 12 hours. This is called "settling the eggs" and lets them recover from the vibrations of travel and eases them into the warmer temperature of the incubator. Start turning the eggs 3 to 5 times a day; do this until three days before their scheduled hatch date.

"Setting the Eggs" the First Step in Hatching

Your incubator will have a tray to hold the eggs (plastic trays allow more airflow and are easier to clean than paper ones; if you use cardboard egg trays, poke holes in the bottom of each holder for circulation). When the eggs are settled and have reached room temperature, put them in the holders with the pointed end down to orient the air sac and yolk for ideal embryo development.

Monitoring the Incubating Eggs

There are two main ways to monitor the egg's progress: candling and monitoring egg weight loss.

Candling tells you if an egg is fertile and helps you monitor embryo growth. To candle an egg, hold a strong light against it so you can see the egg's contents. A fresh or infertile egg lights up like a golden globe. As the egg ages, evaporation forms an air sac at the top. At 10 days a viable embryo is developed enough to see through the shell. Look for vein development and air sac size. When the embryo is halfway developed, there will be vein growth covering everything but the air sac.

Don't let an egg get chilled by being out of the incubator too long, and

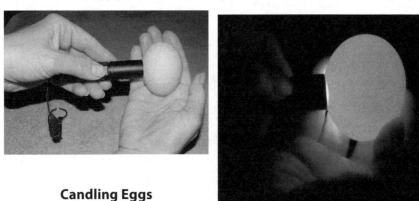

Candling Eggs

Candling eggs is easily done with a flash light. A fresh or infertile egg lights up like a golden globe. At 10 days the embryo is developed enough to see through the shell, and an air sac will have formed.

don't candle one long enough to make it hot. When handling eggs, be careful not to rotate them too quickly because this can rip delicate membranes and damage the embryo.

Weight loss monitoring can be used in conjunction with candling. It gauges embryo growth by measuring an egg's weight loss. From when they are laid to when chicks peck out of them, eggs lose about 15 to 20% of their weight. Ideally, this weight loss is consistent and easy to plot on a graph. You can control weight loss by raising or lowering the humidity in your incubator. If your eggs are losing too much weight, increase the incubator's humidity, which slows evaporation from the shells. Likewise, if the eggs aren't losing enough weight, decrease the humidity to encourage more evaporation.

Keeping Track of Your Hatch

Keeping notes on your hatching experience is worthwhile. It doesn't have to be complicated, but you might want to jot down:

- How many and what kind of eggs you set.

- How many of the eggs are infertile when you candle them:

   on day 10.

   on day 19.

- How many live chicks hatch.

- How many chicks hatch on time, and when or if the stragglers peck their way out.

- Any abnormalities or problems with the hatch.

Most folks leave a notebook and pen by the incubator to jot down observations throughout the hatching process.

The Hatch – Here They Come!

At day 18 (three days before your eggs should hatch) raise the humidity by 2 degrees. This can be done by adding a pan of water, decreasing the temperature by 1 degree, or adjusting the air flow.

If eggs begin hatching a day or two early, decrease the temperature by 1/2 degree. If your eggs haven't hatched on day 21, increase the temperature by 1/2 degree.

Hopefully, you will have a 100% hatch success rate, but this doesn't happen often. There will probably be a few eggs that don't result in healthy chicks. Here are some possible reasons why, some guidelines for general healthy hatching, and what to do about common hatching problems:

Ten Causes of Low Hatch Rates

1. Infertile eggs
2. Old eggs
3. Parents stressed, unhealthy, nutritionally incomplete diet
4. Poor handling of eggs before setting for incubation (incorrect temperature, rough jostling, unclean conditions)
5. Contamination from dirty eggs, cracks, or thin shells
6. Eggs not turned frequently enough
7. Temperature too high or too low
8. Too little or too much humidity in the incubator
9. Improper ventilation or a dusty incubator room
10. Oxygen starvation (too much carbon dioxide built up in incubator)

Three Suggestions for Healthy Hatching

1. Only set fresh eggs less than a week old.

2. Follow your incubator instructions exactly.

3. Maintain the highest sanitation for eggs, incubator, and hatcher.

Four Common Hatching Problems

1. It is hard for a chick to get out of the shell, or the chick feels sticky. This usually means the hatch was not humid enough. The chick doesn't have enough lubrication to rotate inside the egg to peck out of the shell, and the chick is drying out too quickly.

2. The chick seems too large for egg and appears bloated or swollen. Humidity might have been too high during incubation and the chick couldn't lose enough water. Even if an egg lost too much or too little water during incubation, it needs higher humidity to hatch once the chick starts to peck through.

3. *Early hatch.* The temperature was probably too high. Check and recalibrate your thermometer if necessary.

4. *Late hatch.* The incubation temperature was too low. If an egg gets chilled during incubation it will hatch a day or two late. You might have to help the chick out of the egg.

Incubating: The Way of the Hen

If you don't want to mechanically incubate eggs, there is another method: the way of the hen. In our mechanized world of gadgets, gizmos, and mass production, using the old fashioned, one-hen-at-a-time clutch to incubate and perpetuate a flock is becoming a lost art. However, we are in luck; the art of how to incubate eggs is innately obvious to many hens. But there are tricks, tips, and traps that you, as the hen's birthing assistant midwife, need to know.

If you have any of the heritage breeds you might have a hen with broody tendencies in your micro-flock. The Hen Quest chapter contains information about which breeds tend to have chick rearing tendencies.

If you want to get replacement hens, and if you suspect you own a hen that has a tendency towards broodiness, you might be able to get chicks Nature's way. Here's how to start. Place 4 to 6, preferably wooden or fake eggs in a nesting box. This might entice one of your hens to go "broody" and plop herself down on the clutch (group) of eggs. Can you force a hen to be broody? No.

A hen merging into broodiness will sit on a nest a little longer each day until she rarely gets off. Her clucking tone will drop from the high-pitched staccato squawk of an egg laying event, to a lower, vibrant tone of soothing and communicating to her babies. If you remove a broody hen from her nest, she is like a boomerang, returning back after only a short break. She will make it perfectly clear from her body and verbal language that she wants to be a hermit. She becomes guard-dog protective and scrooge-possessive of the eggs she's sitting on. If your hen begins to sit on the nest at night and leaves during the day only to eat, drink, and poop, then she is broody.

Once she has a clutch of eggs — not necessarily her own – she will stop laying eggs for a couple months. This is just about the amount of time it takes to rear the young ones to independence.

Test the Hen's Broodiness. Once a hen shows signs of going broody in the regular nest boxes, put 4 to 6 wooden eggs under her and remove any real eggs. She doesn't realize what you know: her real eggs are infertile because your City Chicks flock does not have a rooster that can legally visit hens within the city limits.

If fertile, real eggs that have not been under the hen for more than a day, or two, they will still be okay to eat. Much longer than that and the fetus shows development which grosses people out, but not pets.

Do not — repeat, do not — leave her to brood in one of the regular nest boxes. Other hens will get into the box with her, laying their own eggs. Even worse, she will take a break and return to the wrong nest letting the eggs you want to hatch get chilled. This causes the developing chicks inside to die.

Even in broody boxes, two hens together will badger and annoy each other endlessly. They might even steal eggs from one another's nest, possibly sabotaging both clutches.

The moral of the story is give a gal a space of her own for the sake of everyone concerned. The colloquial saying, "when mama ain't happy, ain't nobody happy" is true for sitting hens. But in the unhappy mama hen's case, the consequences can be that eggs are abandoned, get chilled, and embryos die. A hen that's going to sit for 21 days to practice her art deserves, and needs, a nest and box that is private, dim, well ventilated, and quiet. Provide her with a new nest in her own broody box. Practice the Golden Egg Rule: treat your hens the way you would want to be treated (if you were a chicken).

The Broody Coop (Brood Box). A brood coop is usually about 3' wide, 2' long and 16 inches or higher and can look similar to a rabbit hutch. Make the space big enough for the hen to get up, stretch, walk around, relieve herself, preen, eat and drink. She also needs room to turn the eggs without stepping on them. A large dog carrier, or crate can provide enough space to serve as a brood coop. I use a medium size dog carrier that is 2' long 16" wide and 20' high. This tucks nicely under my wall-hung nest boxes so that the sitting hen is still with the flock, but has her own mini-coop that the other hens don't have access to. With my micro-flock, I only let one hen brood at time.

When to Hen-Hatch Eggs. With my micro-flock, I'll only hen-hatch 1 or 4 batches a year. I don't want to hatch high numbers of chicks. With time and small batches on my side, I schedule to have the eggs hatch sometime in May, after the cold and rainy season has passed.

Broody Coop Location. My broody coop is a pet carrier that tucks nicely under the wall-hung nest boxes in the main coop. I like having the broody coop in the main coop and with the flock. It keeps the hen with the flock for security, but with the equivalent of an efficiency apartment of her own.

A broody coop can be located just about anywhere. If your broody coop is outside the regular coop, then consider raising it off the ground for predator protection and easy servicing. If your regular coop is large enough, put the broody coop inside, even hang it on the wall to save floor space. You will probably only be brooding one or two hens at a time so you won't need a lot of space.

Brood Coop Flooring. Harvey Ussery uses 1/2 inch hardwire cloth for flooring. He finds the hardwire easier to clean, and it allows for better ventilation. As soon as his eggs hatch he moves the hen and chicks into a low-level hutch that sits on the ground. If you decide to keep the chicks in the brooder box for a while, cover the 1/2 inch hardwire flooring with cardboard or newspaper so the chicks don't get their teeny-tiny legs caught and even broken in the hardwire.

My pet carrier brooder coop has a solid plastic floor that I cover with cardboard, then newspaper and finally aspen bedding.

Brood Nest Box (inside the Broody Coop). Most brood nest boxes are made of wood. I prefer to use a cardboard box with an entryway and threshhold cut along the front side. It's easy to find a 16" wide box that can be cut down to fit inside the pet carrier.

The nest area should be at least 15" by 15" for standard hens. It needs to be large enough for the hen to stand up and rearrange the eggs, rotating the outer-edge eggs to the middle. If the box is too small she will be forced to step on the eggs, possibly damaging them. Make the sides at least 3" tall to be a threshold helping to keep the nesting material in the box. 3" is too high for a new-born baby chick to climb back into the nest, so just before the chicks hatch, stuff some straw or other material against the nest to make a ramp for the chicks to safely come and go.

Nest Material. Nesting material can be clean straw, aspen wood shavings or some other type of organic bedding. One breeder uses chunks of grass sod. NEVER use cedar shavings, and it's not even good to use pine wood shaving for nest box bedding. The off-gassing is toxic in the small space for the long time the hen is setting.

Insect Protection. A hen, sitting so still in one place for so long, is susceptible to mites or body lice (not the human type). Before moving your hen to the broody box, dust her lightly with diatomaceous earth (DE), and put a sprinkle of DE in the bottom of the nest box. This will help protect her from the buggers.

Trim Sharp Toenails. Before moving her to the brood coop trim and file (blunt) her toe nails so they are less likely to crack or scratch the eggs while she rotates them around the nest.

Broody Hen Feed and Water. Always have clean, fresh water and feed available for your hen. You can use the same chick-size waterer and feeder that the chicks will use after hatching. Feed her a poultry maintenance ration. While sitting on eggs a hen doesn't lay more eggs so she won't need the high-calcium layer ration.

Chick Growth Feed. Once the chicks hatch, switch the feed to a high-protein chick ration for both the hen and the babies. Don't let baby chicks eat layer feed. The high calcium in the formula can harm their tiny kidneys.

Collecting Eggs to Set (Hatch)

Here are a few guidelines about collecting eggs that will be incubated.

• Collect only well formed, normal-size, clean eggs. Never set a double-yolk egg; you won't get twins; just trouble.

• Do not wash the eggs; this removes their natural water-soluble protective coating. Slightly soiled eggs can be used.

• The fresher the eggs the better. Don't store eggs to hatch for longer than 7 days. After this fertility drops dramatically. At 3 weeks, viability is about zero.

• Don't shake or jostle the eggs. Sudden movement can damage the delicate membranes inside.

• Store the eggs in a cool area, ideally 55 degrees and 75% relative humidity. Your refrigerator is too cold; a cool basement or root cellar would do nicely.

• Store eggs with the narrow end pointed down in an egg carton. Reposition the eggs daily by turning the carton on one side, or the other.

• Let the eggs warm to room temperature before placing them under the setting hen. The temperature under her bosom is about 100 degrees.

Move the Hen to the Brooder Coop

Once you are sure your gal is dedicated to brooding, move her at night to her special place. By moving her at night, she will be less disoriented and perhaps less cranky. Leave her there for a day or two to let her settle in and make sure she is truly broody. She will go into a sustained, Zen meditative state that is open-eyed, and coma-like. She just sits and sits, maybe thinking, but probably not. She might be having an out-of-body journey or a religious experience. She will position her body to be low and wide in the nest to be as inconspicuous as henly possible.

Give the Broody Hen a Break

In a natural state, hens sitting on eggs take short breaks. During these breaks they stretch, scratch and peck a quick snack, freshen up with dust bath, flap their wings, check the local coop news and, most importantly, relieve themselves.

These breaks are usually when the ambient temperatures are warmer. A full broody-break can last from 5 minutes (if it's cold weather) to, around 30 minutes. The hen seems to have a built in timer/thermometer that allows her to know how long she can be off the nest before the eggs start to chill.

Usually I'll leave the door open to the broody coop in the mid to late afternoon, after the other hens have laid their eggs. If the door is open in the morning other hens will join the brood nest and lay their eggs with those to be hatched. This is bad on many levels and for many reasons.

A Broody Coop

A pet carrier can be used as a broody coop. It is tucked under the regular nest boxes inside the main coop. These chicks have just hatched. Note the cardboard nest box, waterer and floor feeder.

However, sometimes a broody hen is active in the broody coop in the morning. Usually, this is her body language for "I gotta go". As soon as I open the door she will bop out and, within a few minutes, poops. Often a golf ball size dropping because she had not been off the nest for several days.

I've never had to chase and put a hen back into the broody coop. The hen has always returned on her own and hunkered down onto the eggs retuning to her trance state.

Starting Incubation

Begin the incubation, by reaching stealthily under the widely-spread, hunkered-down hen, quietly removing the fake eggs. You will probably get hen-slapped; don't take it personally. Immediately, take a deep breath and gather your courage to replace the fake eggs with fresh, fertilized eggs that you got from a breeder or hatchery – then be as patient as the hen – for 20 to 21 days. Be sure to mark your calendar.

Start all the eggs on the same day and at the same time. One clutch of eggs. That's it for any one brood. Do not sneak in any more eggs under the hen at a later time, no matter how tempting it might be. Once under a setting hen, all the embryos begin developing at the same rate and are on schedule to hatch in 20 to 21 days. Once the pipping begins, the hen's state changes from setting to rearing, and late hatchers can be left behind.

A hen will keep all the eggs at the perfect temperature and humidity if she doesn't have too many under her. Only incubate the number of eggs a hen can easily completely cover under her widely-spread, fluffy body. How many eggs she can easily cover will vary and depends on both the size of the eggs and the size of the hen.

Candle the Eggs

After about 10 days, I prefer to inspect and candle the eggs to remove the infertile, or eggs that show any abnormality, like hair-line cracks. The hen does not like me doing this and makes a drama-queen ruckus. She will display threatening behavior and cuss at me in articulate foul-fowl language that would cause a street gang to run away.

Some breeders just wait until the chicks emerge and then remove the eggs that don't hatch. The problem with this is that if even one of the eggs goes rotten, it can ooze with a ghost-buster sliming that covers everything and stinks. Worse than sliming, is having an egg explode like a hand grenade, emitting a powerful stench that would make all the skunks in the world jealous.

I had an incubating egg that had gone rotten explode in my hand while removing it from an incubator. It took several showers, and a couple days to totally get rid of the smell. That's why I will muster up to a cranky hen encounter and check the eggs midway through an incubation.

Hatch Day

It's birthday time, called "pipping" in chicken circles. The chicks have a special "egg tooth" on the end of their beaks to help them crack and hammer to their freedom from inside. Pipping can be an exhausting journey for the chicks; some don't make the transition and die trying. Some chicks will pip out of their shells earlier than others. There can be as much as a day's difference. The chicks have their yolk sacs to feed on for about 3 days. This allows them to wait for their slower hatching siblings to escape from their shells.

If by the end of day 22 there are any eggs left unhatched, chances are those embryos have died. A way to test viability is to tap, or gently shake, the egg. If you don't hear a "peep" in response, this means the chick has died; discard the egg. Once all the eggs have hatched, or removed, the hen will know to start caring for her babies.

Grafting Purchased Baby Chicks to a Broody Hen

It is possible to get a sitting hen to accept baby chicks that she has not hatched. This has the advantage of not needing to set up an artificial brooder as described earlier in this chapter. Here's a few tips on how to get a successful hen graft.

• A broody hen will probably not accept transplanted chicks if she has only been broody for a few days. You want to graft to a hen that has been setting on fake eggs for 3 to 4 weeks.

• Mixing hatched with grafted chicks can work. Some hens are more willing than others to accept additional babies to their maternal charge.

• Do the graft at night. Remove all the fake eggs and put the live chicks under the hen. The hen will come out of her broody state the next morning when she hears, and feels, the live chicks under her.

• Be prepared to brood the chicks. If the hen won't accept the babies you will have to raise them in a brooder. In severe cases the hen might even kill the chicks regarding them as foreign invaders.

Brooding after the Hen Hatch

Once the babies are hatched under a hen, there are several ways you can manage them. These systems are different than what is described in the previous chapter on artificial brooding.

1. Keep Broody Coop, Hen, and Chicks in Main Coop. If there is room inside the main coop, you can put the brood box on the floor and create a creep space so the chicks can roam around, yet still have access to quick safety and warmth with mama. After few days you can let them free range with the rest of the flock.

2. Keep Hen and Chicks in an Outside Broody Coop. Some breeders keep the hen and chicks in the brood coop for about 5 weeks, allowing time for the chicks to feather out. This requires a larger area than the broody coop.

Brooder Tractor

3. Brooder Tractor. Harvey Ussery, a veteran in using hens for incubation, propagates his homestead flocks solely using broody hens. After hatching is complete, he puts his mama hen and the chicks in what he calls a "halfway house". This is a low, chicken-tractor type structure. His hatchlings

A brooder tractor is simply a broody coop (a pet carrier in this photo) inside some sort of fenced area. In this case, the fenced area is in a deep-mulch chicken run with a section of hog panel fencing around it. The chicks can range outside the fence, but tend to stick close to mama.

At night, and if it is raining or windy the broody coop is carried to inside the main hen house and tucked back into the same spot it was during incubation; under the wall-hung nest boxes. In the coop at night and outside during the day is day range brooding.

and mama live in the halfway house for a few days and then mingle in with the others on pasture.

4. Day Range Brooding. This is the combination using a brooder tractor in your run or garden during the day and put the portable broody coop back in the main coop at night. It's easy to put a temporary fence around a broody box. I use a portion of a bent 3' high pig panel that has smaller openings at the bottom and larger ones on top. This allows the chicks to go through the fence into the main run. Wire fencing, 2" x 4" will also do.

My grown chickens don't bother the chicks. If you want to completely restrict the chicks to the brooder tractor the put chicken wire to surround the area. The advantage to keeping the broody coop in with the main flock is the hen knows where to return to roost and the chicks follow her back into the coop.

Put a step or stone along side the ramp for the chicks to hop up on to follow their mother inside.

Day Ranging Brooding

5. Sell the Chicks. A final way to manage chicks is to sell the ones you don't want to keep. It won't be hard to find people wanting to buy home-hatched and started, local chicks.

After incubating and brooding thousands of chicks, my preference is to use the natural way with the hen. The amount of equipment and

These 8 day old chicks are free to roam with their mama during the day. The chicks follow her into the coop at night to sleep in their broody box. Notice the rest of the flock takes little notice of the young ones. The chicks integrate naturally into the pecking order.

This photo was taken just after a rain storm. It was windy and the air temperature is around 58 degrees. Chicks raised with their mother are much more temperature and weather tolerant than chicks brooded in an artificial brooder.

daily care is minimal compared to artificial incubating and brooding. Natural incubating and brooding uses a 1-watt-hen and has a low carbon foot print. The artificial methods use electricity to keep the eggs at 98 degrees for 3 weeks, and then the baby chicks at 90+ degrees for 5 to 6 weeks.

To hatch and brood massive numbers of chicks for factory farms using hens is not practicable, or economically efficient. But for small-scale breeders and backyard micro-flocks, it is a viable option.

For the record, the hen who incubated and brooded the chicks in this chapter's photos is a 4 year old, high-egg production hybrid bred for the factory farms. She is supposed to have had her maternal instincts bred out of her. She surprised me by going broody, and proving herself by being the perfect dedicated setting and brood hen. In the commercial system, she would have been considered "spent" and put down about a year ago. I bought her from a 4-H auction.

This shows that sometimes, if we give factory animals a chance, they retain more of their innate wisdom and intelligence then we give them credit for. They are not just "dumb animals" without feelings or intelligence.

May all the eggs you hatch grow to be healthy, wealthy, and wise hens.

Hen Home Schooling

Quality Time with Chicks

"A loud voice cannot compete
with a clear voice,
even if it's a whisper."

— BARRY NEAL KAUFMAN

11 Be a Chicken Whisperer

There are horse whisperers and dog whisperers so why not a chicken whisper? I think the term whispering applies because when anybody whispers, the listener has to be quiet to hear. The art of being a chicken whisperer can only be understood by experiencing life from a chicken's point of view. This takes empathy and imagination.

How does a chicken hear? Chickens have a highly developed sense of hearing. Hearing well, they also can produce a wide variety of sounds to communicate. I've read that chickens have about 20 to 30 different vocal sounds, and about 200 distinct "words or sentence noises" they use for communicating.

Chickens seem to have a universal language that transcends breeds and generations. Perhaps someday there will be a Rosetta stone of animal languages and if there is, Chicken will be at the top of the list.

It's hard to describe sounds in writing, just as it's hard to describe tastes. What does a banana taste like? Until you have tasted banana you'll probably say something like "sweet, mushy, aromatic, etc." The same is true with chicken sounds but I'll attempt to describe a few of the more meaningful communications from my feathered linguists. They have their own language, not limited to the common clucking, cackling, or crowing. They also whistle, moan, groan, cheep, sing, coo, growl, and, even "purr".

Hens have their own culture, vocabulary, and their own view of the cosmos. You can learn their culture, language, and world views by being with them and tuning in to their psyches. Your hens can be your teachers if you give them thoughtful observation, respect, and expand your imagination to thinking hen instead of human.

Anthropomorphizing is easy to do and hard to spell. We all project our personal "stuff" onto other beings. For example, I've gotten really angry at my hens for their brutal pecking order. And even worse, for literally pecking a hen to death. In my world, the flock should be protecting her and helping her heal. But that is in my world, not the pecking order world of poultry.

Projecting human values onto chickens is not going to help you understand chickens. You've got to think like a chicken to be a chicken whisperer. Animals communicate through body language and languages of their own. They do their chickenness, or pigness, or goatness, or horseness, just as I do my "humanness". Different codes of existence, different rules of the game, similar pursuit of species happiness.

I'm going to share what I've learned about poultry politics, power, and avian culture. Some of this I've learned from reading books, journals, and listening to conversations. Some I've learned from experiences shared with individual chickens; my feathered philosophers, winged warriors, clucking scholars, and four-toed teachers.

The Chicken Language

Chicken language starts in the shell. Baby chicks begin making sounds while still in the egg. Once hatched, chicks cheep and peep almost constantly when they are awake. They emit a short, clear peep as they scurry about the brooder scratching and running up on each other. If a chick is in distress, the pitch rises and can be a high-pitched frantic falling tone such that even humans can tell something is wrong.

If a chick gets separated from her mama, the broody hen will call it back with a low, fast clucking sound that tells the chick which way to find her. Chicks give a soft twittering if they feel safe and secure in the company of others.

Grown hens cluck gently to themselves when they are feeling good. They can also make a staccato humming sound to themselves while doing a pleasant task, such as spreading mulch looking for insects.

When alarmed, chickens squawk and give a high pitched frantic sound. A chicken will sound an alarm for the others in the flock to be aware. There are different sounds for different predators. A fox, or ground-approaching intruder, will cause a hen to emit a different alarm than for an airborne predator attacking from above. I've herd hens emit low grunting, growling sounds while charging at crows who are trespassing in their run.

Chickens coo and purr. The first time I heard a hen purr was an Auracana that had been severely pecked and wounded while molting. Her back was raw and her tail was a bloody stump when I adopted her. After treating her for a few days and getting her moving around better, I put her in with a few of my hens. A clumsy, large faverolle instigated a broody friendship with the wounded hen. As protection, the wounded hen would tuck herself under the larger bird like a baby chick. Both would do a stutter type of purring that was soothing and melodic. They are family, and one is never without the other, 24/7.

I've heard a Buckeye rooster giggle when he finds food that fits his fancy. He give a quick, joyful hop, scratch and giggle clearly displaying his pleasure.

Trans-species Communications

Between body language and vocal sounds, chickens can be precisely articulate to make their needs known to humans. One story that exemplifies this was told to me by Marsha Heatwole about her sister, Leslie Zambito. It was almost dark when she was startled by hens jumping up to the kitchen window sill and pecking on the glass to get her attention. This was bizarre behavior for her chickens because they should have been bedded down in the coop at this twilight hour. When Leslie went out, she found the coop door had been blown shut by the wind and the only way her chickens were to sleep (and live) safely through the night was to somehow get her to open the coop door for them to gain access to the roosts. Smart chickens.

A similar event happened with my flock. It was post-twilight when suddenly there was such a ruckus that I thought something was killing one of the birds. I ran out to investigate and found about half the flock was still outside. The automatic door had closed too early because I had changed the timing. I opened the pop hole door and they marched in single file, climbed the ladder to their roosts and settled down in their usual locations, chattering all the while about their close encounter with having to sleep outside, vulnerable to night predators.

Egg Proclamations

Before laying an egg, hens have a specific sequence of chatters that announce their condition. Some hens seem to want the world to know they are about to birth an egg and will squawk on long and loud about it. This is helpful if you suspect eggs are being deposited in wayward nests, because you can follow her call. On the other hand, some hens are silent in their egg birthing activity, sitting in composed dignity.

Some of my hens will emerge boldly from the nest box with a proud look, head held high and twinkle in their eyes, as if saying I did it again! They must have a sense of satisfaction and relief that come, from producing such a relatively large object out of their bodies. A hen's egg is usually larger than her head.

A Day in the Life of a Hen

A hen's day starts at daybreak. Some are early birds: up and off the roosts pre-dawn. The saying goes that the early bird gets the worm. In my experience, it's the early free-range bird that gets eaten. There is a time about one hour before dawn, and just after sunset, that is the killing hour. The killing hour is the edge between night and day, light and dark. It's the change-of-shift gray zones when nocturnal critters are just waking up and diurnal beings are headed to bed. It's the time predators, especially foxes, raccoons, hawks, owls, and others are just waking up, and are hungry and on the hunt. It's the time of day you don't want your hens out.

Hens are up feeding and drinking as soon as they can see well enough to hop off the roost. They will be hustling around getting breakfast. Morning is when most of their eggs are laid. By mid-morning chickens become complacent and tend to lounge around, taking dust baths and preening themselves with their beaks. This is the *Gallus* siesta time. Unless something piques their interest, like a grasshopper going by, they will tend to be quiet and passive. In late afternoon the hens will start feeding again to fill their crops with dinner to sustain them through the night.

About half an hour before sunset they will saunter toward their roosts, settling in for the night. Some hens prefer to get to the coop early to reserve their perch space. Others will be out and about until the last bit of light is gone and they can't see. As the late comers jump up and settle down there will be a shuffling of bodies accompanied by protests and squawks, as the higher pecking order hens bang their comrades on the head. If the perches are different heights, there could be trouble every night. This is because any lower pecking-order ranked birds who enter the coop first and get premium places at the top, will be ousted when the higher-order hens arrive on the roost later. Having all the perches on the same level, lets all the hens meet and sleep on the level.

If it's cold, chickens huddle together, some putting their heads under others' wings for warmth. Smaller birds might totally tuck themselves under larger hens.

If it's really hot, chickens prefer to keep a few inches distance from each other. Sometimes panting, they hold their wings away from their bodies to help dispel body heat. Hens pant when hot.

After finding their sleeping places for the night, hens make a gentle soft clucking sound, sometimes a twitter purring. They will bring in their little necks, blink a few times and finally put their heads under their wings for sleep. Loud or unusual noises will cause quite a panic and clatter but if the disturbance stops, they quickly settle down back to sleep.

Chickens are night blind and don't see well without good light. If you want to catch that extra-wild hen, after dark is the time to do it. Even if she runs, she can't see where to go and will tend to hunker down to be easily caught.

Another factor about chicken eyesight is that they don't see far. Being a descendent from the forest where their vision was limited by heavy undergrowth, their vision is developed for small objects 15 feet away and larger objects about 160 feet away. Because of the limited vision most free-range domestic fowl will range only as far as to still see their housing. If you are only two feet tall on your tippy-toes, the vision distance on flat ground is only about 150 feet.

Pecking Order

To understand chicken culture you have to understand the role of the pecking order in chicken society. Not all chickens are equal and they will let each other know who's who by the pecking order. It's called the pecking order because a chicken's beak becomes its intimidation and authority device. The head chicken's creed is to "walk softly and carry a big beak".

Whether male or female, birds fight primarily by pecking. Targets are the neck, head, comb, and face. Sometimes they rush up, wings flapping and loudly squawking, to gain advantage of the opponent's vulnerable parts. Roosters will use their spurs as weapons to bring down their opponent.

Many factors affect how a chicken challenge will turn out. The opponent's appearance is important and chickens are constantly summing each other up. Is the contender proud and self-assured "cocky"? Their posture becomes erect. Fierce eyes flash warning signs. The combs seem larger and brighter red. They become as formidable as possible. Sometimes they're bluffing and sometimes not.

Baby chicks only a few days old display power moves that are a prelude to the pecking order. Little three ounce nurf balls will race up to each other in challenging "what's it to you" defiant stances. At a couple weeks old, the larger chicks will peck at smaller ones, sometimes drawing blood. Once a chick is bleeding, the others might not leave it alone and can ultimately kill it.

Because of this we didn't commingle batches when raising day ranging broilers. As the chicks mature, at about two to three months the pecking order gets serious. Power plays and violent moves that would intimidate the Mafia are the order of the day.

Enter just about any cockerel. If a flock of chickens includes a male he becomes the uncontested top of the pecking order. It makes me wonder if our current patriarchal culture didn't model itself after chickens. Luckily our human culture is finally evolving to empower more women at the top levels of the *homo sapiens* pecking order.

Back to chickens. Assuming there is only one cockerel, the hens will intimidate each other until the order is established. In authority after the cockerel, you don't want to be just a hen: you want to be THE HEN. The highest ranking hen has power over the others. She has first rights to feed, water, pop holes, and perches.

Like the seasons, the pecking order is ever-changing and shape shifts into new centers of power. Here are some of the reasons why the ranking status shifts.

Factors and Influences in The Pecking Order

• *Combs can reflect status - or not.* Have you ever wondered why chicken combs can sometimes be so big and brilliant? Some of the photos of exotic breeds are bizarre. Researching the topic, here's the reason. It's a power symbol.

A large comb indicates power and thus domination by intimidation. The bigger a bird's comb (both male and female) the higher status it might have in the pecking order. Exotic combs serve as power symbols in a similar way that humans use crowns. headdresses, and bishops' miters to display rank and power. But here's a catch with chickens. If a bird's comb looks unusual or weird to the flock, the others will attack her or him as an intruder.

Birds with smaller combs are usually deemed harmless by the flock and tend to be delegated to lower status, unless they have a large body size.

• Size. A standard hen will usually top just about all the bantams in the pecking order.

• *First Place or Bust.* If there are two hens battling for first place, the fight won't stop until one is submissive, or severely wounded.

• *Illness of a Top Bird.* If the #1 bird, or any of the flock, become seriously ill, she or he can become the target of abuse. They get pecked when they are down, and even worse, if they are bleeding.

• *Top Bird Out and Down.* Remove top birds from the flock and they lose their status. There are no holes in the water when it comes to pecking order & poultry politics.

• *Numbers Matter.* Introduce several hens to a flock, and the establishment of the pecking order starts anew.

• *Wimpy Hens of Status.* Any formerly high-ranking birds that have shown weakness (whatever that might be in chicken eyes), will be challenged the most, even by lower ranking birds.

• *Inter-racial Status.* If newer introduced birds are of a different breed with different feather patterns and plumage appeal, the feathers rule over the comb size, shape and, color.

• *Birds of a Feather Rule Together.* If one of the birds with different feather patterns wins a prominent position in the pecking order, then this power is conferred to other birds of that breed or plumage. Now one breed has dominance over the other breed. Sound familiar?

• *Cougar Hens.* Older hens that are more self assured will tend to have a high status. It's the feathered crone factor. Age begets respect. So if you introduce an older hen that had a high ranking in her flock, she will expect, and possibly get, a high position in the new group within a few hours of entering a new flock.

• New Flock Member. If lower ranking hens are put in a new flock they will tend to be timid and expect a lower position. The get what they ask for even though their physical aspects announce that they deserve better.

• *Cock Order.* There is a pecking order code with cockerels that is separate from hens. What's below is probably more than you need to know because urban poultry keepers won't be keeping roosters legally; but it's interesting all the same. It happens often enough that the cute pullet hen you chose turns out to be a young cockerel, or those straight-run chicks have a couple males that slipped in.

Just by their presence, roosters try to assume the highest position. But there are psychological and social variations on this. Cockerels who grew up in a flock without an adult cockerel, or who grew up with aggressive hens, can be intimidated to the point they will have a hard time mating. They have been "hen pecked" and can be poor performers until they gain experience and status. The ladies might even start a fight with the young cock. That's just to show him who has the power, as she then can become submissive and seek his affections. Finally being allowed to mate helps a young fellow's position in the flock. Behind every successful cockerel might be several supporting hens.

During molting season a cockerel tends to loose his interest in the hens and power. He might even tolerate another rooster. Molting doesn't usually affect the top hen at all. In spite of going through the ugly pin-head phase of getting new feathers, she still rules.

How to Help Keep Peace and a Stable Pecking Order

Being in charge, and the highest one on the pecking order, there are several things you can do to minimize pecking order stress and harm.

• Don't overcrowd. Have enough room to roam so that the birds can get away from each other.

• Provide places of refuge for the timid birds to escape or even hide. If you have smaller birds, provide a place they can duck into that larger birds can't go.

• Have enough feeders and waterers. Bully birds can keep underlings from eating and drinking.

• Don't cull a bird because it's the lowest on the pecking order. There will always be a bottom bird.

• If the birds constantly peck and fight, it might be due to poor nutrition and especially not enough quality protein or oils. Check the formulations.

• If one bird is an extreme bully and especially brutal, remove it from the flock for a few days. This might shift it lower on the pecking scale. If, after re-joining the flock the bird's behavior is still harmful, cull the individual. Mean birds do exist and can create a negative flock culture in the whole group. One bad bird can make life miserable for everyone; even you. As the poultry proverb says: "One bad apple spoils the flock".

Broody Hens (Sitting on Eggs)

Occasionally hens will get in the mood to become incubators and sit on a nest of eggs. When this happens, the hen turns inward and tends to remove herself from the flock and flock activities. Their behavior shifts into just sitting with a vacant look in their eyes. A brooding hen tends to cluck in soft and rolling tones. They lose their aggressive tendencies except for defending the nest. If threatened, they will ruffle their feathers and make a rapid twittering sound.

A broody hen's metabolism changes. Since she is no longer producing eggs, she doesn't need to get up early, eat, and produce. Instead, broody hens tend to leave the nest only once or twice a day, usually after mid-day, when the ambient temperature is warmer.

At midday, most of the other hens are taking a siesta, so our mama can take her break unmolested by the pecking order. She will fill her crop as full as possible, drink, and defecate once as compared to the frequent ones of her non-broody sisters. If possible she'll take a dust bath then shake, stretch, and flap her wings to complete her hygiene. If she has enough room to roam, she will not tend to walk directly back to the nest but take an indirect route so that she doesn't lead predators to her treasures. Settling down, tucking eggs under her broody body, she returns to her contemplative state.

She sits and broods on the nest for up to five weeks — or until the chicks hatch. The chapter on incubating and brooding describes how to employ a broody hen to hatch and raise chicks the natural way.

Hens Morphing into Roosters

Here's another interesting factor of the pecking order which is important for City Chick flocks of only hens. Because you will be politically correct and socially sensitive by not having a rooster, that means your highest

Introducing New Members to a Flock

A fence separating new birds from established flock members gives everyone a chance to get introduced and helps determine the pecking order without getting mugged. The ear of corn gives them a common activity and distraction, with a safety barrier protecting the new, smaller members.

ranking hen will take over the role of cockerel. She will tend to seek out food for the others, boss them around, and might even cover the hens (a nice term for mating).

The dominant hen will probably be one of your older hens because, as a hen ages, her female hormones tend to drop. And even more amazing, she might begin to crow. A dike queen in feathers right in your backyard! Nature does have its surprises and ways of coping.

Integrating New Members Into Your Flock

New hens introduced into an established flock will have a hard time as the pecking order gets reshuffled. The hazing and mugging are usually quick and decisive as the social structure gets rearranged; but these fights can also be harmful and even deadly. There are strategies that can help ease the integration. It might take several days for a new hen to figure out her place in the flock. It will help if you can be around as a referee just in case the debates get ugly. How fierce the battles are depends on several factors:

• *Number of New Birds Joining Your Flock.* The more newbies, the less fierce might be the response from the established flock. Introducing only a couple of birds into larger flocks can invoke excessive harassment and possibly result in injury and even death. Don't just drop the hens in with the flock and leave. You might need to stop a merciless gang beating of the new hens.

• *The Size and Age of the Newbie(s).* Older and larger birds will tend to take a place higher in the pecking order.

• *Access to Feed and Water.* With new birds joining the flock access to the feed and water will sometimes be blocked by higher birds. This is a power play. Having more than one waterer and feeder helps avoid this.

• *Give All Birds Space.* Crowding is stressful and will cause problems of its own. Add new birds to an already over-crowded flock and there will be hell to pay. If possible let the birds out to range and have safe places they can run to get away from the bullies.

• *Introduce the New Birds at Night.* Place the newcomers on the perches after dark. If you can, keep the light subdued to make the new birds less conspicuous.

•*Create Distractions*. Spread scratch or table scraps around so the birds will focus on something else instead of the new coop-mates. Even a few drops of an essential oil can help provide an olfactory camoflage.

The best way I've found to introduce newbies into a flock is to make a separate pen using a 3 foot, 2" x 4" mesh fence between the new ones and the established flock. The fence serves as a safety factor for the jousting, and all the hens have the chance to get used to each other. The newbies have their own waterer and feeders. At dusk, I put the newbies in with the established flock and in the morning, if they are still being harassed, then back they go to their own adjoining area for more integrated socialization. After a few days of this, the tension eases and they merge into a unified flock.

Introducing just-out-of-the-brooder chicks into a flock has the additional challenge of feed. The young chicks need higher protein than is provided by layer feed. Even more important is the layer feed is high in calcium that can harm a young chick's kidneys. In this case a chick creep feeder might be useful. A creep feeder, or creep feed area, has some sort of physical barrier that only lets the babies get in. The physical barrier might be a gap in the fence or lower space that the chicks can duck under, but the hens can't.

It's important not to let newbies get into the habit of sleeping in the nest boxes. Initially, the established hens will give newbies a hard time on the perches. The newbie might find peace and quiet in a nest box. But she will also dirty the nest box, resulting in dirty eggs. It's worth a few after-dark visits to the coop to relocate any newbie from nest boxes to roosts and get her properly used to the perches and being side-by-side with her colleagues.

Handling Chickens

With micro flocks, it's easy to get up close and personal with your chickens. This includes handling. One or two hens might protest loudly about being picked up. With time they get used to it as they get to know and trust you. Some of your biddies will become your new best buddy and come running to greet you. They will even cuddle on your lap if you let them, like a cat.

Picking up your birds has another advantage. You can tell how vital they are and if they are losing weight. Feathers can hide how much weight a bird might have lost. They might look normal walking around but when you pick them up it will be obvious how skinny they have become.

Keeping a look out for bullying can also help to know if an individual is being kept from eating her fair share. I've had the lower caste birds so harassed that they lived in a dark corner with heads down. They would have starved if I hadn't separated them so they could eat in peace and gain weight and self-esteem. It gives new meaning to re-cooperate.

If I suspect a hen hasn't had enough food, I'll feel her crop to see how full it is. The crop is in front of the throat area. You will feel a bulbous, but soft (as in stuffed) mass about the size of a golf ball. Ideally this is full as the birds finish their day and digest the food overnight.

If the crop is not full, I'll put the hen in a separate place, like a pet carrier, so she can chow down and drink in peace. If she is really thin, I'll let her eat alone twice a day, in the morning and late afternoon. A few days of this and I can tell when I pick her up if she's gained weight and has regained her vital force. Once her crop is consistently full at night — after being and eating all day with the flock — is when I'll stop providing separate feeding station for her.

Catching Chickens

There is an art to catching a chicken that only comes with experience. The best approach is a slow one. Don't rush up and expect any hen to welcome you. Chickens are easily excitable, and once they are scared, having them running from you can be a very frustrating experience for all involved.

Some breeds are much more docile than others. Selecting and keeping docile breeds can bring more camaraderie with your flock. The Hen Quest Chapter has a table that includes breed characteristics, including docility.

There are ways to encourage chickens to let you catch them consistently. A few of my hens come to me wanting to be picked up. Others I have to corner, or wait until they go to roost to even touch them.

After dark is the best time to handle chickens. They are night blind and relaxed as they prepare for sleeping.

Free range chickens that have not been handled are the hardest birds to catch. Your best approach is to guide them into a corner. Chickens can be herded using your arms or some long object, like a stick as an arm-extender. Once you get close enough to catch the bird make a quick grab aiming for the bird's body, or upper thigh. Once you have a firm hold, use both hands to secure the bird.

Never grab a chicken by the neck, head or a wing, as these are fragile body parts and the struggle can cause permanent damage. A large net is sometimes used to catch birds. I haven't found nets very useful, because the birds quickly recognize the net as something terrifying and will become agitated at the mere sight of it. Then the chase is on.

Leg catchers have a long shaft, usually 4 feet long, with a hook on the end that fits around the leg and traps the foot. These catchers require considerable skill and timing to catch a bird successfully and without harm. The problem is that, like nets, birds quickly recognize the catcher and become adept at avoiding it. I had one that disappeared long ago and I haven't missed it.

Getting hold of a few tail feathers can slow a bird down enough so you can get a hold of its body. Don't jerk back on the tail feathers or they will come out leaving your bird looking odd and without a rudder.

However you catch a bird, don't swing it by its leg, or hang onto flapping wings. This is a recipe for a dislocated joint or sprained tendons. Don't do it; it's better to let go.

Chicken Trust

How Not to Catch or Hold Chickens

Many people are clueless about how to handle chickens. It's as important to know how *not* to catch or hold a chicken. Catching or handling birds the wrong causes more harm than good. It can upset the bird and undo any trust or training you built earlier.

The wrong way to catch a bird is to grab it by the neck, head, one wing, or feet. If any of these body parts are jerked, or abruptly twisted, then permanent muscle or bone damage, disability, and even death can happen.

Luring a Chicken in Close

The logic behind getting more

Monika Eaton has gained the flocks trust by sitting down and letting the flock come to her. They peck grains or pieces of bread from her hand. This is one of the best ways to get chickens to know and trust you. Soon they will be running up greeting you and looking for treats. Eventually even the shyest ones will be eating out of your hand. Some will hop up on your lap.

bees with honey applies to getting more hens with scratch. Think like a chicken to catch a chicken. Take the time to train your chicken to eat things she likes out of your hand. This way she will eventually associate you with something good, rather than scary. All the flock will begin to trust you and come closer and closer for the treats. Preferably, while sitting down or on the ground, and with smooth, subtle movements begin to touch her while she is close and eating out of your hand. Eventually, she will enjoy your touch and will eventually let you pick her up. Sudden movements will scare her and you'll loose her trust, be gentle and quiet.

Another technique to catch a chicken is to scatter some scratch on the ground around you, preferably by a fence. Hens become so distracted by gobbling the goodies that you will probably be able to pick the one you want up.

Cornering a Chicken

Most people catch a chicken by getting it in a corner so it can't get away. This works well in coops, but can be difficult on free range. Don't chase chickens, but rather unhurriedly herd them where they can be cornered and caught.

Once you catch a chicken, don't let the bird go — if you can safely hold one while it is flapping its wings and struggling. Wait till it settles down and then gently let it go. This tells the chicken that you are in control and that you won't hurt it. This will help in the future when you want to catch her again.

Mating Position to Catch a Hen

A hen squats in the mating position when she lowers her head, and raises her hind end and tail feathers. This is the invitation for a cockerel to jump on her shoulders to breed. Hens will often huddle down in submission to their keepers (you). This makes them very easy to catch. If you don't want to pick her up, just give a quick massage on her back and she will get up, shake her feathers, and amble off clucking happily.

You can train a hen to hunker

**Catching a Hen Using
Mating Position**

Hand Position to Hold a Chicken

This highly-paid chicken model illustrates how to place one, or two fingers between a bird's legs so that the hocks (leg joints) don't painfully press against each other as you hold the bird. Your palm supports the chest to give the bird the feeling of support, as is shown by the Plymouth Barred Rock hen to the right.

A Balanced Hen Feeling Secure

This hen is well-balanced and feels safe, and a little silly, by the look on her face. There's no flapping, struggling, or panic.

Holding a Chicken Forwards

Holding a chicken facing forward is the easiest and most instinctive way to hold a chicken. Notice the feet support.

Holding a Chicken Backwards

Tucking a chicken under your arm to carry it allows you to use both hands. Don't squeeze too hard.

down by extending your arm up in the air, and then lowering your arm over her. When she senses your arm low overhead she will tend to squat. What you are doing is imitating her sense that a rooster is coming up behind her. Sometimes a hen will stop abruptly so be careful not to step on her.

Holding Chickens

Now that you have caught a chicken there are ways to make her feel comfortable and not threatened.

Hens feel more secure when their legs and bodies are supported. They, like most animals, respond well to gentle handling, soothing sounds, and slow movements.

Gripping or squeezing a bird does not make them feel more secure, but at risk of being crushed, which can cause a panic attack. Hold a chicken gently, as you would an egg.

One way to make a chicken feel

Chicken Sitting on Arm

Chickens can learn to sit on your arm like a parrot. This Araucana hen enjoys the novelty of being arm candy. Keep the chicken facing you and in close. If the bird faces any other way it will have a greater tendency to fly. Move slowly so that the bird doesn't feel like it's losing its grip or balance.

How to Calm a Chicken

Holding a chicken upside down for a few seconds has an instant calming effect. It also puts them in a submissive mode for discipline and asserts that you are at the top of the pecking order. Never swing a chicken while upside down and don't carry them way very far in this position.

secure is to slide your palm under the chest and back so that at least one finger is between the legs to buffer the hocks from getting painfully squeezed together as you hold the legs together. Keep the bird level and not angled so that it feels the need to stabilize itself by flapping wings.

Getting a Chicken Out of a Nest Box

There are two times you might need to remove a hen from a nest box. The first is at night, and the second if a hen is broody and you don't want her to sit on eggs. If a hen is laying an egg, don't remove her from the nest until she's done. If you handle her during this tender time, or even just pick her up, there is a possibility of cracking the egg in the oviduct and causing trauma, pain, and even death to the hen.

To collect eggs from under a sitting or laying hen, slide your hand underneath her and feel for eggs, pulling them out gently. She will fuss and might peck, but she's harmless and getting pecked isn't painful, at least not for long.

Some chickens tend to prefer sleeping in a nest box at night. You don't want them to sleep in any nest box because they poop in the box and that gets eggs dirty and hard to clean. It's also messy to clean the nest box later.

Manure in a nest box is a tell-tale sign that a chicken has slept there. It pays to check on the hens at dusk and remove any in the boxes onto the roosting perches to join the rest of the flock. Get them trained and they will eventually stop sleeping in nest boxes because they don't want you bothering them as they are sleeping.

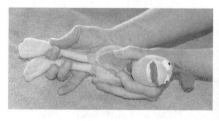

How to Safely Get a Chicken Out of a Nest Box

Removing a determined-to-remain chicken from a nest box can end up in a terrible struggle, possibly putting both you and the chicken at risk. The chicken model shows how place both hands to support and secure the bird. Use this technique and you can gently slide the bird out without a fuss. If a hen is laying an egg, don't move her until she is done. Moving her can cause the egg to break inside her and result in serious injury.

Getting a chicken out of a nest box can be tricky and intimidating. The bird can ruffle it's feathers to look fierce and defiant, truly angry for being disturbed. Hardly anything annoys a chicken more than being moved after it's gone to roost. It will fuss, squawk, growl, and squeal in protest of being disturbed. If you try to just pull the bird out, it will plant its feet, flap wings, and struggle with all its might. Because of the close quarters, this can be dangerous, resulting in sprained shoulders, bruised legs, and twisted necks on the chicken and bruises for the keeper.

The chicken whisperer way to get a hen out of a next box is much easier and safer. Quietly slide one hand under the bird and gather both legs. Slightly pull them backward. This prevents the bird from using its feet and legs to brace against being pulled out.

While moving slowly and firmly, slide your other hand along the side of the hen and tuck one of the wings against her body. Rotate her on her side, supporting her body from underneath with one hand, and at the same time keeping her legs extended behind. Now you can safely and without a fuss remove her from the nest and put her on the perch for the night.

Handling Hens That are Laying

When a hen is in the process of laying an egg it's best to leave her alone until the egg is out of her body. If you pick her up incorrectly, the proceeding egg might break and the sharp eggs shells perforate her oviduct causing serious harm.

Handling Pullets

With young hens it's especially important to handle them carefully. They are at a life phase where their oviduct is developing rapidly. If the hens are mishandled at this tender phase it could lead to a ruptured oviduct and cause misshapen and thin-shelled eggs. A severely ruptured oviduct can cause an egg to be laid into the hen's abdomen. This is fatal. Here are some precautions to avoid damage to the young hen:

- Don't lift hens off nest boxes when checking for eggs. Feel under them to gather eggs.

- Catch and inspect your hens in the nest box, which lessens the risk of handling a hen with an egg passing through the oviduct.

• When holding a hen put your hand under her breast bone to support her body weight.

• Don't squeeze the hen.

Hens as House Pets

There are stories about people keeping hens as pets inside the home, in high rise apartments, and even in travel trailers. I don't have any experience with inside chickens, and it seems at first glance that it would be inappropriate, stinky, and messy. But there are some people who have kept chickens inside. Below is a summary of what's been discussed on chat groups and blogs.

• Chickens, as inside pets, have desirable qualities. They are personable, curious, friendly, fluffy, they don't snarl, bark, bite, or require daily walks. They will snuggle on your lap (use a towel), and greet you at the door (of their pen). They eat food scraps and give back eggs.

• Chickens can get along amicably with dogs and cats.

That's the upside, all of which can be appreciated with outside chickens. Here are some other downside considerations.

• Chickens are messy, both with their manure and spreading bedding around with scratching. You have to be prepared to manage the waste products.

• Chickens need a relatively large space to live and get exercise.

• Choose smaller breeds such as bantams. They lay small eggs and make smaller messes.

• Chickens tend to poop at random, without warning, wherever they are, and in the moment. Unlike a cat, they rarely confine poop to one spot like a litter box. It's best not to let them free range inside unless newspaper is spread abundantly around.

• Chicken poop can be large and smelly.

• A chicken often selects a favorite place to perch, like the back of a silk couch.

• Chickens gobble down dog and cat food if they have access.

• Some smarter hens have learned to navigate pet doors.

• Chickens have a natural curiosity and will explore just about every nook and cranny in your house, and lay eggs there, if you let them.

The bottom line on hens as inside pets is that it has been done, but is a questionable practice from just about every angle. Keeping a hygienic indoor coop, with clean bedding and no smells is not as easy as with a pet such as a cat or parakeet. Think thrice before you decide to co-habitate with a chicken as a roommate.

Therapy Chickens

There are therapy dogs, cats and horses that interact with some folks in ways that humans can't. So why not therapy chickens? I have a few in my flock that love people; especially one named Oprah Henfrey. She is so named for her highly developed social skills, empathy, communication and entertaining talents. She is a certified therapy chicken; tried and true. Not all chickens can therapy chickens. They have to be handled and specially trained. They have to trust humans, and even like humans.

When I take Oprah Henfrey, or any of the therapy chickens, to nursing homes or schools for a show-and-tell there is always "eggcitement" among the residents. I bring the bird out sitting on my arm, like a large parrot. Everyone in the room always perks up (or wakes up) and gives us their attention. We talk about their experiences with chickens. Usually, about 75% of the senior citizens had chickens growing up. Seeing and holding a docile chicken brings back childhood memories and lots of stories to share. One fellow was catatonic until he saw Henfrey; he wanted to hold her in his lap. He rhythmatically stroked the hen in a steady 1-2-3 waltz tempo; she enjoyed it and hunkered down in his lap as if she were settling on a nest. It was the most responsive he had been for a long time. Usually there are one or two folks who don't want to hold the chicken, and even the hesitant ones ended up stroking the bird after they see that it's safe from others experience.

While lap-sitting, I always have the bird on a towel to catch any

Therapy Chickens

Seeing, petting and holding a docile chicken, with its soft downy feathers and gentle clucking, can be a multisensory delight. The therapist in these photos is Oprah Henfrey, a Chantecler hen so named for her highly developed people and social skills of her namesake, Oprah Winfrey.

This fellow was unresponsive until the chicken was put in his lap.

Reach out and touch — transgenerational and trans-species.

Both these lovely chicks are pleased to meet each other. They seem to share the same enchanted expression;

A discussion of something important .

Just about everyone will pet the chicken.

spontaneous rear emissions. There are chicken diapers specifically made to keep chickens inside without any droppings on inappropriate places.

With each visit, I pass out hatchery catalogues that have colored chicken photos to everyone who wants them. I often hear from the activities director that many of the residents bring the catalogue down to meals. For several days afterwards, residents and the staff are still talking about the chicken visit.

Dealing with Aggressive Chickens

Some chickens seem to be hostile, aggressive, and downright hateful. Perhaps they came into the world and were hatched that way. There are a couple maneuvers you can try to modify hateful acts and egregious behavior.

- If the chicken attacks you, or is aggressive to other birds, catch it and, holding it properly with your finger between the hocks, hang it upside down until it stops struggling. This calms the bird and lets her know you are in charge. Afterward you cuddle the bird for a while until it relaxes. Then gently put it down. If it has that look of being aggressive again, then do another upside-down/cuddle round.

After a few times the aggressive chicken will decide that you are really in control and the discomfort isn't worth it. This lesson might even take a few weeks, but do it every time she shows any aggression towards any human. She will eventually understand humans are higher on the pecking order than she is.

Don't swing, toss, or otherwise terrorize the hen, as this inhumane treatment will backfire on you. If everything else fails, invite her to dinner.

- If a chicken is hurtfully aggressive with another bird, catch it and do the upside-down hold. Let the two birds be face-to-face with the bully hanging upside-down. This puts the aggressor in a compromised, vulnerable position and the under-bird (as in under-dog) gets the advantage. The bird on the ground might take a chicken version of a punch by pecking at the bully. This upside-down, showdown might change the pecking order and put the combative bird in a lower slot, and in theory, less aggressive. This doesn't always work.

• Separate a bully hen for a few days. By removing it from the flock the power pecking order will shift leaving no holes in the water. If possible, put the aggressive bird in a pen where it can still be seen by the flock. Once back in with the tribe, the bully will have to reintegrate the flock. Usually, it will merge back into the flock as a kinder, gentler, and politer poultry member.

Seven Outrageous Chicken Tricks

There are a few tricks you and your chickens can perform together to truly amuse and amaze your audience. These dare-devil acts take skill, courage, trust, and practice by all involved.

Be aware that what you practice and achieve in private might be thwarted by chicken stage fright in public. Chickens tend to be shy and private about

1. Hen Jump Up Trick

This is an easy trick to train a chicken. Get your bird comfortable to eating out of your hand. Then keep raising the treat up so they have to jump. Bread is a good treat to use; just not too much..

2. Chicken Conversation Comedy

This is a "Birds and the Bees" comedy act with the delightful, dynamic duo, Paige and Minnie Pearl (the hen).

performing. <u>YouTube.com</u> might be a resource for you to get ideas and refine your skills. People will regard you, and your chicken, with new respect.

These outrageous chicken tricks are listed in the increasing order of difficulty.

1. Jumping Chicken Trick

This is the easiest to train a chicken to do. Just as a dog is trained to sit up, chickens can be trained to jump up. Start with getting members of your flock to eat out of your hand. After they are efficient at hand eating, hold some bread with your fingers and get them comfortable with taking the bread. Over time, raise your hand so the chickens have to hop up to get the treat. Eventually, a few of the more athletic ones will be able to jump up to several times their height to snatch the treat with deadeye accuracy from your fingers.

2. Chicken Conversations Comedy

Just about any chicken who likes to be held and is comfortable with people can be the star of this act. This is something like the ventriloquist dummy acts of Howdy Doody and Lamb Chop, except that your bird can participate in the conversation.

Hold your chicken, preferably sitting down, and just start a conversation with her. Ask a question and wait for her answer. A gentle squeeze or turning her can invoke a verbal response. You can create sitcom-like comedies, or sympathetic empathy discussions to fit the situation. You choose the topic and take the lead; the chicken will follow.

3. Funky Chicken Act

This comedy scene is not for every chicken, as some will find it offensive. Sit with a very tame hen in your lap, one that has a sense of adventure and humor. Don't exaggerate the movements too much or the chicken will get frightened and the fun is over. Be

3. Funky Chicken Act

considerate of the chicken's feelings and dignity.

Start by making weird chicken sounds as an introduction. Here are the lyrics. The tune you can hear at YouTube.com.

> *"Put your left arm out.*
> *Your right arm too.*
>
> *I'm going to tell you what*
> *you've got to do.*
>
> *Flap your wings, feet start*
> *kicking,*
>
> *Now you know, you're*
> *doing the funky chicken..."*

Chicken Hypnosis

These next outrageous tricks are done through hypnosis. There are several ways to hypnotize a chicken, but the easiest way is to place your subject on the ground with the legs extended backward. Hold a stick directly in front of the beak/eyes and draw a quick line on the ground directly away from the head. You might have to repeat this several times. The chicken will stare at the line and be hypnotized. Just using you finger in front of the chicken's beak can (sometimes) put it into trance.

Next is a variation of the hypnotized chicken, except you put it on its

4. Lap Chicken Stunt

5. Dead Chicken Trick

6. Gallos Gynecological Exam Act

back, still holding the chicken with one hand gently stroke its throat with the other. After a short while you can let go holding the chicken. Now put one hand over its head which will put it deeper. Softly and quietly back away.

It will wake up shortly. A gentle nudge will bring it out of its reverie faster. I don't like to use loud noises to bring the bird back as it rudely scares the bird, making it an unpleasant experience. You want to keep your chickens calm and trusting.

4. Lap Chicken Stunt

Lap chicken trick, you get a bird to be on its back and then hypnotize it by stroking its throat and talking it down. The head might need support to keep it feeling safe.

5. Dead Chicken Trick

Unlike the "dead dog" trick with a canine, you can't just say "dead" or "bang" and expect the chicken to drop. This less dramatic trick requires you to hypnotize the bird, preferably on its back and on the ground. After the chicken is in an altered state, to enhance your act, kneel down beside the chicken and tell the bird, "Don't be afraid, I'm here to help".

6. Gallos Gynecological Exam Act

Jim Bates showed this outrageous act at a dinner party starring his trained parrot. This is a variation of the dead chicken trick.

Hypnotize the hen on her back with her legs extended. You might be able to combine the dead chicken trick with the gynocological exam in one act if the bird is deeply hypnotized.

Place a white washcloth or a paper towel over her lower body as is done in gynecological exams. This stunt always draws loads of laughs, guf-faws, snickers, and shaking heads from your audience. It's either a real crowd pleaser or will cause people to leave in disgust.

7. Hen as Platter Act

7. Hen as Platter Act

This is one of the advanced, and potentially most dangerous tricks. Heavier birds require some strength to hold up. If you're not very strong, work with a bantam. This trick requires the chicken to trust you and feel totally safe and relaxed.

Until you (and your bird) get experienced, start this trick sitting down. Position the chicken on its back with its legs extended toward you. Stroke the bird until it is still and hypnotized, supporting its head if necessary. Slide your hand underneath the chicken's body so you can lift it up.

After the chicken is hypnotized, slowly, very slowly raise the chicken in your hand like a platter, keeping it level. The chicken might appear awake but if you don't make any sudden moves it will stay in your palm. Only hold it in this position for a short time. Usually the bird will do this trick once or twice and then will have had enough.

I've seen a video of a chicken fetching. It was dressed complete with a cape and diaper for indoor entertainment. My chickens tend to just race after a treat and then keep on going.

That's it folks! Come back soon!

12 Profitable Home Eggri-business

A backyard flock of 3 hens can provide enough eggs for a family of 4. A micro-flock of 20 can produce enough extra eggs to sell. Once you master the basics of chicken keeping, the next step for entrepreneurial-minded folks might be to start a small business selling eggs, compost, extra pullets, or value-added products, like cooked foods or Easter Eggs.

A chicken cottage industry could create a small additional income stream for a college fund, chicken facility remodel, fun money, etc. I'm a fan of the "multiple streams of income" approach; it has gotten me through some difficult times.

Now is a great time to begin a small urban egg business. Consumers are learning more about commercial food supply systems and often don't like what they hear. Many are willing to pay more to purchase local, organic, or free range foods from small farmers because they are concerned about several aspects of industrialized egg, meat, and produce production. Some consumer concerns are:

Your Eggs-traordinary Product

The Nutritional Value of Commercial Eggs vs Home Grown Eggs

Eggs produced by free-ranging hens that eat a variety of healthful foods are more nutritious than those from industrialized hen houses. Cheryl Long and Tabitha Alterman report that an egg-testing project conducted by *Mother*

Earth News in 2007 compared the nutritional profile of eggs from 14 free range flocks (from different parts of the country) to the USDA's nutrition data collected from "conventional" (commercially raised) eggs. The analysis was performed by an accredited laboratory and concluded that – compared to industrially produced eggs – free range chickens' eggs generally have:

- 1/3 less cholesterol
- ¼ less saturated fat
- 2/3 more vitamin A
- 2 times more Omega-3 fatty acids
- 3 times more vitamin E
- 7 times more Beta Carotene

The complete results of this study are available online, at www.MotherEarthNews.com. *Mother Earth News* isn't the first institution to come to this conclusion. Their data is supported by several previous studies:

- A 1974 study conducted by the *British Journal of Nutrition* determined that eggs from pastured hens had 50% more folic acid and 70% more vitamin B12 than eggs from industrial environments.

- A study completed by Animal Feed Science and Technology in 1998 concluded that eggs produced by free-ranging hens had higher levels of omega-3s and vitamin E than those from their caged counterparts.

- Barb Gorski, of Pennsylvania State University, found in 1999 that eggs from pastured birds contained 10 percent less fat, 34% less cholesterol, 40% more vitamin A, and 4 times the omega-3s than commercially raised birds.

- In 2003, Heather Karsten (also of Pennsylvania State University) compared eggs from 2 groups of Hy-Line variety hens. One set lived in crowded factory farm conditions, and the other had access to mixed grass and legume pasture. Both groups of eggs had similar fat and cholesterol content, but the pastured eggs had 3 times more omega-3s, 220% more vitamin E, and 62% more vitamin A.

The differences in egg nutrition are probably a result of the differences in hen nutrition; hens allowed to free range can choose from a variety of foods, including greens, seeds, bugs, composted food, layer pellets, and scratch. They can pick the diet that is best for them, instead of trying to survive on cheap rations with high amounts of corn, and soy.

Unfortunately, the egg industry has a much bigger marketing budget than small eggri-businesses. They have developed a series of arguments to debunk the claim that free range eggs are more nutritious than commercially produced ones. First, they muddle the definition of a "free range hen".

The American Egg Board's Web site (http://www.aeb.org) reports, "True free range eggs are those produced by hens raised outdoors or that have daily access to the outdoors". The USDA's definition of free range echoes that of the egg industry; they say a free range chicken is one that is, "allowed access to the outside". These are fine definitions in theory, but they are exploited in practice. Producers are allowed to label their eggs as "free range" as long as a small door is left open so chickens can (in theory) leave the giant shed and stand outside on a small plot of grass or dirt patch. This is not the type of range that consumers imagine. To the producer's relief, few, if any chickens will learn how to exit using the door.

A more ethical and accurate definition of a free-range chicken would specify that birds are allowed and encouraged to range outdoors for several hours a day on a pasture that includes several types of green plants and the accompanying insects, earthworms, weeds, and seeds.

The egg industry also claims that egg nutrition is not affected by a hen's environment. The statement on the Egg Board's website goes on to say,

> *"The nutrient content of eggs is not affected by whether hens are raised free range or in floor or cage operations . . . Barring special diets or breeds, egg nutrients are most likely similar for all egg-laying hens, no matter how they are raised."*

The red flag in this statement is "special diets" and begets the question: how is an animal's diet not an essential part of "how it is raised?" Interestingly, the Egg Board does not include on their website any links to scientific studies supporting their claims.

Of course, the Egg Board's position is supported by most other commercial egg producers, including the American Council on Science and Health (an industry-funded nonprofit), the Iowa Egg Council, the Georgia

Egg Commission, the Alberta (Canada) Egg Producers, Hormel Foods, CalMaine Foods, and NuCal Foods ("the largest distributor of shell eggs in the Western United States"). However, the faith of many consumers in "big business" has been shaken recently, and they are less likely to believe all the claims a food manufacturer makes.

Food Safety and Food-borne Diseases

Beef, spinach, and peanut butter are a few of the recent victims of food recalls in the U.S. A quick scan of the U.S. Food and Drug Administration's website (www.fda.gov) this morning revealed that there were 22 food recalls issued today. And it is only 10:30 AM. Warnings about potentially contaminated food have been fast and furious in recent years, and folks are legitimately concerned about salmonella, botulism, mad cow disease, and countless other strains of food-borne bacteria. Consumers are realizing that factory farmed food is not necessarily safe just because it was produced by a large company. They are realizing that produce and animals cultivated on small farms are fed and cared for carefully, handled by fewer people, and transported shorter distances. These factors (and many others) make them more resistant to disease, less likely to be infected by toxic bacteria or viruses, and generally healthier and more nutritious.

The Unnatural and Unhealthy Lifestyles of Commercially Farmed Hens

Numerous exposés in recent years have increased public awareness about how factory farmed animals live. Bestselling books such as *Fast Food Nation* by Eric Schlosser, *The Omnivore's Dilemma* by Michael Pollan and *Animals Make Us Human* by Temple Grandin illustrate the inhumane and unhealthy ways that most commercialized animals are raised, housed, managed, and slaughtered. These books and reports make it harder for consumers to purchase industrialized meat in good conscience.

Agribusiness's Dependence on Oil

America's "oil addiction" has gotten a lot of press lately, and statistics regarding the food industry's oil consumption are now widely reported. Commercial production and distribution often uses two calories of energy to produce, process, and transport one calorie of food.

The Dubious Nature of Many "Organic" Claims

As the demand for organic food increased, industrial organic producers began to lobby for looser definitions of "organic" and more flexibility in labeling. For example, a granola bar that advertises it is made from organic granola can include non-organic ingredients, and the list of pesticides that can be used for "organic" farming gets longer every year.

Test for Fresh Eggs

Eggs can be stored in a variety of ways and for varying amounts of time. Most consumers want fresh eggs. There are a few ways to check eggs for freshness:

The Water Test

You can easily test the freshness of an egg by putting it in a bowl of water. Eggs that sink are fresh because the air sack has not yet formed inside the shell. A floating egg is an older egg, and rotten eggs swim to the top. Be careful; rotten eggs can explode.

The Candle Test

Another way to check for freshness is to shine a strong light through the egg with an egg candler to see the egg's interior. Put the large end of the egg against the egg candler's light hole; the small end should extend downward at a 45 degree angle. Quickly twirl the egg to check for blood spots, dark colored goo, and bloody or watery whites. A large air sac indicates the egg is old – don't sell those to your customers. Feed eggs of questionable age to other pets.

The Yolk Test

A third way to ensure freshness (chefs often do this) is to fondle a raw yolk. A fresh yolk from a healthy hen will not break when you hold it up and even pinch it between two fingers.

Eggs for Cooking

Most of your egg buyers will be folks who plan to cook with them. You can generate repeat business by educating your customers about how to properly cook eggs. They are an adaptable and nutritious food. Chefs adore them, children will usually eat them, they are versatile and economical. Eggs are used in an incredible variety of dishes: they create elegant soufflé presentations, hold meat loaf together, keep oil and vinegar from separating in mayonnaise, and form crystals in candies. They spin magically into meringues and thicken smooth custards. Eggs build cake batters by providing structural framework and produce finely grained ice creams, enrich soups, and glaze rolls and pie crusts.

How To Properly Cook Eggs

Many culinary schools preach that "you can tell a great chef by the way they cook eggs". Great egg dishes are not to be rushed. Cook them at a low to moderate temperature. High heat sets the protein too rapidly and causes it to shrink quickly. Whether you are scrambling, poaching, baking, hard-boiling, or frying the egg, cooking it fast will make it tough and rubbery.

An egg cooked too quickly in a sauce will not hold the liquid properly and your sauce will curdle with small, tough lumps of egg. An egg hurried into a soufflé, meringue, or angel food cake will not have time to expand its protein and the finished dish will lack volume and lightness.

Hardcooked (Hardboiled) Eggs

Hardcooking eggs makes them easy to carry for snacks and egg salad. Instead of boiling, we prefer to steam our eggs. This seems to make the shell easy to remove and a more tender hard boiled egg. Put some water in the bottom of a steamer cook for about 20 minutes. Once the eggs are done, dunk them in ice water until they are cool. This makes them easier to peal.

To hardcook eggs, cover the eggs with about 1.5 inches of cold water. Bring to a rolling boil. Turn off the heat letting the eggs cook in the hot water for 15 minutes. Rinse under cold water until the eggs are cold.

If you use eggs that are at least a week old, or older, the shells will be easier to remove. The shell peel-ability is affected by the egg's pH. A fresh egg has a pH of about 8; a three-day-old egg has a pH of about 9.2. That's why some cooks put salt in the water – it raises the pH to a more basic level. Another way to make hardcooked eggs easier to peel is to put them in cold water immediately after taking them off the heat.

Green yolks form when iron in the yolk reacts with sulfur (hydrogen sulfide) in the white to form an unappetizing green layer of ferrous sulfide (FeS) around the yolk. Heat speeds up this chemical reaction so the longer an egg cooks, the greater chance you'll get green yolks. Watch cooking time and temperature carefully to avoid the yolk discoloration.

A technique to get the yolks centered for deviled eggs is to store eggs on their side rather than with the large end up. This helps center the yolks. Just store the egg carton on it's side in the refrigerator. Another way to center the yolk is to spin the egg for several minutes, then lay the eggs on their sides while cooking.

Not sure if an egg is hard-boiled? Just put it on the counter and twirl it. Hard boiled eggs will spin like a top. Uncooked eggs will wobble like Humpty Dumpty.

Fertile or Infertile Eggs

Because your City Chicks flock won't have a rooster, your eggs will be infertile and you won't have to worry about embryo development. Customers don't like to crack open an egg and find a bloody mess where they expected a golden yolk. In certain markets, fertile eggs bring a higher price because some folks think fertile eggs have more vitality and strength, but these markets are few and far between. Your infertile eggs have plenty of health benefits without the drawbacks of possible fertilization.

Hens produce the same number of eggs irregardless if a rooster is around. If they could vote, they would probably elect to not have that cocky guy bossing them around and dominating each and every hen. The advantages of producing infertile eggs, by not having a rooster are:

- The eggs do not hatch.
- Egg will not develop into a chick (and gross out your customers).
- They will not have unappetizing blood spots on the yolks.
- They ship better and are easier to preserve.
- They are slower to decay.

Income from Raising Pullets for Replacement Hens

If you decide to brood chicks, you can sell extra bantams or pullets that you decide not to keep, or brood two batches at once and keep one for you and one to sell. As the City Chicks movement expands, the market for ready-to-lay pullets and hens will increase. Plus, your friends are watching you enjoy your micro-flock and are bound to want their own chickens – you might make some money by being the neighborhood source for pullets and hens.

Value-added Goods

A "value-added good" is a raw product that is modified, changed, and/ or enhanced to turn it into another product worth more money. Possible value-added products for micro-flock keepers include quiches, breads, cakes, or cookies made with homegrown eggs, painted Easter Eggs, artistic photos of chickens, or feathered jewelry. The opportunities are limited only by the micro-flocker's imagination and available time. Folks who already know you and enjoy buying your eggs are a "captive audience" and are primed to purchase value-added goods.

Where to Sell Your Products

Many people think they are too shy to be good salespeople. If you are one of them, start slowly by telling your friends and relatives about your products. Then branch out to business associates, folks in your church, and so forth. Many excellent books exist about how to improve your sales abilities. Zig Ziglar is a great sales trainer; check out some of his tapes or books and start practicing. Being a good salesperson will also help you in other areas of your life. Salesmanship is worth studying.

Neighbors

Neighbors are probably the easiest folks to sell your eggs to. Delivery is not an issue, they can see how well your chickens are cared for, and they will probably be thrilled to support your business venture. You can use neighbors to practice your sales pitch, and they can spread the word about your products to their friends and co-workers.

Social Networks (Workplaces, Book clubs, Church, Exercise Classes, Etc.)

A friend of ours has a micro-flock and sells her eggs to the members of a swim class she teaches each week. Her students have standing orders and she brings the eggs to the pool. This is a great system: it is consistent income, and she doesn't have to make an extra trip to deliver the eggs. You could develop a similar arrangement with any group of people you see regularly. If your workplace has a refrigerator big enough, you could pick a day (Friday seems to be a good one) to bring eggs to work and collect payment. Bible studies, committee meetings, book clubs, or kids' sports practices could all be venues for egg sales.

Food Co-ops and Community Farms

Group sales are another avenue. Marlin and Christine Burkholder have a CSA (Community Supported Agriculture) farm just north of Harrisonburg, Virginia. The Burkholders work with local farmers who sell produce, eggs, and meat. They take orders from subscribing customers, get the products from participating farms, and handle delivery to customers. Every Monday evening they deliver about 50 bags of food to their drop-off point in the city. The Burkholders generate sales, get a commission for brokering the orders, and make money for themselves and their fellow farmers.

Restaurants

There is a recent proliferation of restaurants that use locally grown ingredients for some or all of their cooking. They plan each day's menu around the food farmers have available. Meals at some of these restaurants are pricey affairs, and others are sandwich-type joints. They might be interested in purchasing eggs or related products from you, but often need large quantities of eggs and might expect a discount for the volume they want to purchase.

Grocery Stores

Grocery stores could be a consistent customer because they carry eggs year-round. The problem is they keep a large chunk of each sale; usually about 50%.

Farmers' Markets

Some farmers' markets are barely hanging on, particularly ones in small rural communities that only generate a few dozen customers each day they are open. You probably won't have many sales at these markets, but there are plenty of buyers in large, inner-city or suburban markets. These can attract hundreds or thousands of shoppers each day. They are usually a great venue for the sale of value-added goods, too.

Barter

The ancient custom of bartering (trading goods of approximately equal value) is becoming popular again. Classified websites like craigslist.org now have sections specifically for bartering. Bartering possibilities are limitless; I have a friend who trades eggs for horseshoeing. You can trade eggs for bread, meat, beer, milk, cheese, guitar lessons, you name it. Not everyone knows about the joy of bartering, so don't be shy about asking people if they would like to trade their goods or services for your fresh eggs. The worst thing they can do is give you a funny look and say no.

Bed and Breakfasts

Bed and Breakfast (B&B) establishments have become fashionable again, as travelers and tourists get tired of sleeping in matching chain hotels. Local

B&B proprietors might be an excellent source of business; after all, eggs are a popular breakfast dish, and serving local eggs and produce adds to the charm of cozy accommodations.

Cooking Schools and Specialty Shops

If you live in a city big enough to support a cooking school – either a professional program or one that offers hobby classes – they might be interested in purchasing your eggs and related goods. Teachers and students in a cooking school will appreciate the nutritional and flavor superiority of homegrown eggs. Some cookware stores offer demonstrations or evening classes; perhaps you could donate eggs for these events in exchange for an advertisement in their flyer, website, or on their bulletin board.

Egg Share Cooperative

Some neighborhoods have egg-share cooperatives, in which several households on a block work together to take care of a flock and share the eggs. All participating households should collect their food scraps and yard waste to feed to the hens. The actual costs and labor time should be recorded and summarized so no one feels cheated or taken advantage of.

Internet

If you have consistent access to high-speed Internet and are proficient with computers, you might be able to sell some eggs or goods over the World Wide Web. Perhaps one page of your website could feature an order form that people could fill out and email their egg orders to your inbox. If nothing else, it is helpful to have a website where customers can log on and learn about you and your chickens.

How to Advertise Your Products

If you have a hard time selling your products, the problem probably isn't that people don't want to buy from you; they probably don't know about you. We live in a county with 25,000 residents and it amazes me how many people don't read the local newspaper or watch television news. We've had numerous articles written about us in area newspapers and have been featured on TV, yet almost everyone I meet for the first time says they've never heard of us. Most of our business comes from word of mouth, which is great, but only works as long as our reputation is sterling.

It might take some time to build your core group of satisfied customers who will spread the news about you. You need to find a way to tell potential customers that you are in business and have great products for sale. Then get the products to them efficiently and make it easy for them to buy.

Word of Mouth is Usually the Best Way

Nothing beats word of mouth advertising. It is free, spreads like wildfire, usually comes from sources folks trust, and builds community and friendships. Don't be afraid to ask your customers to mention your business to their friends. If they send you a customer, thank them with a gift of extra eggs, a loaf of bread, or a chicken painting. Anything that expresses your gratitude (and maybe gives them a sample of a new product) is worth it.

Samples

Samples can be a good way to get your foot in the door. We delivered gift baskets with farm brochures to each quality restaurant and B&B in our area so they could taste our (naturally superior) products. Discerning chefs have no trouble detecting the quality, and can turn into faithful buyers if you make the purchasing process easy. Send them a price and availability sheet by fax or email weekly. It also pays to call them regularly to get their order and make sure they are happy. Any time you have a new product, give it to the local chefs first so they can introduce it to the community.

In other cases, a business card is a lot cheaper than giving away samples and will get the point across just as well. Consider donating goods to silent auctions or for use in fund raising events. Sometimes it is a good idea to hand out taste samples at farmers' markets or food events.

Educate Potential Customers

To make your eggri-business profitable, you will probably have to charge more for your eggs than the local supermarket does, which means you have to convince consumers to pay a higher price. You can do this by educating and communicating. You have to communicate the information you learned in this book and explain to customers why homegrown eggs are better for people, chickens, local economies, and the planet. Here is a review of why consumers should pay more for free range, homegrown eggs:

- Better nutrition
- Happier chickens

- Stronger local economies
- Less chance of food-borne illnesses
- Healthier environment

Give Talks and Slide Show Presentations

Volunteer to give talks and Power Point presentations about chicken keeping and products at community meetings, 4-H events, book clubs, or grocery stores. Consider having an "open coop" day in which you invite existing and potential customers over to nibble on complimentary food (made from your hens' eggs, of course) and visit with the chickens. Show your guests what you do with your birds and explain why you do it.

Prepare Business Cards, Flyers, and Brochures

Printed advertising materials can be simple and cheap. Most computer software includes a program to help you prepare these documents on your home computer. Make sure you include your contact information, location, and briefly discuss your business philosophy. Then include any other information you think is necessary. Remember that people love pictures.

Establish an Online Presence

Designing and maintaining a website gets easier every year. If you aren't interested in creating an entire site for your chicken venture, consider developing a blog (they are free; try blogspot.com) or signing up for a page on popular networking sites like facebook.com or myspace.com where you can advertise your hen business to your online local friends.

Advertise in Local Newspapers or Classified Magazines

Running a classified advertisement is an inexpensive way to promote your products, but the success of classified ads varies with each region. Several years ago we lived in North Carolina and generated lots of business by running classified ads in our local paper. But here in Virginia our ads didn't even lead to enough business to pay for themselves.

Direct Mail

Advertising your poultry products by direct mail probably won't pay off. It is expensive and, in our experience, doesn't generate enough customers to make it worth it. Posting brochures and signs in places where food-conscious people congregate is more efficient; we have had moderately good response from tear-off flyers in food co-ops and health food stores. Try email advertising instead; check with your local provider about the marketing services they offer.

Competition

Your poultry product sales will probably fill a small niche market because it won't make much sense to compete for large-scale business. You will be selling to a small customer base that appreciates the quality of your products and likes working with you. Even if your neighbors get micro-flocks and start selling their own eggs, they will have other friends and family to sell to, and probably won't affect your sales volume much. Or you can consider joining forces with other small egg sellers and taking turns selling your collective products at farmers' markets and other sales venues.

Mind Your Own Business

Home-based egg businesses can use the same accounting and management principles outlined in the City Chicks Micro-Flock Poultry Project Record Book. The worksheets help beginning eggri-entrepreneurs understand the microeconomics of keeping chickens for profit. Another good resource is *Backyard Market Gardening: The Entrepreneur's Guide to Selling What You Grow* by Andy Lee and myself. This book describes how to efficiently market locally produced products from your micro-flock.

People who run small businesses (like micro-flocking and curb-side lemonade stands) don't use accounting. They buy what they need for the business and when anything sells, the money goes in their pocket; it is quiet income. But to really know if you are making a profit, you have to keep data on loans, repayments, interest, income, labor, and other expenses. This lets you know if the poultry production efforts are worth your time financially, or if your flock-keeping should be relegated to strictly hobby status. Most gardeners, family farmers, and eggri-entrepreneurs ignore or underestimate the time they spend on production. They end up losing money and subsidizing the products they sell. You must produce profit/loss and cash flow statements to know if your enterprise is worthwhile.

Having a small business can be a wonderful tax write-off. Hobby costs become legitimate tax deductions. Losses from a small business can defer taxes from other income, especially for 1099 wage income earners. My accountants, tax advisors, and business courses I've taken taught me how to make almost every expense in my life legally tax deductible. Books on this subject are listed in the Resources section.

Here are worksheets to help you track your business data by hand, on electronic spreadsheets, or in accounting programs such as Quicken (Intuit), or Microsoft Money.

The Basic Eggri-business Chart of Accounts:

ASSETS (cash, things & stuff)	
Cash on Hand (or Petty Cash)	Asset (Current)
Checking Account	Asset (bank)
Accounts Receivable (people who haven't paid you yet)	Asset (Current)
Undeposited Funds (Money you received but hasn't been entered in petty cash or checking accounts yet)	Asset (Other Current)
Liabilities (money you owe)	
Accounts Payable (money you owe to vendors)	Liability (Current)
Deposits (money people gave you in advance, like pre-paid egg subscriptions)	Liability (because you haven't provided the eggs yet)
Owner's Equity (Capital stock or money you have invested in your business)	Equity
INCOME	
Income (eggs you've been paid for)	Income
Income (other – perhaps you sold some chicken feed or compost)	Income
EXPENSES	
Chicken Coop expense	An Asset because it's something you own long-term.
Feeders, waterers, brooders, fencing	
Hens, hatching eggs, or chicks purchased	Expense or Cost of Goods Sold

Transport expense (gas, auto, bike expense to pick up feed, supplies, delivery eggs, any thing or task in your business that requires transport)	Expense or Cost of Goods Sold
Feed and supplements	Expense
Supplies (egg cartons, leg bands)	Expense
Dues and Subscriptions (poultry club dues, subscriptions to poultry magazines)	Expense
Interest Expense (interest on loans, charge card interest)	Expense
Licenses and permits (city chicken license)	Expense
Labor (paid to others, as a business person, you will figure your time and its market value later)	Expense
Postage or Shipping	Expense
Printing and Reproduction (sales flyers, news letters, etc)	Expense
Repairs and Maintenance (fixing the chicken coop, fencing expenses, whatever to repair or maintain fixed assets in your business)	Expense
Taxes (Federal, State or Local)	Expense
Phone and Internet	Expense
Training (job related courses or seminars you take to maintain or improve your business)	Expense
Meals, related business expenses, travel & entertainment. For example, write off poultry shows where you could pick up potential customers or sell your goods.	Expense
Professional fees (accountant, legal, etc.)	Expense
Expenses, misc. (all other business related expense you don't record elsewhere)	Expense

Customer Service

Sales is a customer-service based business. Your happy, loyal customers are your most important asset, especially for word-of-mouth recommendations.

Complaints

When you turn potential customers into paying customers, you begin the most important part of your business: keeping them. It is vastly easier to keep an existing customer happy than to find a new one. Good customer service is generally the same in any business; think about what it means to you and deliver that to your clients. Clichés like "the customer is always right" are often true; promptly address any complaints or questions, apologize if necessary. But don't grovel. Some people can never be satisfied. Fire those customers – they are not worth the aggravation. Remember that word of mouth can work against you as fast and powerfully as it can work for you.

Delivery

Part of good customer service might involve delivering your products. If you do deliver goods, make sure that your delivery is on time, and the product is neat, clean, and in top shape when you hand it to your customer. But be careful about how or if you make delivery routes; delivery can be costly, both financially (gas) and time-wise (driving). If possible, have customers pick goods up from you, or arrange a pick-up/drop-off time and date that works for both parties.

Challenges in Egg Production and Sales

Like any other business, selling eggs and related products will present problems to solve and challenges to overcome. Here are some things we have dealt with in our poultry business:

Seasonal Differences and Cycles in Egg Production

Although it's hard to tell by constantly stocked supermarket shelves, food production is a seasonal endeavor. Fruits, vegetables, meat, and even eggs are subject to seasonal and physiological cycles that affect their quality and availability. Food production has a natural ebb and flow. It's important for egg producers to understand how these cycles affect their hens' physiology and egg production.

People have disposed of perfectly fine hens because they didn't know about hens' natural cycles. One hen owner put her bird down because she didn't know about the molting process. The hen looked pathetic while losing feathers and the owner thought the bird was ill and suffering.

Yolk color also fluctuates with the seasons. We've had customers comment on how pale our free range egg yolks are. We explain to them that when fresh grass is available, the yolks are a deeper yellow. When the hens don't have as much access to fresh greens, the yokes are paler but the egg whites taste better. Education is the key.

True chefs understand the seasonal differences of eggs and cook dishes that optimize egg use during different parts of the year. For example, summer eggs are better for deviled eggs and hollandaise sauces. Winter eggs are great in dishes like angel food cake.

The Hen and Egg Maturing Cycle

The egg production cycle begins with a rapidly growing baby chick. The young hens (pullets) begin laying when they are about 20 weeks old. Because of their immaturity, they lay small eggs, aptly called pullet eggs. The eggs will get bigger as the hens' reproductive organs grow. When pullets are about 26 weeks old, their eggs approach a consistent standard size. If you have a young flock of pullets, expect very small eggs and tell your customers that the eggs are little because they came from young hens. We advise our egg buyers to "Take two; they're small".

Egg Production and The Molt Cycle (Changing of the Feathers)

The molting cycle of a hen affects her egg production. Hens lay for about a year before they go through a molt, at which time they lose feathers and stop laying. Protein previously used to build eggs is used to make new feathers. This feather-creating cycle lasts about a month. When newly dressed, the hen will start laying eggs again. Like other growing patterns, post-molt egg production isn't just switched from "off" to "on"; it will take a while for the hen to produce her pre-molt egg size and quantity.

Egg Production and Temperature Cycle

Temperature also affects a bird's ability to lay eggs. A chicken has a high metabolism. In cold temperatures, she uses more calories to keep warm. This doesn't leave much energy for egg laying, so her production drops. Too-high temperatures also discourage hens from laying and cause premature molting.

Day Length Cycles

Light stimulates the glands that cause chickens to lay eggs. As days get shorter, egg production drops. Natural egg production is optimal on the longest day of the year, around June 21st (the summer solstice). Conversely, egg production is at its lowest around December 21st, the shortest day of the year (winter solstice), and starts to increase as the days grow longer. Eggs are best in late May or early June, before really hot weather sets in.

Egg factories use artificial lights to stimulate production. How this affects the long-term hen health and egg quality is a question that I haven't seen good data on. I suspect extreme artificial lighting takes a toll on the birds by interfering with their sleep rhythms. I use some low-wattage, full-spectrum lights in my hen house during the evenings of shorter days, but they are mostly for my convenience. If there isn't much sunlight it's a good practice to give your hens additional vitamin D3, either by feeding cod liver oil or by adding a supplement to their rations. I suggest you also take some.

Consumer Demand Patterns

Egg producers often notice an interesting customer demand pattern. Egg consumption (and with it egg buying) peaks in November and drops dramatically in the spring. This is exactly opposite to a hen's normal production pattern.

Egg Storage & How Long Eggs Keep

How long will an egg keep? Longer than you think. Most factory farm eggs are about 2 to 3 months old before they are eaten by consumers. Here's a summary of the pipeline and egg freshness.

- Egg producers have 30 days to get their eggs to groceries.

- The stores have another 30 days to sell the eggs.

- The FDA suggests consumers keep eggs a maximum of 5 weeks in their refrigerator before discarding eggs.

- Eggs properly stored rarely spoil, they are more likely to dehydrate and dry up.

• If an egg is spoiled, you will know. It will smell like a rotten egg (the hydrogen sulfide odor), and you instinctively know not to eat it.

• The sailing ships of old kept eggs on board for months while at sea as a protein source. This includes Columbus's ships in the 1400s.

• Eggs have a natural water-soluable coating that seals the eggs from bacterial invasion. This is sometimes called "the egg bloom". Just by putting eggs in water, you wash away that protection. If you expect to keep the eggs for long, don't put them in water.

The longish shelf life of eggs is one of their selling points, and they can keep for varying lengths of time, depending on how they are stored. One long-term storage method requires only a few half-gallon jars (with lids that screw on) and a material called water glass. (Some folks call it soluble glass.) It is a mixture of potassium and sodium silicates and is sold as a thick syrup or powder. Combine one gallon of water glass with nine or ten gallons of water. Boil the water, add the water glass, and mix the solution thoroughly. Put the eggs in the half-gallon jars (you should be able to fit about 14 eggs in) and pour the solution over them; there should be at least one inch of solution over the top layer of eggs. The water glass method works best for clean eggs with strong, thick shells, preferably stored the same day they were collected. Don't wash the eggs prior to storage. Washing removes the egg's natural protective coating that is water soluble. Screw the lids on tight and store the jars in a cool place, but don't freeze them. Remove as needed for immediate use; it has been reported that eggs will keep for up to a year when stored this way.

The Pure Food Cookbook, written by Harvey Wiley in 1915, reports, "Repeated tests at various experiment stations have demonstrated that eggs properly packed in water glass after three and one-half months still appeared to be perfectly fresh. In most packed eggs, the yolk settles to one side (a sure test of an egg not fresh laid), but when packed in water glass, the yolks remained in their original position. They lost no weight, 'beat up well' for cakes and frostings, and would keep four weeks after removal from the preservative solution. In other words, water glass adds no flavor to the eggs, and takes away no flavor from them".

Dr. Wiley also wrote that the shell of an egg preserved in water glass is apt to burst when placed in boiling water, but pricking the shell carefully with a needle solves this problem. I haven't had any trouble with eggs breaking if they are cooked in water that doesn't reach the boiling point.

Table below has guidelines about how long eggs keep.

- Locally produced, refrigerated, in carton.   3 months or more
- Commercial, refrigerated, whole, in carton   5 weeks after purchase
                                                   from grocery.
- Refrigerated, whites, in tightly sealed jar   4 days
- Refrigerated, yolks, covered with water   2 days
- Refrigerated, hard cooked in shell   2 weeks
- Refrigerated, hard cooked, peeled, in water   1 week
- Hard cooked, peeled & pickled   6 months
- Water glass   6 months at 34°F
- Frozen   12 months at 0°F

Legal Concerns

You should consult your attorney for advice regarding the legal logistics of your eggri-business. You might need to obtain a business license, and incorporating your business might also be helpful. *Inc. and Grow Rich* by Allen, Hill & Kennedy is a good resource and should be available at your library. A few books written about this subject are listed in the Resources section.

In summary, being an urban producer, even on a micro-flock level, is a multi-discipline profession. Everything you learn about chickens, business, nutrition, farming, carpentry, and sales can be useful. This is not a "get-rich-quick" venture; the scale is too small to generate a large amount of cash flow. Hopefully, after all is said and done, you'll have some "egg money" profit in your pocket from an additional income stream.

Animals are such agreeable friends -

they ask no questions, they pass no criticisms.

— GEORGE ELIOT

13 Children and Chickens

Children are fascinated by animals. It's only natural that kids will want to help tend the chickens. By handling chickens, children learn how to be gentle, responsible, how to care for pets, and where eggs come from. Pets teach children about life cycle processes, as the passing of a pet might be a child's first experience with death. Children can form legitimate bonds with chickens; one of the benefits of having a micro-flock (as opposed to a larger one) is that, when given attention, hens develop individual personalities and can be affectionate – a real treat for kids. Small flocks tend to be safe for children to handle; they are generally calm, gentle, and attached to their keepers.

How children are introduced to poultry makes a big impact on their impressionable minds. You want to make it a delightful and educational experience. Children have told me stories about how snippy hens pecked their little hands or the hateful rooster that attacked them. My neighbor grew up afraid to collect eggs because the hens pecked her so viciously. A broody, puffed-up hen can be cranky enough to scare a child.

And, of course, there are stories galore about mad roosters attacking everything in sight: feathers flying, spurs slashing, leaving bloody gashes on hands and legs. Such negative experiences can cause kids (and adults) to become anti-chicken for life, such that the only chicken they like comes in nugget form.

If you combine pleasant experiences with chickens and gardening, you might begin the development of wholesome interests that last a lifetime. Here are some tips on how to introduce children to chickens and vice versa:

1. *Keep Children Quiet.* Kids have a tendency to be loud, which upsets hens. Be firm about using "inside voices" and keeping the volume down.

2. *Don't Let Children Run or Chase Birds.* The faster a hen is chased, the faster she will run way. Make children walk slowly and quietly around the flock. Preferably, children should sit down and let the hens come to them; hens are naturally curious and will eventually come over to check them out. Have children hold some grain and hens will probably eat out of their hands.

3. *Make the Encounter Simple.* Keep the conversation at the child's level and answer questions as they ask them. Suggest they watch the hens for awhile and tell you what they observe.

Soon, both children and chickens will get comfortable around each other. When kids have learned how to behave around poultry, and neither party is afraid of the other, it is time for a hands-on, up-close-and-personal experience.

Children Handling Poultry

Handling poultry – especially baby chicks – is exciting for children. As the adult-in-charge, you must pay attention to behavior of both species at all times.

I once let a 5 year old hold a baby chick. Before I could stop him, he grabbed the chick and held it by the neck. The chick was frightened and fluttered, scaring the child, who

Child Holding Baby Chick

Sarah is sitting on the floor with a towel in her lap to hold a baby Black Australorp chick. This is a good way to introduce chicks to kids. They learn to hold the chick gently, and never, ever to squeeze hard.

instinctively jerked his hand up. The chick died a few minutes later. The kid had no idea what happened; his parents were mortified; the chick was dead. Not a good experience.

Here are a few suggestions how to avoid an unpleasant scene:

1. Always stay with a child holding poultry. Be ready to rescue the bird from being improperly held, squeezed, thrown, or dropped. Sometimes it's better if you hold the chick and let the child pet it.

2. Have the child sit on the floor or in a chair. This does two things:

   • If dropped, the bird has less distance to fall.

   • The child will be less distracted if sitting.

3. To start, put a towel over the child's lap and have him or her put one hand flat in their lap so the baby chick can sit on it. The hand and the towel give the chick footing and sense of stability. Have the child use their other hand to gently hold the chick in place. Emphasize that they must not squeeze the bird tightly or it will get hurt. If the bird is struggling or the child starts to get scared, take the bird away quickly.

If you are using an adult hen, wrap her in a towel with her legs back. Having the legs extended backwards calms the hen and she can't get a foothold to stand up. The towel restrains her wings so flapping won't scare the child or get a wing tip in an eye.

Child Holding an Adult Hen

This White Leghorn hen is wrapped in a towel with her legs extended backward and the wings covered. This prevents her from struggling and keeps wings from flapping, possibly hitting the face or even worse, an eye. Take the hen away from the child before she gets up.

4. Be ready to take the bird before the child gets up, or to rescue the bird if necessary.

5. Don't let the child stand up while holding the bird. The movement might frighten the bird, or the child might lose balance. Either will result in a mishap.

Children Helping to Tend Poultry

With supervision, there are many chores children can do to help with flock care. Here are a few ideas, depending on the child's age:

• *Collecting eggs.* Kids love to collect eggs. When children come to my house, they always ask to see the hens and collect eggs. If I know they are coming, I'll leave the eggs for them to fetch; it can be Easter any time of the year. I never tire of collecting eggs.

• *Filling the feeder.* Kids can use beach buckets and shovels to scoop feed and fill feeders. Expect some to spill, but the hens will clean it up later.

• *Feeding table scraps.* Feeding table scraps to enthusiastic hens is fun for both species.

• *Opening and closing the coop door.* If your system has a pop hole that requires daily opening and closing, this might be a good task to assign to a responsible child.

• *Clean the waterer.* Waterers need a swish every so often to keep the water clean and algae growth down. This is an essential task, so make sure it's done correctly and routinely.

Hatching Eggs as a School Project

Unfortunately, many children in urban areas don't have a clue where food comes from. The not-so-far-from-the-truth joke is that city dwellers believe milk and eggs come from cartons. In England, a company called Living

Eggs addressed this gap with practical knowledge. Mark Hunt (owner of the company) has developed a 10 day educational project that leads students and teachers through the incubation, hatching, and brooding processes. The program is delivered in kits by licensed Living Eggs providers.

The project is timed so that the package arrives on Monday and the eggs hatch two days later. The eggs are pre-selected for quality and have a high viability rate – usually 100%. The chicks themselves are the lesson plan; they provide opportunities for discussions about birth and growth. There are structured exercises and hands-on exercises that, for some kids, are the first time they have held a baby animal. The lesson objectives for the project are for children to:

1. Gain an understanding of the life cycle.

2. Learn how to care for other living beings.

3. Stimulate interest, inquiry, and communications skills.

The Living Eggs service is wonderful because it takes all the guesswork, expertise, and timing out of classroom hatching. The incubation time is long enough for the kids to understand there is a living being inside the eggs, yet not long enough for them to get bored and take the eggs for granted. The chick's rapid growth and feather patterning keep them interested in the development process.

Learn more about the Living Eggs program at www.LivingEggs.co.uk.

Chicks as a Learning Experience

Baby chicks and poultry can serve as learning opportunities and special projects. Chickens can be the subject of, and participate in, many science studies, including biomass recycling and compost creation.

Hens and Environmental Studies

Living Eggs' system is an excellent way to teach children about growth and development, but the academic contributions of chickens could be much greater. Chickens would be an excellent way to teach children about recycling, composting and soil production, gardening and food production.

Increasingly, American schools are developing educational gardens in the curriculum. One class room garden is the Roots & Shoots Project at Herrington Waddell Elementary School in Lexington, Virginia. Dirck and Molly Brown are the coordinators of this innovative program that serves over 300 children from Head Start through the 5th grade. The name is a metaphor. Young children are known as "Shoots".Older community volunteers are called "Roots". This weaves together an inter-generational school garden to grow vegetables, flowers, and herbs.

The outdoor classroom serves as the laboratory for science lessons. The garden program is year round. After spring planting, the harvesting is done throughout the season and into the fall. Before putting the gardens to bed for the winter, the students dig potatoes, pick pumpkins to make jack-o'-lanterns, and make other garden crafts. The children and volunteers "work together as partners making friends of plants, insects and one another".

Jane Goodall's Roots & Shoots program sponsors positive community change for animals and for the environment. Active in almost 100 countries, the Roots & Shoots network connects youth of all ages who share a desire to create a better world. Young people identify problems in their communities and take action. Roots & Shoots members take turns caring for the plants. Gardening projects teach children about plant growth, the importance of fertile soil, and where food comes from.

Incorporating chickens, along with composting some of the food from the lunch rooms in school gardens would accomplish several more life-lession objectives:

1. Show children the art and value of composting.

2. Enrich the garden soil and improve crop yields.

3. Make gardening more fun because of the cute animal factor.

4. Illustrate the process of biorecycling and provide an opportunity to discuss the crisis of overflowing landfills.

5. Decrease the amount of waste produced by the school; chickens can eat leftovers from the cafeteria, then the residuals can be composted.

6. Teach children how to care for hens. Classes could take turns caring for the small flock.

The subject of environmental responsibility is getting more emphasis in class syllabi, as well it should. Today's children face unprecedented environmental challenges and need to understand the precarious state of our planet's health.

But it is just important to teach them things they can actually do about it, instead of overwhelming them with bad news. The mentality of, "I don't know what to do – so I won't do anything," can be replaced with a mind set of:

> *"Composting and food production is important, and I*
> *know how to do it; I learned it in school. Now my family*
> *composts in our backyard!"*

Having a micro-flock of hens working in school gardens is a great way to show kids how to make small – but meaningful – improvements in our planet's health. Here's how it could get started:

1. A local chicken-aficionado (trust me, it will be easy to find one) will volunteer to help school officials incubate eggs or purchase peeps.

2. The shop classes could design and construct chicken housing, and build fencing to keep chickens where they are supposed to be.

3. Eggs are incubated or peeps are brooded in a classroom. Children take turns caring for the chicks and learn about nutrition, growth, and development along the way.

4. When ready, the chickens move to the school's small coop and garden beds, where they will help with weed and insect control and soil fertilization. Classes will take turns doing daily chores.

5. Students could do science projects researching some part of the recycling/food production system. Student research might focus on

quantifying soil fertility, composting biology, soil fertility and plant growth/food production, chickens' effectiveness in the roles of fertilizers, pesticiders and herbiciders.

6. Install food residual collection containers in the cafeteria where children can separate leftovers from their lunches (after carefully removing any non-food materials). Some of these residuals could be used for ongoing compost demonstrations using different types of composting techniques.

7. The resulting compost is put back onto garden beds to raise more food, some of the food served back to the students.

8. Students learn how to save seeds and the value of seed saving for next year's harvest.

Introducing Adults and Chickens

It's just as important for adults to have a pleasant experience when first introduced to chickens. When I took Attila the Hen to the Lexington City Council meeting, the members were all impressed with how calm, pleasant, clean, and warm she was. She looked them in the eye and was not afraid to be stroked and have her comb and wattles fondled by strangers. Her personality shone through and bridged the inter-species good will gap.

When a dinner guest asks questions about my chickens I will often invite a hen to dinner – as a guest. I get one of my hens that especially likes people and let folks hold her. She likes to snuggle down in a lap and drift off while being stroked. Just to be safe, I'll put a towel over the lap of whoever is holding her. Just about everyone wants to hold the hen once they see how gentle she is. Usually she is passed around from one lap to another.

There is something about chickens that can put a smile on faces of all ages. Even grumps soften at a hen's downey feathers, comical looks, and awkward antics. Close encounters of the hen kind can create memorable and magic moments.

Kids of All Ages Enjoy Interacting with an Amicable Chicken

Marie Foreman, 90 something, chats with an Araucana chick – 2 about weeks old.

Kaia and a Partridge Rock pullet.

Duke chuckles at a chicken's joke.

Chris, Duke, and Agnus, a Black Australorp chick.

Rebecca and Tasha take turns cuddling Q-tip, a White Rock hen.

14 Poultry Clubs & City Chicks Projects

Keeping City Chicks doesn't have to be a solo journey. The communication networks provided by the internet and telecommunications can connect lonely poultry people from around the world or down the street. City Chicks groups, blogs, and activities are popping up everywhere. This chapter describes a few ways chicken folks can connect with each other and learn more about keeping chickens.

Poultry and Garden Clubs

Combining poultry clubs with garden groups is a marriage made for food production. We started The Shenandoah Valley Poultry and Garden Club. Our club is for chicken owners (and wannabes) who keep poultry for pets, eggs, meat and show. It is also for gardeners who are growing food (or want to grow food) for themselves, and perhaps others to generate income, for barter and/or to participate in local farmers' markets. We collaborate with the 4-H poultry club with lectures, workshops, field trips and participating in poultry shows.

We exploring how chicken "skill sets" can be employed in our local "Chicken Have-More Plan". We talk by turning kitchen, yard and coop bedding into compost. We explore how to integrate chickens as pesticiders, herbiciders, and insecticiders in our gardens. We talk about the local lymes epidemic

We discuss how chickens are bio-recyclers and can be employed in our area as clucking civic workers in solid waste management systems to save thousands, even millions of our precious tax payer dollars. We "eggstaticly" and "eggcitedly" meet every month. Think about starting a poultry and garden club in your area. We offer support and out-reach assistance to other groups wanting to get started with the "Chicken Have-More Plan".

American Poultry Association and the American Bantam Association

The APA and ABA have strong youth programs and sanctions poultry shows and club meets across North America. They both have strong youth programs and educational out-reach.

4-H Poultry Clubs

4-H is sponsored by the USDA Extension Service. 4-H Poultry Clubs are for ages 5 to 19 and members usually participate in the poultry chain that gives 50 to 100 chicks to members for brooding and showing at local fairs. At the fair livestock auction, each club member give 10 of their birds to be sold to keep the revolving poultry chain fund going.

Poultry Blogs & Websites

Many City Chick groups publish blogs and websites that communicate useful and time-saving information, including:

- Meeting and event dates
- Member contact information
- Where to find replacement hens
- Poultry rescue contacts & resources
- Information about hen keeping and coop construction classes
- How to connect egg buyers and sellers
- Poultry sitters
- Local ordinances and applications for chicken-keeping permits
- Poultry feed and supply vendors
- Links to other poultry resources
- Resources for round up of unwanted chickens.

It's easy to start a blog, online discussion group, or club website. If you want to know more about what's happening with hens in your area, don't be shy about networking.

Poultry Rescue

The issue of abandoned chickens usually comes up in discussions about whether or not to allow hens in city limits. Chickens are often abandoned. We've contacted our local humane society and told them we will take any unwanted poultry. Most animal control facilities are not designed to keep chickens.

A local chicken rescue network, linked to the local poultry club website (if there is one), would be a valuable community service and solve a lot of problems. If the rescue network had a clear plan of action, abandoned fowl – even roosters – could be adopted with a few phone calls between club members. We have a local zoo that takes orphaned chickens as free-ranging zoo animals or, as a last resort, dinner for a captive carnivore.

Local Abbattoir

An abbattoir is a killing and processing place for livestock. Abbattoir is derived from the French verb abattre, meaning "to strike down". Usually processing chickens within city limits is illegal. But a meat purveyor that takes unwanted chickens, and lawfully converts those chickens to food, even if it's pet food, is a valuable service for poultry people. The meat vendor takes problems and turns them into meals. Candidates for a trip to the abbattoir could include roosters, old hens, and stray birds that no one will adopt.

Having a local chicken processer would be in the spirit of the "Grow Local, Eat Local" movement. Many chicken breeds are "dual purpose" meaning useful for eating and laying eggs. They are sometimes called "table birds" implying the dining room table.

Many urban chickens will be treated as pets. Even so, having a quick and humane final exit to their lives is worth pondering. Most chickens are kept mainly for egg production. As hens age, their egg production drops dramatically. To supply yourself, and others, with eggs, you might want cull (remove) older hens for younger ones. With reverence for life and respect for death, deal with removing individuals in your flock humanely and within the local laws.

Some people won't eat a pet. But the idea of *someone else* eating it, instead of disposing of it in a landfill, is consistent with the spirit of local food production. Either way, it's part of the life cycle and the natural food chain.

Master Chicken Keeper Certification

The Master Gardener program has inspired the creation of Master Bee

Keeper and Master Composter classes, sponsored by some USDA Extension programs. I will not be surprised if someday there is a Master City Chicks course. It would be wonderful if these Master Series were offered throughout the country so that gardens, bees, and chickens could support local food production. Food supply safety is a national security concern, and I think our government should strongly support and subsidize local production.

City Chick Coop & Garden Tours

Joining a City Chick tour is a super way to meet like-minded poultry people. Annual chicken coop design and gardening tours are increasing in popularity.

Seattle, Washington, Madison, Wisconsin, and Portland, Oregon, sponsor annual City Chick tours as fund raisers. These tours offer workshops and insights into the world of City Chicks. The tours are usually self-guided and give folks a chance to talk with local chicken owners about building coops, selecting bird breeds, managing eggs, and the trials and tribulations of keeping City Chicks. Coops featured in the tours range from simple coops that were built in a day to converted sheds, spaces in garages, and exotic, architecturally-inspired hen chateaus that took weeks to plan and months to build. Some of the high-end coops are designed to blend with the housing style of their neighborhoods.

Poultry Club Blogs & Chat Groups

City Chicks blogs and chat rooms are popping up everywhere, helping hen keepers share tips, support, and vacation coverage. Some urban poultry clubs host classes about hen house building and hen keeping 101.

Local Poultry and Garden Clubs

Although there are a plethora of online hen discussion groups, meeting locally with other folks and talking face-to-face about fowl subjects is not only fun, it's informative. I've had the pleasure of meeting some of the most interesting people in my life because of chickens. Who would have thought?

Inter-disciplinary Poultry and Garden Clubs

Inter-disciplinary clubs gain strength through collaboration. Gardening groups that include hens as garden helpers have more dimensions than just a gardening club or just a poultry club. Supporters of the "Eat Local" and "Eat Clean" movements would also be good people to network with.

Bee keepers, bakers, equestrians (they have lots of manure for composting), vegetarians and vegans, and local restaurant owners might also be interested in collaborating with "clucks and shoots" chicken/garden clubs.

State and National Poultry Associations

State and national poultry associations usually focus on large-scale commercial poultry, meat, and egg production.

The American Poultry Association (APA) hosts poultry shows all over the US and Canada, including an Annual meet and a Semi-Annual meet. The APA publishes the *American Standard of Perfection*. This book has complete breed descriptions and sets the standard for judges.

Individual members earn points at these shows toward becoming a Master Exhibitor. The APA publishes a quarterly newsletter and an annual yearbook. The yearbook features articles, membership listings, advertisements, APA licensed judges, and master exhibitor information.

International Poultry Clubs

The Internet has just about made a global village for poultry keepers. There are discussion groups in many countries that you can join. These connections can give you contacts and advice from abroad with different perspectives.

The different clubs' blogs and websites can be a rich source of information about a particular breed and what works in their country.

You can meet fellow poultry lovers on line and it's a genuine treat to meet them in person when traveling abroad. With adequate advance notice, you even might be invited to be a guest speaker to talk about poultry at one of their meetings.

Import laws won't allow you to bring home live chickens or eggs from another country.

In summary, there are many ways to connect with other like-minded and special interest poultry people. Chicken owners are increasingly important players in local food production because they provide eggs, compost, and garden produce. City Chick keepers are doing great things for their neighborhoods and the environment. Chickens can serve as a catalyst and a focal point to bring communities together at many levels and for many purposes.

15 City Chicks Project Work Book

The *City Chicks Project Workbook* was developed with a grant from the Gossamer Foundation. It is adapted specially for use with micro-flocks. Much of the focus is on conservation because users are encouraged to record the amount of food and yard biomass "waste" they divert from landfills. This "waste" diversion is to feed their chickens and create compost they produce with chicken manure. In the spirit of public service and local self-sufficiency, you are free to copy, use, and distribute the *City Chick Project Workbook*.

The *City Chicks Workbook* can be downloaded for free in an 8.5" x 11" format from www.GoodEarthPublications.com. Changes and suggestions about how to expand and improve the *Workbook* are welcome. Updated editions will be published on the website.

Customized project, city, or organization workbooks containing specific phone numbers, dates, names, and policies can be developed by the author and Good Earth Publications. Contact the publisher for more information.

This workbook focuses on projects that involve different knowledge and skills necessary for the management, and humane care, of micro-flocks. The questions, forms and data you collect will help guide you through the many considerations, benefits, and unintended consequences of being involved with a poultry project. Keeping a journal of your experiences throughout the project can give you valuable insights and hindsights about your project. Have fun with your journey. Your hens will be your most valuable teachers. Listen, observe and respect what they tell you about how to care for them; what works and what doesn't.

City Chicks
Project Work Book

*

Name:_____Years in Poultry Projects: _____

Address: _____

Phone: (H) _____ (C) _____

email: _____ @ _____

Date of Birth: _____ Age: _____

Poultry Project Goal

Describe below what you want to accomplish with your chickens. Write down your goal now and review it at the end of your project. Be as specific as possible. Someone once said, "An unwritten goal is an unmet goal". You might find that as you learn, and do more, your goals will change.

Learning & Skills Development Objectives

You have to define what you want to learn about chickens before you can gain the skills you need to be a good hen keeper. This knowledge and skill set will expand during your years as an urban hen keeper. Beginning to learn a new subject is exciting, but can be daunting. Just take it one step at a time and build your experience one day at a time. Some suggestions for learning objectives are:

1. To understand humane ways of keeping chickens and develop pride in and responsibility for chicken ownership.

2. To acquire information and an understanding of hen management by keeping records of my flock.

3. To learn how to collect, process, and store eggs safely for human consumption.

4. To develop integrity, showmanship, cooperation, and communication skills by giving demonstrations, talks, and participating in local fairs and competitions.

5. To learn and use chicken skill sets to help the environment. This includes egg production, gardening with hens to create a local food supply, biorecycling food and yard waste to divert biomass from landfills, and using chicken manure for compost and soil enrichment.

Write in the space below your own goals and desired poultry skill sets.

This record book works for all aspects of your poultry project. By keeping records up-to-date, you will be able to see how much progress you make as you set goals and work to accomplish them. Begin keeping data at the beginning of your project. Write clearly and keep notes about everything you do with your chickens. Make a commitment and discipline yourself to follow through. You might want to develop spreadsheets and keep data on your computer. Keep a record of all the money you spend on or generate from this project. Also, track your time; it's your most valuable asset. Take photos from the first day to build a slide show or power point presentation. If you need help, ask. Add pages as you need more space.

At the end of your project you should be able to answer these questions:

1. How did you house your chickens to keep them safe and healthy?

2. What did you learn by doing this project?

      About chickens

      About yourself

      About food production and biorecycling

3. How did you research your project? List references including books, websites, people, seminars, field trips, etc.

4. What was your biggest challenge?

5. What did you like most about chickens and your project?

6. What did you like least about chickens and your project?

7. How could you have expanded your project?

8. Describe how keeping chickens could be beneficial to the environment. List ways they might help decrease pollution.

9. Describe at least 5 ways chickens can help with local food supply.

10. Would you consider doing this type of project again next year? Be specific about why or why not.

11. Review your original goal from the beginning. Did you accomplish your goal? Why or why not? What other unintended goals did you accomplish?

City Chick Micro-flock Worksheet and Information Tracking

This section provides worksheets to document how you obtained and managed your chickens. It tracks project progress from hatching eggs, raising day old chicks to pullets, or buying adult hens.

Source(s):_____ Breed (s): _____

Table 1: City Chicks General Information

Identi-fication (leg band, markings)	Date of Hatching or age of hen at purchase	Date of Purchase	Purchase/ value at start of project	Value at end of project if sold
			Total purchase price	Total value at end of project

Comments and observations:

Table 2: City Chicks Feed Information

Record all feed your chickens consume during this project. Include kitchen food waste.

Date of purchase	Brand or type of feed used. Include on a separate page the formulation if you made your own.	Pounds of feed	Cost of feed	Pounds of kitchen waste given to hens.
		Total pounds of feed	Total cost of feed	Total food waste fed or composted

Describe your feeding program in as much detail as possible. List any supplemental food and/or vitamins/nutrients that you gave your chickens. Describe all brands and types of feed you used and the reasons for choosing one type of feed over another.

Did you have any food-related diseases during your project that supplements or food management could have prevented? Describe the morbidity and mortality (sicknesses and death) that might have been caused by nutrition.

Table 3. City Chicks Other Expenses

Other expenses include the costs of housing, waterers, feeders, bedding, fencing, leg bands, storage cans, nest boxes, transportation, fair fees, veterinarian bills, medications, etc.

Date	Item	Cost
		Total Cost

In any venture, there will be unexpected expenses that can increase the cost of the project. In real business, often these miscellaneous expenses are overlooked and not included in accounting. Not including such expenses makes the project look more profitable than it really is.

For tax purposes, any business-related expense is deductible. Keeping receipts and recording expenses can dramatically reduce federal and state tax obligations.

Time & Task Tracking

In any project or business venture, the amount of time spent is often greatly discounted. To really know what your project might earn, it's important to include time as an expense. This will help you estimate your pay/hour. One way to estimate how much time you spend is to take about 5 typical days and accurately record the time (start to finish), then total the time by task. Typical tasks include:

- Daily care (feeding and watering)
- Collecting, cleaning, sorting, and putting eggs in cartons
- Cleaning the hen house
- Hen health care, giving medications, special treatment etc.
- Getting feed
- Having fun with the flock

City Chicks Project Daily Time Log

Start & End Time	Duration	Activity Description	Task

Observations about time and activities.

Time Log Summary

Task	Day 1	Day 2	Day 3	Day 4	Day 5	Average
Total						

You can use the results of where you spent your time to see where your operation could be streamlined. For example, would adding an automatic pop hole door keeper save enough time to justify the cost?

Once you know your total profit (or loss) you can calculate your pay/hour as if you were working for someone else by taking your total net profit and dividing by your time spent generating that profit.

Income

Micro-flocks can be a source of multiple streams of income. The most obvious is selling eggs. There might also be a market for hatching eggs that you get from breeders, selling newly hatched chicks, ready-to-lay pullets, grown hens, compost, or feathers for fly fishing and costumes.

Table: Income from a City Chicks Flock

Date	Kind (eggs, chicks, hens, compost)	Amount	Price (per dozen, per bird))	Total Income

City Chick Production Indicators

An indicator is a ratio that tells you about the status of something and how the status has changed from an earlier time frame. Whether you are keeping hens as pets to provide eggs for your household, or as an egg business, you might want to track your flock's status. Useful indicators for City Chicks include:

- Average egg production per day/week/month/season/year
    = Number of eggs layed/time frame
- Egg production per number of hens
    = (Number of Eggs/week)/total number of birds in flock

- Average age of your flock
    = Average Age of Hens in Flock/Total Age
- Cost of Egg Production per Dozen
    = Total Costs of Keeping Hens/Number of Eggs Produced
- Profit per Dozen Eggs
    = Total costs of egg production/Income generated

These indicators are useful when making management decisions. For example, if you notice the egg production for the entire flock has dramatically dropped, there is a reason for it. It might be that it's winter and egg production has naturally decreased because of the shorter days. As your flock grows older, they will lay larger eggs, but there will be fewer of them. The flock might be ill or infested with worms, lice, or mites.

Data collection doesn't have to be complicated. It takes only a few minutes to record egg production and track flock status. I hang a calendar in my kitchen and mark how many eggs I collect each day. The cost of feed is easy to track if you keep your sales receipts. Tracking indicators can be very basic and provide valuable information.

Summarize Your Project

Record the total income, if any from your project   _____

Subtract the total expenses from your project: _____

Total profit or loss   _____

Describe factors that affected your accounting that might have resulted in a profit or loss.

Describe what your indicators told you and how you might use the indicators in flock management.

Keeping chickens is not just about money. The most important things in life don't have a price tag. The most important things cannot be bought and are priceless. List at least 5 intangible, valuable benefits you received from doing your poultry project.

*It is a wise man's part, rather to avoid
sickness, than to wish for medicines.*

— THOMAS MORE, UTOPIA

*The greatest mistake in the treatment
of diseases is that there are physicians
for the body and physicians for the soul,
although the two cannot be separated.*

— PLATO

16 Poultry Primary Health Care

To manage the health of my micro-flock, I use what Western (allopathic) medicine would call "alternative" therapies and preventative practices. I use a combination of traditional (folk) treatments and Functional Medicine methods to care for my birds. Functional Medicine is a way of practicing medicine that uses symptoms as clues to how a body is functioning (or not) as a whole. Dr. Jeffery Bland describes Functional Medicine as "a systems biology approach to health". This approach was developed to treat humans but applies to all animal species. It emphasizes the comprehensive manner in which all the components of biological systems interact with each other, and with the environment, over time.

In other words, what works for one sick chicken might not be appropriate for the entire flock. Even at the chicken level, I believe in personalized medicine. Rarely does my entire flock need treatment for an infectious disease.

The development of Functional Medicine is based on three primary questions:

1. How are the body's physiological systems linked together?

2. How are those functions influenced by environment and genetics?

3. Is there something the body lacks (nutrition), or has in excess (toxins)?

Western medicine has delegated each body function and organ to specialists such as Ear, Nose, Throat (ENT), oncology (Cancer), endocrinology (glands), cardiology (heart), etc. This type of compartmentalization is the opposite of the Functional Medicine model, which views the body as a web of interconnections of physiological processes and biochemical pathways. More robust processes and cleaner pathways result in healthier animals. Each of the body's functions enhances or hinders the other functions. For example, the immune system begins in the digestive tract and is influenced by the endocrine glands. To alleviate allergies (an immune system malfunction), the digestive tract must be healthy – a connection that most Western physicians have not been trained to address. Only by looking at our bodies holistically can we understand the complex symptoms that humans and animals (and chickens) present. Functional Medicine is a dynamic approach to preventing and treating disease. You can find a list of Functional Medicine physicians at the Institute for Functional Medicine's website: www.FunctionalMedicine.org.

Functional Medicine emphasizes healthy digestive systems, with good reason. About 70% of our immune function ability begins in the GI tract, which is a concept that is just now being introduced in mainstream medicine.

Every animal's digestive system works by breaking down nutrients in the stomach and absorbing them in the intestines. Some reasons for a malfunctioning digestive system are:

- An overload of toxins
- Infestations of parasites
- Overgrowth of "bad" bacteria and fungus
- Inability to absorb nutrients properly
- Leaky gut (a breach of the intestinal mucosa)

The Functional Medicine "4R" model represents the process of re-establishing digestive balance to enhance well-being and immune system functioning. It works for humans and hens. The four R's are for:

- *Remove* pathogenic yeast, bacteria, and protozoa from the GI tract and stop the intake of foods and additives that result in disease (which could also be thought of as "dis-ease") or allergic responses.

• *Replace* beneficial digestive enzymes and other factors that support a healthy GI function.

• *Re-inoculate* the gut with beneficial bacteria to establish a healthy balance of microflora. This can include probiotics or prebiotics.

• *Repair* the mucosal layer so the gut isn't "leaky". A number of specific nutrients and phyto-nutrients can help repair the GI lining.

The 4R model has important repercussions; if your hens' digestive systems are out of whack and they can't properly digest and absorb nutrients, supplementing their diet will not help. The old saying "an ounce of prevention is worth a pound of cure" applies to poultry as well as people. Almost all chicken ailments (except old age) can be avoided with proper diet and nutritional supplements, abundant fresh and clean water, suitable housing, and a low-stress environment.

Leg Bands

Leg bands tell you who's who in your flock. This hen has numbered plastic bandettes on each leg and a spiral leg band on the right leg.

Chicken Identification

Being able to identify the chickens in your flock is important, especially when you are treating a particular bird or trying to catch a hen with a bad habit, such as flying over fences or eating eggs.

Even in micro-flocks, birds of the same breed look so much alike, it can be difficult to know who is who. Using leg bands to ID your entire flock is usually not necessary, but it is useful to know the birds' ages and recognize a hen with vices, such as excessive aggressiveness or an egg eating suspect.

Colored and numbered leg bands are a great way to identify chickens. Each color represents the year a hen joined your flock, and the number on each band corresponds to each chicken. For example, a yellow leg band means the chicken was born in 2009, a green banded hen was born in 2010, etc. Most chickens don't live more than 4 to 6 years, so you can re-use the plastic leg bands as hens come and go from your flock.

I put the chickens' primary ID band on their right leg. If I'm tracking a particular hen, I'll put a colorful marker band on her left leg so I can identify her in the crowd, or on the roosts.

You will need to get leg bands in a variety of sizes to fit each chicken in your flock. Like rings on people's fingers, sometimes a particular size of band fits perfectly, and at other times in life it doesn't. Leg band sizes range from size 4 for baby bantam chicks to 14 for big adult birds. The band sizes and diameters are below:

Size 4 = 1/4"   Baby Bantam Chicks

Size 5 = 5/16"   Chicks

Size 6 = 3/8"   Month old chicks

Size 7 = 7/16"   Bantams, Hamburgs, growing chicks

Size 8 1/2"   Larger bantam hens, some bantam roosters

Size 9 = 9/16"   Most standard size hens, some larger bantam roosters

Size 10 = 5/8"   Heavy breed hens, some roosters

Size 11 = 11/16"   Rock and Wyandottes hens, smaller roosters

Size 12 = 3/4"   Large Rocks, Reds, Roosters

Size 14 = 7/8"   Jersey Giants, Large roosters

Keep a keen eye on banded birds to make sure the bands don't constrict or irritate the leg. I put an expandable, standard size 9 bandette on a heavy breed hen when she joined my flock. About 3 months later I noticed she was limping. The band had become so tight it was digging into her scales and causing her leg and foot to swell. I removed the band immediately, and after a couple of weeks her leg healed. I put a larger, size 12, bandette on the hen and she has been fine ever since.

Here are few things to consider when choosing and using leg bands in your flock:

• Plastic bandettes and spirals are removable and can be used for years. Metal bands can only be used once.

• Monitor band tightness. Young chicks and even full grown birds can outgrow the bands. Tight leg bands can be disabling and grow into the shank. Just because a spiral band is "expandable" doesn't mean it expands on the bird's leg as it grows larger. If you see a chicken limping, check the leg band. It might be cutting off circulation or irritating the leg with a rough edge.

• Stretching new expandable plastic bands makes them easier to put on the bird and, once on the leg, they will expand easier, if necessary. If a band looks tight on a leg, remove the band and stretch it so it's not as tightly wound.

• Spiral leg bands are cheap, and made of thin plastic, about the thickness of spaghetti. They have sharp ends that can pierce the leg shank which can hurt the hens when you put them on or take them off. Spirals can also irritate a leg if there is a plastic rough spot that rubs, causing a blister. Trim the sharp ends of the band to make it blunt before putting it on your chicken.

• Metal (aluminum) leg bands have their own applicator. They are not reusable and are hard to remove. Make sure the edges are smooth so the leg doesn't get irritated.

• In very cold climates there might be a danger of metal bands getting wet and freezing to the leg.

What You Can do to Prevent Health Problems

It's much easier to prevent health problems using proper flock management than it is to cure disease. Below are three suggestions for keeping your flock in optimal health.

1. Clean, Abundant Water as Preventative Health Care

So many health problems can be avoided by simply providing access to clean water at all times. Hens require ample water to form eggs and dehydration can kill a chicken fast.

How clean does water have to be? The rule of thumb is to ask yourself if you would drink the water you give your flock. If the answer is no, then the water is not clean enough. How much water should be available? Enough so it never runs dry. Never underestimate the importance of water in your flock's health care plan.

2. Personal Hen Hygiene as Health Care

Keeping chickens, their housing and run(s) clean is non-negotiable for healthy chickens. Part of the daily chicken chores include adding more bedding inside the coop to cover droppings, or moving the shelter to fresh ground to keep manure from accumulating into hazardous waste levels. If you are not willing to do this consistently, then don't keep chickens.

Bathing Dirty Chickens

There may come a time when you need to give a hen a bath. It might be to prepare for a poultry show, or because she got exceptionally dirty, or to treat her for infestations of mites or lice. A clean chicken is usually a healthy chicken.

The last hen I bathed was a wounded and filthy "underdog" hen that was given to me. She was in desperate need of tender loving care including de-bugging and cleaning. Her wounds would heal faster if she was clean and started eating clean, nutritious food. She also had to have a clear bill of health and hygiene before introducing her to my flock.

The good news is that hens are easier to bathe than cats. Here's one way to do it.

Fill a bowl with enough warm water (never hot) so the bird can stand on the bottom with her chest and back above the water level. A chicken floats, but not as well – or as willingly – as a duck. Usually birds will calm down when their toes touch the bottom of the tub. That's when they look at you with questioning eyes, "What did I do to deserve this?"

If your bird has dirty feet, clean them first, using an old toothbrush to remove the dirt from between the toes. A full pedicure (trimming the toenails) is a good idea. Do the towel wrap to secure the chicken while you do the pedicure.

After washing particularly dirty feet, you might have to refill the tub with clean water before you wash the bird's feathers. Once the feet are clean enough, lower the chicken into the water by putting one hand under her chest for support and the other over her back to keep wings from flapping. Let her stand in the water and gently work the water and soap onto her skin. After letting her soak for a minute, scoop water onto her back and wings, making sure she is rinsed thoroughly. If she is really dirty you might have to do another wash cycle. Some chickens relax during their bath, as if they are getting a spa treatment. They start playing with rubber duckies in the water, splashing and making giggling sounds.

When your hen is clean and has had enough of the spa, wrap her in a towel to soak up most of the water. If you really want to give her a coiffure, use a hair dryer on the cool or low setting. Hold it about 18 inches away from her skin to blow her feathers dry and style her to her best advantage.

A final dusting with diatomaceous earth (DE) completes the makeover. DE kills lice and other parasitic visitors. Keep her confined until she is

Chicken Bath & Foot Wash

A chicken is calmer during a bath if it can stand on the bottom of the tub and it's chest remains above water. Use an old tooth brush to clean extra dirty feet. This is also a good time to trim toe nails.

completely dry. If you let your squeaky-clean but wet hen out right away, her daily dust bath will turn into a mud bath: good at a spa, but bad for a clean chicken.

3. Heat and Cold Regulation for Healthy Chickens

Chickens don't have sweat glands. In order to stay cool they open their mouths and breathe quickly to dissipate heat (much like a dog panting), and hold their wings away from their bodies. They will drink the coolest water available to fill their crops and create a mini-radiator system so that blood flowing through the arteries in their necks cools as it passes the crop.

Heat stroke and dehydration are common causes of hen mortality. This is especially true when hens live in Chicken Tractor housing with galvanized roofs that turn into ovens on hot summer days. When it is hot and humid it is essential – absolutely essential – for hens to have an ample supply of clean water and shade.

Cold temperatures can also kill hens. The normal body temperature of chickens is between 103.6 to 110.5 degrees Fahrenheit (39.8 to 43.6 Celsius). Their body temperature is warmest around 4 PM and coldest around midnight.

Frostbite

Most chickens can tolerate cold temperatures, but even the most winter hardy breeds can suffer frostbite in severe cold. Providing ventilated, but draft free housing helps the chickens retain body heat. Proper ventilation keeps the air moisture lower. Lower humidity helps the chickens retain body heat better. In other words, frostbite is more likely to happen in drafty, damp housing.

When it's cold, hens sleep with their heads tucked under their wings, and they huddle down on the roost to cover their feet. Don't use metal roosts as this can promote frostbitten feet. Roosters don't tuck their heads under their wings and are more likely to get frostbite when night temperatures plummet.

Chickens can get frostbitten on the exposed parts of their body which are combs, wattles, ear lobes, and toes.

Toes Lost to Frostbite

At freezing temperatures (in the single numbers and below 0° C, or 32°F), blood vessels close to the skin constrict to preserve core body temperature. The decrease in circulation can result in frozen tissue as cells form ice crystals and erupt. The damage can be permanent. Frostbitten areas become discolored and turn black as the tissue dies. Nerve damage is painful, blisters can form, and feeling can be permanently lost in frostbitten areas.

Frozen combs and wattles look pale before they thaw. After the tissue has thawed they will be red, hot, and swollen and can be very painful to the bird. After the swelling goes down the skin can peal, scab, ooze pus, and eventually fall off. If a frostbitten body part turns black, the tissue is dead and there is a possibility of gangrene setting in. This tissue should be surgically removed.

Frostbite Prevention. You can avoid frostbite by selecting chicken breeds that have small wattles and combs that lie tight against the heads; the less tissue available to freeze, the better, particularly if you live in a cold climate. Coating combs with petroleum jelly provides some protection from frostbite.

Some cold climate poultry keepers dub (cut off) the combs, wattles, and ear lobes of their adult birds to prevent frostbite and the accompanying problems. Many folks say it is best to do it at night, before the new moon, when the birds are quiet and not moving around. Dubbing requires a pair of razor sharp, clean, and sterilized scissors, rubbing alcohol to sterilize the scissors, and blood-stopping powder in case there is excessive bleeding. Usually the comb is dubbed to about ¼, inch and the wattles and earlobes are cut close to the head. The dubbed area will heal within 2 to 3 weeks. *Backyard Poultry Magazine* has a detailed article about dubbing on their website: www.backyardpoultrymag.com.

My suggestion is that if you live in a severe winter zone, choose a winter hardy breed that has pea combs and short wattles. See Chapter 7 on Chicken Quest for more information.

Frostbite prevention for toes begins with the roosts. When a bird roosts at night, it hunkers down and covers its feet with belly feathers. If a roost is too narrow, the bird's toes are exposed below the feathers, like having your feet stick out underneath a blanket that is too short. Roosts must be wide enough so toes don't hang down. Usually 2x4 lumber with the 2" side up is sufficient.

Build hen housing to be draft-free and keep it that way. Hens can handle the cold but wind is hard on them, especially at night. Keep the hens dry; wet hens will be cold hens. Wet feathers can't maintain body heat very well, as they lose some of their insulation value. The fine feathers of some breeds, like Silkies, make them especially vulnerable to getting wet and chilled.

Frostbite Treatment. There are several approaches to treating frostbite, depending on the severity. One approach is to dub the affected area. It takes less time for a chicken to heal from dubbing than for dead, frostbitten tissue to fall off. Swollen wattles and ear lobes can be lanced to let the fluid drain, and then removed.

You can also treat your chicken's frostbite the same way human frostbite is treated. Move the affected hen to a warm location. Soak her in warm (not hot) water and warm her comb by contact with the skin of warm person. Continue until she regains sensation, but remember that the first sensation is painful as the nerves thaw. Never rub, slap, or shake the stricken region, as ice crystals in the frostbitten skin will damage surrounding tissue even more. Keep the frostbitten areas from refreezing, which worsens the damage.

Administering Medicine

Giving medicine to a chicken can be challenging, whether it is an eyewash, drench (oral dose), or an injection (shot). In order to effectively treat a hen you need to secure her so she can't flap about and, especially, bob her head around when giving oral medicine or eye treatments. Hens have remarkably flexible necks and you don't want to try to treat a moving target.

The Towel Wrap to Secure a Bird

Keeping a bird quiet and still while providing treatment is critical. The towel wrap technique works very well to calm and secure a chicken for

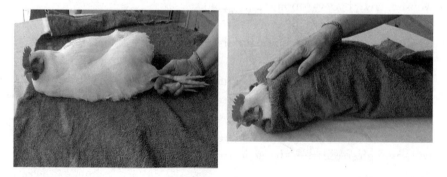

The Towel Wrap

Wrapping a chicken in a towel has a calming affect and keeps it from thrashing around while you administer medical treatment. Don't wrap the towel so tightly she can't breathe.

medical treatment. Lay the hen on her side on top of a towel. Gently pull her feet behind her; this will calm her, and she won't be able to use her legs to struggle. Wrap the towel around her securely but not tightly – like a burrito – so that her head and/or feet are showing, depending which end you need to treat.

Oral Dose (Drench)

Drenching is a good way to administer colloidal silver, probiotics, oral re-hydration, nutritional supplements, or antibiotics.

Use a needless syringe to give oral medications. For a dose of antibiotics, use a 1 cc to 3 cc syringe. To give oral re-hydration, a larger (5 or 10 cc) syringe will do. Pharmacies will usually give away 5 cc syringes used to dispense children's doses of oral medications.

With your patient secured in a towel wrap, place her on her side with her head resting on your knee or a flat surface. When the hen's head is secure, give the oral medicine. Gently hold the top of her beak – being careful not to cover her nostril – and carefully pry her mouth open. Place the tip of the drench syringe at the side of her mouth.

Put only a few drops of the liquid in and release her head so she can swallow. Repeat until the desired dosage is complete.

Giving a Chicken an Injection

Injections are used to administer antibiotics and vaccines. It's best to use a sterilized needle. If you use the same needle to inject more than one bird, dip the needle in alcohol, or pass it through a flame. If you are giving a live-virus vaccine, don't use alcohol to sterilize the needle.

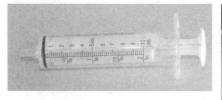

Giving a Chicken Oral Medicine

To give oral medication or rehydration to a chicken use a dosing syringe (used for kids medicines) like the one shown above. These are available at pharmacies, often free.

Syringes can be purchased for animal use at farm supply stores.

Use a syringe big enough to give the entire dose and the smallest needle possible. To treat a single hen, sometimes a 1 cc insulin syringe (the type used for insulin shots) will be large enough.

Subcutaneous Injections

SC or SQ injections are easy and safer to administer than injections in a muscle (intramuscular or IM). The dose will last up to 2 days, but it takes longer to get into the bloodstream and circulate through the body than an oral dose does.

Give a subcutaneous injection where the skin is loose, such as at the breast or nape of the neck. To give the injection, pinch a bit of skin to raise it up and away from the body. Insert the needle at an angle so it slides just under the skin. Be careful not to prick yourself, which can happen if the hen struggles at the sudden pain from the injection.

Intramuscular Injections (IM)

IM injections are given in the meaty portion of the breast or thigh. IM injections are more difficult because as the needle goes into the muscle it can pierce a nerve or bone, or inject the medicine (and air bubbles) directly into the bloodstream, which can kill a bird.

Insert the needle into the muscle at a right angle. Before injecting the drug, pull the plunger back slightly. If blood appears, then you have the tip of the needle in a blood vessel. Don't give the injection there; move to a different site. IM injections distribute medicine throughout the body quickly (in about an hour) and last about 8 hours.

Infectious Diseases in Poultry

Just like people, chickens can get sick from bacterial and viral infections. The condition of their immune system determines how ill they get; usually, well-nourished, free-range birds with access to sunlight are resistant to infectious diseases.

Giving a Chicken an Injection

Injections for chickens are usually given just under the skin (subcutaneous or SQ) on the breast or thigh.

Avian Influenza (AI)

AI is the chicken version of the common cold. It is caused by airborne viruses and causes flu-like symptoms, similar to those that people experience. There are many strains of AI virus and they mutate all the time. One strain is called the "fowl plague" and has a high mortality rate. Other forms of AI are less virulent and result in few deaths.

Low-Mortality Avian Influenza (AI)

A chicken with the low-mortality strain of AI might have a slight swelling of the head and neck, a stuffed-up nose with nasal discharge, and will sneeze. She will have her head down, eyes half-closed, and her feathers will droop in the "I feel like crap" position. If she could talk, she would probably say that she has a headache and aches all over.

There isn't any treatment except supportive care: keep the bird warm, give her plenty of fluids, and maybe add an oral re-hydration solution and vitamins to her water. There is a product (similar to Vicks VapoRub) called VetRx that might help relieve symptoms. Antibiotics are not indicated, as influenza is a viral disease, and antibiotics treat bacterial infections. Generally, an otherwise healthy bird will get well within one or two weeks.

High-Mortality Avian Influenza (AI)

Symptoms of the high-mortality AI strain include diarrhea, rasping breath (rales), excessive fluid leaking from eyes and nostrils, decreased appetite, swelling of the head and face (including cyanosis in the wattles) and combs turning purple. Mortality can be up to 70%. An accurate AI diagnosis involves sending a live or dead bird to a laboratory, where the virus is isolated from tissue samples.

AI in Humans

There has been a lot of press about the possible crossover of AI from birds to humans that could result in a pandemic. But the numbers of actual mortalities don't support the threat. Even in Asia, where millions of birds were infected and slaughtered, only a few people died from infection, and those that did lived in very close proximity to their birds. This is what the Centers for Disease Control had to say about avian flu:

> *Usually, "avian influenza virus" refers to influenza A viruses, found chiefly in birds, but infections with these viruses can occur in humans. The risk from avian influenza is generally low to most people, because the viruses do not usually infect humans. However, confirmed cases of human infection from several subtypes of avian influenza infection have been reported since 1997.*

> *Most cases of avian influenza infection in humans have resulted from contact with infected poultry (e.g., domesticated chickens, ducks, and turkeys) or surfaces contaminated with secretion/excretions from infected birds. The spread of avian influenza viruses from one ill person to another has been reported very rarely, and has been limited, inefficient, and unsustained.*

THE CENTERS FOR DISEASE CONTROL (CDC)
WWW.CDC.GOV/FLU/AVIAN/GEN-INFO/FACTS.HTM

Many public heath officials and organizations – including the agricultural group GRAIN – maintain that small flocks of chickens are not a public health threat; in fact, they are part of the solution to managing the potential AI problem.

Chickens raised in commercial farming environments are more likely to contract any disease, including AI, because their immune systems are compromised from routine exposure to antibiotics and stressful, overcrowded environments.

If micro-flocks are properly maintained and protected from contact with wild birds, they tend to be resistant to infectious diseases, including Avian Flu.

Cocciciosis

Coccidiosis is a serious infectious disease. Luckily, once identified, it is easy to treat. Coccidiosis is an intestinal disease caused by microscopic parasitic protozoa organisms called coccidia. There are at least 8 different types of poultry coccidiosis and they are universally present in poultry operations. Coccidia are found in the intestines under normal circumstances and are relatively harmless. But if a chicken's immune system is compromised, it becomes more vulnerable to the parasites, and an infection can cause extensive damage to the intestinal mucosa.

A bird infected with Coccidiosis will stand around, looking lethargic, hunched over with ruffled, dull feathers puffed out. The bird will feel light when you pick it up, will not be eating or drinking regularly, will have anemic-looking combs, and will get progressively weaker.

To diagnose Caecal coccidiosis, gently squeeze the abdomen, and blood or bloodstained feces will probably be expelled. Blood in the feces of any animal is a cause for concern. Keep in mind that if your chickens have been eating beets, their feces will have a bloody look caused by the betacyanin compound in beets that can also turn people's urine and feces reddish.

To prevent Coccidiosis, keep water and bedding clean. The disease occurs only after the ingestion of large numbers of oocysts (parasite eggs), which live in dirty water and bedding. Oocysts can be killed by freezing or applying high heat.

Drugs for treating coccidiosis include: amprolium (Corid), chlortetracycline, oxytetracycline, sulfadimethoxine (Di-Methox/Albon), and toltrazuril (Baycox). See the Poultry's Pharmacy for description of these products.

Roup

Sometimes roup is a vitamin A nutritional deficiency and can be classified as "nutritional roup". Symptoms in nutritional roup include drowsiness, poor growth, sparse feathering, nasal and ocular discharges especially during weeks 1 to 7 of brooding.

The treatment for nutritional roup is vitamin A in the drinking water. Prevention includes supplementation with vitamin A, antioxidants and feed formulation with quality base materials.

Roup is also the term used for an infectious poultry disease characterized by inflammation of, and mucous discharge from the mouth and eyes. The eyelids can be inflamed and adhered shut. The comb and wattles are usually pale. Differentiate roup from infectious coryza, chronic fowl cholera, infectious sinusitis, and other poultry diseases.

Treatment: colloidal silver eye wash plus 3 to 5 cc oral dose twice a day until clear.

Colibacillosis (Colisepticemia, Escherichia coli infection)

Colibacillosis is commonly known as an E. coli infection. E. coli bacteria exist naturally in almost all animal species. An E. coli infection can usually be traced to environmental stress or earlier infectious condition.

Systemic infection (septicemia) occurs when large quantities of pathogenic E coli enter the bloodstream from the lungs or intestines (via leaky gut syndrome). The bacterial infection spreads to other organs and membrane surfaces, causing widespread inflammation and even death.

Preventive measures include keeping water and bedding clean, environmental stress low, and nutrition quality high.

Because of the widespread use of antibiotics in animal feed, most infectious E. coli bacteria are resistant to tetracyclines, streptomycin, and the sulfa drugs. This is also true for E. coli in humans. Keep an infected hen warm and give her colloidal silver to make her more comfortable; if she is dehydrated, give an oral re-hydration drench and electrolytes in the drinking water. How to make colloidal silver is discussed in the Poultry's Pharmacy chapter

Fowl Cholera

This is a contagious disease caused by bacteria and spread by feed, water, and ground contamination. The main symptom is profuse greeninsh-yellow to white diarrhea. To prevent Cholera, keep bedding and water clean, rotate chicken runs, and don't overcrowd your birds. Cholera can be treated with sulfadimethoxine.

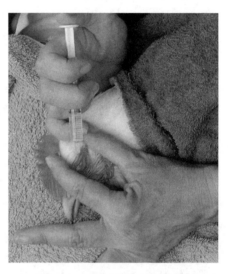

Eye Problems

Chickens can have eye problems: infections, cornea scratches (caused by pecking or fights), and foreign objects (dirt, bedding, etc) lodged in the eye are the most common afflictions. Eye problems can also be caused by the common cold (avian flu) or by a buildup of ammonia fumes from dirty bedding.

Giving a Chicken an Eye Flush

Secure the bird in a towel wrap so you can keep the head steady. Have a pre-filled oral syringe ready. Very gently, open the eye and squirt the wash so that the eyeball is covered. Don't let the syringe tip touch the eye.

Ammonia fumes cause lung problems as well as eye problems, so if you ever smell ammonia, it's past time to clean the coop and create some compost.

Just about any eye problem will benefit from an eyewash. With larger, stronger birds this can be a two-person job, one to hold the bird and the other to administer the wash. But one person can do it alone if they secure the chicken in a towel. We use the same treatment for all eye problems:

1. Sit down with the bird wrapped in the Towel Wrap method. Position her so her feet are behind her and she's lying on her side with the affected eye upwards.

2. Get the eyelid open and remove the deposit if it is crusted over. Sometimes a hen with an eye problem will wake up with her eyelid sealed shut from dried drainage material. There might also be a white, waxy deposit that can be removed, but removing the deposit doesn't solve the underlying infection. Help the hen open her eye by gently pulling on the upper and lower edges of the eyelids to get them apart. The lids and eye are extremely fragile, so be extra-super-gentle. If the lids don't separate easily, get a warm, wet towel and put it over the eye to soften the crusted fluid. It could take several attempts to open a badly crusted eye.

2. Rinse the eye with colloidal silver or normal saline eyewash to rinse the eyeball and give her some symptomatic relief. Use one hand to gently press and hold her head on your leg. Carefully open the eye with the fingers of your other hand. Squirt the wash in the eye using a pre-filled 1 cc or 5 cc syringe (without the needle). Don't put too much pressure on the syringe plunger; a too-powerful squirt can harm the eye and is rude to the chicken. If there isn't a foreign object lodged in the eye, just put a few drops in to get the eyeball wet. Let her blink and regain her dignity before repeating the procedure, if necessary.

3. Follow up with an antibiotic eye ointment if there is an infection, applying a thin ribbon to the eyeball.

4. Keep an eye on the eye and continue treatment for at least 3 days, or until the problem disappears.

Sometimes, a single eye wash will correct the trouble. The worse thing you can do is ignore an eye problem. Chickens can lose their eyesight from minor eye problems that aren't given simple care and attention.

Worms (internal infestations)

Pale or shriveled combs, loose stools, or a drop in egg production might indicate a worm problem. It is good to rotate wormers to avoid resistance. Don't confuse internal worms (bad) with earthworms (good). You want to have earthworms in your garden and yard soil. There are a few facts about parasitic worms that can help you control them.

Parasitic worms cannot live in:
- Very dry environments for a long time
- Temperatures below 10 C or above 35 C
- An oxygen-free environment

Worm Eggs are destroyed by:
- Very high temperatures, such as in compost
- Prolonged freezing conditions, as during northern winters
- Direct sunlight
- Long droughts

Worm eggs can survive:
- For many months in litter or soil
- In an intermediate host, such as snails, fleas or mosquitoes
- The application of most disinfectants

Worm Precautions:
- Avoid grazing continually on contaminated yards
- Let land rest at the end of the season
- Keep grass cut short so direct sunlight and heat reach worm eggs
- Keep litter dry and clean. Compost bedding
- De-worm with a worm medicine
- Worm especially when the temperatures are below freezing because the cold will kill the eggs

The easiest way is to treat your entire flock is to put the wormer in their water. Piperazine is a type of wormer that can be added to the water.

Diatomaceous Earth (DE) is a fine powder that consists of tiny phytoplankton skeletons that are microscopic and razor sharp. When parasites come in contact with the tiny particles, their exoskeleton is pierced and the insect loses fluids, dehydrates, and dies. The Chinese have used Diatomaceous Earth to rid animals and fowl of intestinal parasites for more than 4,000 years.

Wounds

If I see blood on a chick or adult hen I treat it as soon as I can get my hands on the bird. Usually wounds from pecking are on the tail or back, especially if the bird is molting. It's best to treat them at night when they are easier to catch.

Ointments like Bag Balm or petroleum jelly work well for scrapes and cuts. You can make a healing comfrey ointment for quick wound healing. How to compound the comfrey ointment is described in the Poultry's Pharmacy chapter.

Ointments stay on wounds better than creams. Add a drop of an essential oil to the ointment – like rosemary or lavender – to divert the other chickens' attention from the wound.

Feather Loss

Feathers are a good indicator of hen health. Healthy hens have smooth feathers with a sheen you can see in the sunlight. Dirty, broken, and soiled feathers might be a sign of nutritional problems, parasite or other infestations, or poor hen management.

Feathers also tell you about a hen's lifestyle. Broken tail feathers might indicate that the nest box is too small, or the roost is too close to a wall. There are two primary reasons hens lose feathers: a natural feather replacement cycle called molting, or the presence of disease, mites, or lice.

Molt or Molting = new dress of feathers

Just like our clothes, feathers become worn, old, and broken. Every year hens grow a new set of feathers. Molting, as alarming as it might look, is a natural and normal process that takes from 6 to 12 weeks.

The first two molting cycles take place as the chick is growing. When a baby chick is about 6 weeks old, its down is replaced with juvenile feathers. This is when that cute little chick goes through the "ugly ducking" phase.

The juvenile feathers are replaced by adult plumage when the chicken is about 5 months old.

Every year thereafter, your hens will molt to replace worn-out feathers. Shortening days trigger molting, so in late summer or early autumn the hens begin sprouting pinfeathers, which blossom into new feathers. A hen doesn't lose all her feathers at once. Nature allows for a succession of feathers to be lost and replaced so the bird is never totally nude. Feathers drop and re-grow in a specific molt sequence:

> Feathers at the head and neck are the first to go.
> Then the feathers on the breast, back and body.
> Wing feathers are usually last.

A chicken looks pitiful and pathetic during the molt. It will appear disheveled and scruffy while molting, as if it fell briefly into the plucker. Egg production decreases and often stops; this is called being "out-of-lay". The hen's body reabsorbs the energy and material normally used for egg production and uses it to make new feathers. Her body weight can decrease. I've seen birds look truly butt-ugly during molts.

I know my hens' personalities well enough to know when they are embarrassed by their molt feather transition. They act depressed, dejected and tend to stay away from the flock. I have no doubt the chicken ego and sense of self is affected as they are stripped of feathers. If I were running around with pinfeathers and pink skin showing I would be embarrassed and depressed as well. Of course, their entire biochemistry is shifting, which might add to the personality and PMS-like mood change.

It would be interesting to study how the pecking order is affected by molting. Do feathers represent a kind of armor to hens? Do the more aggressive hens – like Sampson with a haircut – lose power as their feathers drop away, leaving them to face the world wearing only pink skin as outerwear?

Mites Causing Feather Loss

Mites are very small bugs – less than 1/25th of an inch in diameter. Some can be seen only with the aid of a microscope. They get nourishment from blood, tissue, and feathers. Most mites live in nooks and crannies of the hen house or a bird's skin. Some live in the feathers, lungs, liver, or other organs, but these types of mites are relatively rare.

Good hen management requires constant vigilance for mites and swift action if they are found, as a mite infestation can kill chickens. Signs of infestation include feather damage, increased food consumption with decreased egg production, retarded growth in young birds, hens sleeping away from their usual roosts, and anemia.

Red Mites. Red mites are one of the most prevalent and harmful external parasites. They are tiny, blood-sucking critters so small that you will feel them before you see them; they look like red dots from a ballpoint pen (except they will be moving). They often become active at night and will suck the blood of any animal: chicken, human, goat, or cow.

Red mites live and breed in cracks, crevices, and under dry crusts of manure. They especially like wood or cardboard nest boxes and perches. They tend to be dormant during the winter in unheated hen houses and thrive in hot weather and dirty conditions. These dot-sized arthropods are why we prefer metal or plastic nest boxes and plastic nest box liners; mites tend not to breed as well in metal or plastic.

Several steps are needed to get rid of red mites:

1. Disinfect all perches, scrape off any droppings, and coat them with oil.

2. Sweep down the inner walls and remove litter from nests and the floor.

3. Bathe the birds with a flea or tick shampoo, being careful to rinse them completely, and dust them with diatomaceous earth when they are dry.

4. If the red mite infestation is widespread, treat the entire hen house – including the roof and floor – with a good mite-exterminating product. A power spray pump or powder sprayer will help penetrate crevices. Be sure to follow the instructions, as sprays can poison hens if used incorrectly. Mitecides include diatomaceous earth (DE), pyrethroids, organophosphates, carbamates, vegetable and mineral-based products (both liquid sand dusts), and citrus extracts.

5. Fill in obvious cracks in hen house walls to eliminate mite-hiding places.

We had a mite infestation while using plywood nest boxes and had to burn the boxes because there were so many nooks, cracks, and crannies where mites could hide from the pesticide spray. We watched what seemed like millions of mites running out of the boxes as they burned.

To prevent mite infestations, change the bedding often. If you use cardboard or newspaper in the bedding, you can put the soiled bedding over your garden beds, where it will become manure-enriched mulch, eventually decompose, and turn into compost.

Scaly Leg Mites. Scaly leg mites burrow under a chicken's leg scales and the resulting detritus buildup makes the legs look rough and dry. They cause irritation and swelling, and eventually the scales break off. Leg mites can make it painful for a bird to walk and can even be disabling.

Leg mites travel slowly from bird to bird and do not affect humans. We've purchased a few older birds with scaly leg, but it healed quickly with treatment and did not spread to others in the flock.

If more than one bird is affected, it might be wise to treat the whole flock. This is easy if you have only a few chickens, but impractical for larger flocks.

The usual treatment is to coat the legs with an ointment, such as petroleum jelly, to smother the mites. A more aggressive treatment is to make a 5% to 10% sulfur-based ointment; the ointment immobilizes the mites and the sulfur kills them. Sulfur also works as an antiseptic and an antifungal for conditions like ringworm. A 10% solution soothes the skin, but higher concentrations turn into sulfuric acid and burn. Sulfur also kills burrowing mites (scabies) and lice. The formula and directions for making sulfur ointment is in The Poultry's Pharmacy chapter.

Put the ointment on daily or every other day for about 10 days. Repeat weekly if the scaly leg returns, or until the leg scales look normal.

Another treatment for mites, lice, and worms is Ivermetin, a broad-spectrum antiparasitic medication. Give bantams a few drops and larger birds ¼ ml (1.25mg) by mouth, using a eye dropper or oral syringe (without a needle).

Lice. Chickens take dust baths to keep their feathers clean and rid themselves of lice, but new birds joining the flock, wild birds, or rodents can still introduce lice to a micro-flock. Brooding hens are the most susceptible to lice, because they sit for hours on clutches of incubating eggs and can't scratch themselves or get up more than once a day for a dust bath. If you see a bird scratching and pulling out her feathers, check for lice by looking at the fluff feathers around the vent and under the wings. Lice move quickly, so look fast.

Chicken lice are not the same as the human head variety. Sometimes you might find one or two lice crawling on you, but they are harmless. Some types of lice live in the cracks and crevices of the coop and only climb on the birds at night, while others can't survive for more than a few hours off the bird.

If you find lice in your flock, dust all the birds under the wings and around the vents with an insect power or DE. Putting DE where your birds nest and dust bathe helps control parasites. Cedar shavings help keep lice away, especially in the nest of a brooding hen.

Laying Problems

Egg Bound

If your hen is trying hard but still can't produce an egg, she might be "egg bound," or unable to lay because the egg has become stuck in her oviduct. She will make frequent trips to the nest and show signs of distress with her body language. If a hen could cuss, she'd do it while being egg bound. A hen laying an unusually large egg might show similar behavior; hens can lay astonishingly large eggs (10 ounces) without being harmed. The biggest danger of egg binding is that the egg can break in the oviduct and the sharp eggshell fragments can cut the oviduct tissues, causing bleeding, infection, and even death.

Occasionally you'll see an egg with red streaks, caused by strained and torn laying muscles. Sometimes a hen will pass an egg but will be left with a bloody vent or a partial egg hanging from her birdie-butt. This puts her at risk to cannibalism and infection.

Egg binding and egg laying problems have several possible causes:

1. A change in environment and surroundings
2. A change of feed
3. Obesity
4. Not enough exercise

If your hen is egg bound there are a few things you can do to help her. Pick her up, lifting her very gently; if the egg breaks in her oviduct it can be disastrous. If you are up to it, lubricate the vent with some oil or a lubricating jelly to help the egg pass. If you can feel the egg inside the vent, then very gently insert a lubricated finger and manipulate the egg towards the vent and put the hen back in her nesting spot.

If an egg breaks inside your hen, then do your best to remove the fragments and set up a prayer circle.

If the egg never passes, then something more serious might be happening. The "something more serious" might be internal laying, a condition in which a hen is totally unable to lay the egg. The egg might have missed its passage into the infundibulum and wrongly passed into the peritoneal cavity. There it becomes infected, a condition known as egg peritonitis. Little can be done for a bird with this condition, and she will die a painful death unless she is put down. This is a situation for the Hen's Hemlock Society to allow her a dignified, quick transition.

Prolapsed Uterus

Uteruses don't prolapse often, but when they do, it's alarming and life-threatening for the hen. You must act quickly to save her. A portion of the hen's lower oviduct has turned itself inside-out and was pushed through the vent as she tried to lay the egg.

The earliest sign of a prolapsed uterus is when a hen looks like she is having a hard time laying an egg, or is running to the nest box throughout the day. Young hens are especially vulnerable to a prolapsed uterus when they lay their first eggs. After about 7 months of age, the probability of a prolapsed uterus diminishes. A hen that is too fat or one that lays unusually large eggs can also be at risk for a prolapsed uterus.

If you discover a prolapsed uterus, immediately remove the ailing hen from the flock. The blood and fleshy protrusion from her rear will attract the pecking beaks of other hens, who will peck the vent tirelessly and pull out the sick hen's entire oviduct and intestines. A painful death is guaranteed unless the wounded hen is isolated.

Treatment of Prolapsed Uterus. Be very gentle while handling a prolapsed hen. There is probably an egg in the oviduct. Start by making sure the oviduct is clean. This is two-person job; one should hold the hen while the other plays doctor. Only use warm water and clean cloths. Once clean, apply an antiseptic cream.

The next step is to remove the egg trapped inside the oviduct. Lubricate your fingers with petroleum jelly and gently probe to find the vent opening. Don't apply any squeezing or pressure, lest you break the eggshell somewhere inside.

The oviduct will probably be stretched over the egg, but you can slide it back with gentle manipulation and ample lubricant. Don't be alarmed if there is some bleeding. In most cases the egg is visible, making it easy to manipulate out. If the egg breaks, gently remove all the shell pieces – every piece. Like broken glass, shell fragments cause internal serrations and bleeding that can lead to infection.

When the oviduct is clean and the egg removed, guide the oviduct back through the vent. A hemorrhoid ointment (such as Preparation H) will give the hen some relief.

Over the next few days, chances are high that the oviduct will protrude again. This means you'll have to clean it and put it back in place. Use a hemorrhoid cream each time because it helps shrink and tighten the oviduct so it will eventually stay put, and the hen can lead a normal life.

Keep the hen isolated and quiet until the oviduct remains in place for at least a week. If you put her back with the flock they will peck at the vent or popped-out oviduct, causing more damage. Feed a low protein diet to keep egg production low.

If this treatment grosses you out, a veterinarian can replace the uterus and will probably put a stitch in the vent to keep it from popping out.

I've only had two cases of prolapsed uterus and both hens recovered to live productive lives.

Foot and Leg Problems

A chicken's feet are critical for its health. Their feet and toe nails affect the bird's ability to navigate and forage, scratch for food, and get up and down to roosts. There are three common conditions that are easily cured and prevented. These are scaly leg, long toe nails and toe balls.

Scaly Leg

Occasionally, we've obtained birds that had scaly leg, usually

Scaly Leg and Long Toe Nails

This chicken's feet and toe nails need attention. The scaly leg can cripple a bird and is easily cured. Long toe nails cause the toes to turn painfully sideways and impede the chicken's ability to scratch and find food. Use nail clippers to correct this problem.

orphans that we've been given. Scaly leg is not a cause for alarm and is not contagious to humans. It is contagious to other birds by direct contact.

Scaly leg is caused by a burrowing parasitic mite, *Knemidocoptes mutans*. The mite lives under the scales in the bird's legs, causing the scales to raise with a white crusty appearance. Scaly leg is irritating and can result in lameness.

The mite spends its entire life cycle on the birds and is usually spread by direct contact.

Scaly leg can be cured by applying a thick coat of an oily substance, such as petroleum jelly, vegetable oil, or a commercial chest rub. The oils gum up the mites and suffocate them.

A 5% sulfur ointment is very effective against leg mites. How to make a sulfur ointment is described in Chapter 16, The Poultry's Pharmacy. Eventually, with treatment, the raised leg scales will drop off and be replaced by new scales and the entire leg will regain a healthy look.

It's not necessary, but if you want to remove the loose crusty scales, just soak the afflicted bird's legs in soapy water for a few minutes. Scrub the encrusted areas gently with an old tooth brush. The scales that are ready to come off will detach. Probably, not all the scales will be ready to release. Don't pull the scales off that are still firmly attached as this can cause the leg to bleed. After the leg soak, apply a thick layer of the sulfur ointment or petroleum jelly.

Completely curing the legs of the mites will take multiple treatments that last over several weeks. You'll want to be on the lookout for new, developing cases of scaly leg on other birds legs and treat accordingly.

Toe Balls on a Young Chick

Poultry Pedicures

Chicken toe nails that have grown too long will cause the toes to be crooked. Worse yet, long toe nails

Toe balls form when mud and crud collect on the end of toes, making it hard for the affected bird to forage and walk. If not treated, toe balls can get so big they cause the toe to break.

impair a chicken's ability to scratch and forage, resulting in malnutrition and nutritional deficiencies.

It is impractical for commercially kept chickens to have toe trims, so this is rarely mentioned in books on commercial agriculture.

It's easy to trim the toe nails with clippers. Clip hen's toe nails as you would trim your own, being careful not to trim too much and cause bleeding. You can see where the blood vessels (and nerves) start and the toe nail ends on the feet of some chicken breeds. Put on your pro-health checklist to check periodically for long nails and keep clippers handy.

If you notice that a rooster is leaving marks or wounds on a hen's back, a quick clipping and filing his toe nails can help solve the problem.

Toe Balls

Toe balls are formed by the mud, crud, and crap that get on a hen's feet. The buildup makes foraging and sitting on perches difficult. Toe balls can get so big and heavy they cause lameness and toes to break.

One of my feather-footed ladies developed toe balls. It was rainy and she got mud caked on her feet and leg feathers, which hardened into near concrete; she not only waddled when she walked; she limped.

The best way to avoid toe balls is to change bedding often and keep the birds out of their excrement and off mud as much as possible. The poop acts like glue and attaches mud to their toes. As they scratch, new layers get stuck on the toe ball, which eventually encases the nail.

The cure for toe balls is a foot soak. Warm water works best. Sometimes it might take several soaks to soften the ball enough so you can gently pry or cut it away from the toes.

Long Leg and Foot Feathers

Leg feathers can be beautiful. However, they can get so long that it's hard for a bird to walk. One foot steps on the feathers of the foot behind, like a person stepping on their shoestrings and tripping. Long foot feathers can also contribute to dirty eggs, making cleaning the eggs more of a chore.

In feathered-footed birds, (ones that won't be going to shows), you might want to trim the leg and foot feathers so they can't collect mud or interfere with the chickens' mobility. I was given a Faverolle, whose career as a show bird was over. We called her Phyllis Diller, as she had a wacky-looking feather-do from head to toe.

Phyllis was culled from the flock because they didn't want to keep her over the winter, and her usefulness as a breeder and show-girl was declining. By joining my flock, she could free range, but she had a terrible time walking and getting around. This meant that she couldn't get away from the bullying birds in her new flock. She'd hunch, waddle, stumble, and hunch again as the blows from the other chickens rained down on her. It was a brutal initiation.

Trimming Phyllis's leg and toe feathers gave her a new gait and a new life. Right after the feather trim, she would pick her feet up very high, like a standardbred horse with weights just taken off its front feet. In a few days her coordination improved, she gained confidence, and she even learned to climb up to roost gracefully. Trimming her long leg feathers was like removing shackles. With crew-cut legs she gaind a new image. She might eventually want a Mohawk and red feather dye as the punk rock biddy.

Crop and Gizzard Problems

The crop and gizzard are essential parts of the poultry digestive system. The crop is a food collection sack where food is softened before it passes to the gizzard and the rest of the digestive system. Chickens don't have teeth; instead, they have a gizzard, which is a muscular pouch that contains grit to grind food before it gets to the stomach.

There are three somewhat common digestive problems: sour crop, pendulous crop, and impacted crop (or gizzard).

Sour Crop

Sour crop is also called poultry thrush or moniliasis. It happens when the contents of the crop have begun to ferment and produce a sour or rancid smell. A bacteria called Candida Albican is the culprit. To diagnose, get close (nose to beak) to your hen and gently squeeze her crop. If she has sour crop, you will smell a rancid gas. Sour crop can be a symptom of – or precursor to – the more serious conditions of pendulous crop and crop impaction. Birds with sour crop are depressed and have little appetite. Eventually they can become emaciated and die.

To prevent sour crop, keep the waterer and feeder clean. Don't feed more mash than your chickens can eat in one day so it won't sour. Don't feed rotten foods.

If caught early, sour crop can be treated with some success. The contents of the crop may become semi-liquid and will drain out if the bird is held upside down. Hold her upside down, with one hand gripping both legs, and gently massage her crop with your other hand to drain out the sour contents. Give the hen a break and repeat until the crop is empty. A hen can't vomit, so this is as close to throwing up as she can get.

Once the crop is emptied, isolate the bird and medicate her water with an antifungal agent, such as mycostatin or nystatin. If these drugs are not available, try aloe vera juice and/or tea tree oil. Both can be helpful given as a drench or diluted in water.

Pendulous Crop

Pendulous crop is a condition in which the crop becomes enlarged and hangs down noticeably in front of the bird. It is caused by excessive consumption of water or food. I once saw an entire flock of breeder broiler chickens with pendulous crops. They were on a limited ration so they wouldn't gain too much weight to make breeding difficult. At feeding time, they were so hungry they would gorge themselves, eating as much as possible. The result was low-hanging crops that waddled when they walked.

Stretched or weakened muscles can also cause pendulous crops. Whatever the cause, the crop is unlikely to return to normal size and can't completely empty its contents into the stomach, so the hen will have chronic sour crop. If a hen's crop becomes oversized and pendulous, consider culling her from the flock, as there is no cure for this condition.

To prevent pendulous crops, always make sure your birds have ample clean water and food available. If they are abnormally hungry, feed a little bit at a time until they are sated and don't try to gulp their food.

Crop Impaction

Don't confuse crop impaction with sour crop. In both cases the crop will be distended, but an impacted crop feels solid, not squishy like a sour crop. The distention is caused by a blockage at the exit of the crop that causes feed to press against the crop wall. Chickens who eat long, partially dried grass, feathers, bedding, or any other hard-to-digest matter are vulnerable to impacted crops. If the impaction is severe, the hen will die of starvation.

To prevent crop impaction, don't give birds access to long, coarse grass when you first let them on pasture.

To treat it, gently massage the crop to help dislodge and move the impacted material through the digestive system, or manipulate it up and out through the mouth. A teaspoon of olive or mineral oil can help move the dislodged material to the lower esophagus and on to the gizzard.

Give an oral dose of mineral oil to help move the binding material. Surgery is an expensive option, but hens have recovered from the operation.

Gizzard Impaction

Although similar to crop impaction, gizzard impaction is caused by inadequate access to grit, sand, or crushed granite. Birds need about 1 ounce of grit per month. If birds have ample grit, impacted gizzards are rarely a problem.

Impacted gizzards can also be caused by litter, grass, string, nails and even plastic bits wadded up in the gizzard.

To prevent gizzard impaction, give chicks access to fine sand after they have their first fill of starter feed. This will help develop a strong gizzard. As the birds grow, make increasingly coarser (larger) grit available.

There is no effective treatment for an impacted gizzard.

Disappearing Chickens & Eggs

A missing chicken is probably a dead chicken; the ultimate health problem. Rats, minks, snakes, and feral ferrets can squeeze through unbelievably small holes to steal eggs and kill chickens. We had this problem for a while and solved it by borrowing a live wire trap to catch the criminal; ours turned out to be a feral ferret. You can ask your local vet, Humane Society, or wildlife trust if they know of anyone who can lend you a trap. Bait it with some cat food in gravy; the smell is irresistible to murderers and burglars of every species.

Snakes tend to be active during the evening and lay low during the day, so check under the coop around noon if you suspect you have a reptilian thief afoot. A snake wouldn't eat adult chickens unless you live in a place that has pythons. Even then, snakes usually fall asleep near where they last ate a large meal, so you'd find the snake in the chicken house if they did snag a whole chicken.

Does your chicken pen have a roof on it? Owls are very adept at carrying off chickens, although they won't steal eggs. Weasels can also get into hen houses to eat their fill of eggs, but it's doubtful that one could remove an adult chicken, since it would have to fit the chicken through the small spaces in the wire. Another possibility is that your chicken/egg thief is human; a homeless person could be taking them to eat, or kids might be stealing them for kicks.

Rodents as a Health Problem

Rodents are one of the top hen health problems because of the damage they do to housing, equipment, feed, and the birds themselves. The combination of poultry houses, feed availability, litter, and the hens all attract these furry freeloaders.

In urban areas, neighbors might be quick to blame your hens for attracting rats and mice. You can nicely point out that other neighborhood rodent attractions include warm and soft compost heaps, feed put out for wild birds, and garbage awaiting pickup. Rubbish around or in buildings can also become rodent condos.

Chickens are not necessarily the only attraction for rodents on the block. But if you have poultry, it's only a matter of time before the nude-tailed, large-incisored, furry bundles arrive for dinner in your hen house.

Rodents can ruin sheds, feed bins, and electrical wiring. You will have to fix patches of gnawed wood, deal with the fire danger of chewed electrical wires, and keep your hens safe from them. Rats will chew on hens as they sleep at night. A friend of mine had his bird's legs and toes bitten off by rats. He was mortified and had to totally redesign (and plug holes in) his coop to keep the rats out. Because he was in an inner city area, rats were constant company.

Rodents can get into almost any building or coop. Rats can squeeze through a crack less than an inch wide and mice can get through even smaller crevices. If they can't get through an existing crack, they make their own entryway by digging and chewing.

Feed sacks and plastic storage bins are like dental floss to rodents. They chew up the plastic and spit it out in rodent-sized bits. I once bought a 50 pound bag of scratch from our farmer's co-op and as the salesman lifted the bag into my Prius, we saw a neatly chewed hole in the bag with grain streaming out. We looked back across the parking lot and saw a trail of grain that Hansel and Gretel would have been proud of. A rodent's dream home is dark, warm, and has an easy food supply. That's why poultry houses are such rat magnets; they love the sacks of chicken scratch, feed in feeders, available water, and crevices in walls and roofs that protect them from predators while they rear their young. Below are a few tips, tricks, and traps for rodent control:

1. Raise your coop or hen house off the ground by a foot or more so the space underneath is light and open and cats, dogs, and other rodent

killers can better protect your coop. If your coop is high enough, the underside space can double as shade for the chickens in the summer.

2. Use metal feed bins. Plastic trash cans and paper feed bags will not deter razor-sharp rodent teeth.

3. If you feed your chickens outside, bring the feeders in at night. Leaving them out is like running a rat buffet.

4. Fill in wall cavities in the coop where rodents can hide and nest. If you live in a severe climate and want to insulate your coop, only install insulation around the roost area. Consider making the insulation removable so you can take it out during warmer weather and get rid of open spaces behind the walls.

5. Use ¼ inch hard wire for screens and doors, as rats can get through chicken wire easily.

6. Keep holes and crevices covered or repaired.

7. Block rat footpaths between coop roofs and rafters that offer refuge from predators.

8. Be tidy. Clutter and junk attract rodents. That old rusted barbecue sitting in the corner could be a cozy nesting place for raising a rodent family.

9. Use dogs and cats as rat and mice security. I have two mixed-terrier dogs from the Humane Society that go with me to the coop at night and are very good at this job. One pounce, bite, and shake of their heads will send any rodent to the big hen house in the sky. I noticed the rodents were escaping from the dogs via a hole in the wall, so I plugged it and the hunting success of my pooch poachers soared. My cats patrol the coop at night and often leave tokens of their devotion (in the form of rat corpses and body parts) by the door or in the drive. They have learned not to bring their gifts in the house; I don't appreciate animal parts in the bathtub.

10. A late night surveillance of your coop is a good way to detect rodents. When you walk in they will scatter as if you had caught them drinking moonshine and playing poker.

11. Think like a rat. Stand back and twitch your nose while looking at your operation from a rat's perspective. Where could burrows be created? How accessible is feed? Are there predators around?

12. Black snakes are talented ratters and mousers if you don't mind having them around. They are not poisonous and can be semi-tamed to handle. My uncle had commercial greenhouses and kept two black snakes as part of his pest control. Black snakes are egg eaters, but the price for an egg or two is a cheap price to pay for rat and mice control. They will also eat the faux eggs which will kill them. If you want to have the benefits of black snakes around; don't use wood or ceramic eggs in the nests.

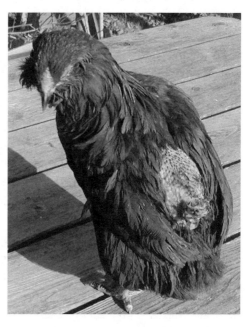

Poultry Vices

Poultry vices are included in Hen Health because most of them can be avoided with proper nutrition, preventative health care, and healthy housing.

The most common vices are pecking and cannibalism, egg eating, and roosting outside of the coop.

Pecking Wound

The back feathers of this unfortunate hen were pecked while molting. Her tail was a bloody stump and the pinfeathers were pecked off. If left untreated, she would have been killed by the other chickens. Her owner was going to put her down. The good news is that we adopted her and, with treatment, she healed and is a happy, healthy, productive hen.

Pecking and Cannibalism

Pecking can lead to cannibalism and causes many time-consuming problems for poultry keepers. Pecking is easier to prevent than control. Pecking and cannibalism can happen in micro-flocks if the birds are bored, stressed, or hungry. By far, the biggest causes of pecking and cannibalism are:

1. Crowding
2. Stress
3. Malnutrition

Often pecking is a first sign of protein deficiency. Adding quality protein, fiber, and dark leafy vegetables will often stop the problem quickly.

Not enough salt in the diet can lead to birds attacking each other, especially in hot weather. Offer the flock a low salt solution to drink. Have two waterers available, one with salt and the other regular water. A salt solution of not more than 1 tablespoonful/gallon is enough.

Trimming the beaks of hens to minimize their ability to peck is standard procedure at commercial farms; the process is called "debeaking". In my opinion, debeaking is not the way to prevent pecking. It is a cover-up of poor animal management, cheap and inadequate protein in the feed, and/or overcrowded conditions. Birds use their beaks the way we use our hands to eat, communicate, defend, clean, and preen themselves. Cutting off the beak is a form of mutilation.

A hen is also susceptible to pecking after she lays eggs, as her pink vent area is exposed; this is why nest boxes ought to be dark and private.

A bird being pecked is a pathetic sight. She will hunch down and give up while suffering the bloodthirsty pecks of the other birds. Chickens can be heart-wrenchingly cruel to each other. Pecked areas encourage more pecking and it's best to remove the bird being pecked.

Blood on a bird – especially on the back or tail – can lead to more pecking. If you spot a pecking victim, remove her from the flock immediately and treat the wounds with a healing balm or anti-pecking ointment. One commercially available salve is called Black Salve; it has a petroleum base and ingredients like methyl salicylate, aloe vera gel (soothing), salicylic acid, thymol, and oil of eucalyptus (chickens don't like the scent). Chapter 17, The Poultry's Pharmacy Chapter has a recipe for a highly effective, comfrey-based healing ointment you can make in your kitchen.

I adopted an Araucana hen that was badly pecked on her back and tail.

She had scabs across her back, and her tail was totally plucked and bleeding; it was raw and ugly. The breeder was going to put her down but I wanted to give her a chance. She cuddled and shivered in my arms as I carried her to the car.

Once home, her new abode the first few days was a dog carrier with a small log perch. She got quality feed, probiotics in her water, colloidal silver as an antibiotic, and lots of tender, loving care. I lavishly spread the comfrey healing balm across her back. With quality food, care, and a healing environment to recover in, she perked up and was able to walk around outside in the yard in 3 days. Within a week she was clucking happily along. That's when I had no doubt she would live.

I loaded comfrey healing balm on her wounds and put her in with the flock. As usual, the hens re-shuffled the pecking order. By the end of the day, she was bleeding again. I had forgotten to put a drop of an essential oil on her feathers.

Essential oils, sprinkled on the feathers of a pecked victim, deter further pecking. I compound a mixture using equal parts of clove, lemon, cinnamon, eucalyptus and rosemary oils. This formulation known as thieves oil. It's called "thieves oil" because 19th century French grave robbers protected themselves from the plague by using these aromatic oils.

The essential oils protected my Araucana hen until one of the larger birds adopted her and protected her from there on. They became inseparable; the Araucana snuggled under the large hen like a baby chick. The larger hen loved her new role as a caregiver and was incredibly solicitous with the smaller hen's needs. The Birdie Best Friends Forever (BBFFs) would roost together side-by-side, on the perch like bookends. Hens have deeper emotions, fiercer loyalty, and more feelings than people realize.

In summary, the best way to control pecking and cannibalism is through management. Providing your flock with an environment and life-style that has variety, space, private nest boxes and complete nutrition will prevent, and deter pecking problems.

Egg Eating

Egg eating is a bad habit that is hard to stop after it begins. It usually starts when an egg cracks in the nest box. A hungry or bored hen will take a bite, develop a taste for raw egg, and spread the habit through the entire flock like wildfire.

Egg eating is another form of pecking with some of the same solutions as listed above. There are ways to prevent chickens from eating eggs and some things to try to stop the eating habit after it's begun, but before it becomes an epidemic.

Ways to Prevent Egg Eating:

1. Remove the eggs as soon as possible after hens have laid them; late morning or early afternoon is the best time to collect eggs.

2. Use soft nesting material so that eggs won't break easily. Thoroughly clean any nest with a broken egg and remove all the eggshell fragments.

3. Have enough nest boxes available so that hens don't have to crowd (stepping on eggs) or lay where eggs are likely to get broken.

4. Don't let your hens get so hungry they go looking for eggs to eat.

5. Make sure your hens have enough calcium, protein, and grit. If their eggshells get thin or soft, add more grit or oyster shell to your chickens' diet.

6. Behaviors like egg eating, feather picking, pecking, and cannibalism all increase when hens don't have enough vitamin D. Add cod liver oil to their feed.

7. Don't feed hens eggshells that have any resemblance to a whole eggshell. Make sure the shell is very finely crumbled or broken into itty-bitty bits.

8. Put something hard in the nest that resembles an egg but won't feel pleasant to a hen if she pecks it. Wooden eggs work best, but I've also heard of folks using golf balls and ceramic eggs.

9. Bored hens are more likely to peck at eggs. Keep them occupied scratching and pecking at food bits instead.

10. Install roll-away nest boxes that roll the eggs to a safe area where they are not available for pecking.

Once egg eating starts, it can spread quickly to all your hens, especially if you haven't been leaving out enough feed for them. Egg eating is serious and entire flocks have been put down because of it. Act fast; time is of the essence.

Ways to Stop Egg Eating Once Started

1. Identify the Guilty Hen. If you find a hen eating an egg, put a leg band on her so you can identify her.

2. Close Down the Nest Box Where the Broken Egg was Found. Take the nest box or area where you found the broken egg out of service. If you use metal nest boxes, remove the bottom from the nest, or put something in the box to block the entryway. If the hen is laying outside of a nest box, make that place inaccessible or unattractive; for example, put a bucket on the nest. This seems to break up the eating pattern. In Hen Neuro-Linguistic Programming (NLP) this would be called "changing her state".

3. Feed Quality Protein and Have Grit Available at All Times. An egg eater might have a nutritional deficiency of protein and/or calcium, or she might just be hungry. Constantly available food can deter hungry beaks from seeking snacks. If grit isn't available, eggshell is the next best thing. Enough high quality protein and grit access can deter chickens from craving eggs and eggshells.

4. Decoy and Fake Eggs. Put decoy eggs in the offender's nest. These decoy eggs can be hard boiled or filled with something that will disgust the hens, like mustard.

5. Darken the Nests. Covering the nest box openings with a cloth makes the nesting area darker so it is harder to see the eggs. Chickens can't see in the dark and what they can't see, they won't peck.

6. Bored Chickens. Chickens confined in small areas will sometimes peck at eggs because there isn't anything else to do. Letting them out to free range and have access to scratch can deter egg eating by keeping them busy and full.

7. *Isolate the Egg Eater.* Remove the guilty hen(s) from the flock for a few days, then reintroduce her. This changes her patterns and possibly the pecking order – including the order NOT to eat eggs (excuse the pun).

8. *Trim the Upper Beak.* Use a finger nail clipper to cut the top beak back so it's blunt. This limits the chicken's ability to crack an egg. Don't cut the beak back too much or you will hit nerves and cause permanent damage and disfigurement. Blunting the tip of the beak is different from debeaking done in commercial poultry houses. The beak is shaped, not cut off.

9. *Find the Egg Eaters a New Home.* Sometimes relocating egg eaters stops the habit.

10. *Capital Punishment.* If none of the above stops the egg eating, then putting the offender down might be your last resort.

Sometimes young pullets will lay soft-shelled eggs or even eggs without shells at all (nude eggs); these are very fragile and tempting to eat. If the young gal gets enough pellet and grit, the problem will probably disappear.

Roosting Outside

Occasionally, birds prefer to roost in trees or on garage tops and can be difficult to get down, leaving them vulnerable to owls and hawks. If your birds are not willing to go in at night, there might be a problem with their roost or roost area.

Some possible problems with roost areas that cause chickens to nest outside are:
- Irritating red mites breeding in wooden areas
- Rats causing disruption
- Dirty quarters, especially with ammonia odors

If the roost area seems fine, you might have to retrain your birds to roost. Do this by locking them in for a few days and then letting them out for about 30 minutes before dark – just long enough so they can get outside and range a bit, but not so long that they can go far from the hen house. Progressively let them out earlier in the afternoon until they resume their normal schedule.

How to End it All (Euthanasia)

Probably the easiest, most humane, and least gruesome is the neck dislocation method. Hold the bird's feet in one hand so it is hanging with the breast towards you. Hold its head in the palm of your dominant hand. Firmly jerk the head downwards and simultaneously rotate your wrist so the neck stretches tight and the head pulls back to break the neck. All this happens very fast and, after you make the motion, there's no going back. You will feel the neck dislocate and might even hear a crack as the vertebrae separate and break. The body will go into violent spasms for a short time, less then a minute if you did the deed correctly. The body and legs twitch until the bird is dead.

A variation of this method is to step on the head to hold it on the ground while jerking up on the legs to disconnect the neck.

There are other ways to euthanize a chicken but all of them involve more trauma and blood. Emotionally, it is very hard to put an animal down, but it is harder to see one suffer.

17 The Poultry's Pharmacy

The life style of chickens in small flocks helps to avoid many of the infections and behavior problems that are so feared in factory farms. Micro-flocks don't have the same stresses and pressure for high production as do commercial flocks.

But as good a life as your hens have, occasionally, one or all of your birds will need health care. The Poultry's Pharmacy describes treatments for most common poultry primary health care that you can do at home inexpensively. I've heard lots of stories about how a $20 hen had a $300 veterinarian bill. The economics of professional health care for chickens is a consideration, especially since hens have a relatively short life-span. At 4 years old, a hen is considered past her prime. An 8 year old chicken is a senior citizen and at 12 years or more the bird is downright ancient and deserves an award for longevity.

With a sick or injured hen the question will always come up: is it worth the money to take this feathered friend to a veterinarian? It's a tough question and an emotional one. I've pre-answered that question for myself. My answer is "no". I've yet to take a chicken to a veterinarian primarily due to the cost. This is a double standard in my household, because if a dog or cat needs veterinary care, I take them in and pay the bill. I'm grateful to have them with me a little longer. The "go" or "no-go" to the vet question is one you'll have to answer for yourself.

The Poultry's Pharmacy describes over-the-counter medical treatments for the common diseases that might affect your hens. It also includes alternative treatments and home remedies you can make in your compounding kitchen to effectively and safely treat your backyard flock. If you have had first aid training, much of it can apply to a chicken. Below are suggested items for a poultry first aid kit.

Poultry First Aid Kit Check List

[ ] Q-tips, cotton balls, cotton pads
[ ] Paper tape (to keep gauze in place for wounds)
[ ] Disposable gloves
[ ] Measuring spoons
[ ] Syringes with needles, 5 cc or larger syringe
[ ] Saline or colloidal silver solution for eye infections and wound wash
[ ] Scissors, preferably blunt
[ ] Flashlight
[ ] Old clean towel to wrap bird so it is quiet while you apply treatment
[ ] Paper towels as there's often a mess of some sort
[ ] Nail clippers for trimming toe nails
[ ] Healing salve containing comfrey (that you can make yourself)
[ ] Diatomaceous Earth (DE)

The Poultry's Pharmacy is categorized by treatment class. For specific health problem identification see the previous chapter on Hen Health. At the end of the chapter are common systems of measuring and conversion for calculating dosages.

Consult your veterinarian or the USDA extension for assistance with the diagnosis, treatment, and control of infectious diseases for parasitism. Label any products you make "for animal use only".

Antibiotics

Antibiotics can be life saving if the right one is administered in the right dose, at the right time. If one or more of your chicks or hens has an infection, and an antibiotic is necessary to save its life, then use your good judgment. Very rarely have I used oral antibiotics to treat members of my flock.

Antibiotics are available without prescriptions (over-the-counter) for use in farm animals. These antibiotics are NOT approved by FDA for use with chickens. The labeling reads: "For animal use only...do not administer to chickens or turkeys producing eggs for human consumption".

There was not information about the withdraw time for eating the eggs of a hen that had been treated.

The USDA Food Animal Residue Avoidance Data Bank (US FARAD) position is never to consume the eggs from laying hens that have ever been dosed with antibiotics. US FARAD would not even comment on the use of any of the antibiotics. This is what they replied to my simple dosing inquiry:

> "According to the Animal Medicinal Drug Use Clarifica-
> tion Act (AMDUCA) extra label drug use is only permit-
> ted under the supervision of a veterinarian when there is
> a valid patient-client-veterinarian relationship. Further-
> more, FARAD provides withdrawal intervals on a case by
> case basis and therefore, we cannot make blanket recom-
> mendations. Dose, age and disease status as well as other
> variables can affect our evaluation and the data that we
> use to make recommendations are continuously updated
> which may affect future recommendations. I am sorry
> that we cannot be of more assistance".
>
> — US FARAD

There you have it, no comment on antibiotic dosing schedules for chickens from any of the experts. It is just as well you don't use antibiotics and give alternative treatments instead. In my opinion, antibiotics are too often used as a shotgun, practically placebo approach in both animals and people. The undesirable consequences these drugs can have on the gastrointestinal (GI) track ought to be taken into consideration before using a shoot-from-the hip treatment. We have never used antibiotics in our home flocks and here's why.

About 70 percent of the immune system functionality starts in the gut. That's where the T-cells generate. Every being's GI tract is loaded with good bacteria that is necessary for digestion and energy production. The word "antibiotic" means against life (anti = against and biotic = life). These anti-infectives work by killing the "bad" bacteria. Unfortunately, they also kill the good bacteria and often cause an unbalance in the GI tract.

Improper use of antibiotics can pave the way for even more bad bacteria and fungus to repopulate the gut. It also breeds antibiotic-resistant bacteria. There is so much inappropriate antibiotic use that trans-species (factory farm animal to human) infections are growing more common.

If you decide your poultry patient requires an antibiotic to survive, the general rule of thumb is that if the bird doesn't show improvement after three days, the antibiotic is probably not effective against the invasive agent and you might want to try a different treatment.

Neomycin

Broad spectrum antibiotic commonly used in the treatment of intestinal disease referred to as "greens", "enteritis", and "mud fever". Neomycin is also used to control colibacillosis (bacterial enteritis) caused by Escherichia coli (E. coli).

Neomycin is also used topically for staphylococcal skin infections such as bumble foot and infected breast blisters. For topical use, neomycin comes in powder form or as an ointment (Neosporin®).

Neomycin Dosage Calculations

Comes as a packet with powder and as an oral solution.
Dosage = 15mg/kg given orally, twice a day for 5 days.
A dose for an 8 pound bird would be 55 mg of neomycin.

Tetracyclines

Both Chlortetracycline (Aureomycin® & Chlortet-Soluble-O®) and Oxy-tetracycline (Terramycin®) are tetracycline drugs used in poultry. Both these forms of tetracycline are broad-spectrum and have similar uses and dosing. Indications for treating include:

• Chronic respiratory diseases (CRD) and air sac infection caused by mycoplasma gallisepticum, Escherichia coli (E. coli).

• Infectious synovitis caused by Muyoplasma synoviae.

• Foul cholera in growing chickens caused by Pasteurella multocida.

• Coccidiosis caused by a protozoa.

Both tetracyclines are available in packets as soluble powder to put in drinking water.

Tetracycline Dosage Calculations

Commercial use of tetracycline antibiotics are put into the flock's water supply. The dosage on the label is given as number of packets per gallon. Each packet contains 10 grams of pure tetracycline in a filler that totals 181.4 grams.

The Pfizer Animal Support line said that one teaspoonful (3.2 grams) of the Terramycin® soluble power contains a 200 mg dose of oxytetracycline (if only that had been printed on the label). They also said that birds will drink the medicated solution and therefore dose themselves according to their body weight. But the dosing for an individual chicken is still murky. They said the dosing depends on how much that bird will drink. Here's the recommended amount, from the label, to put in the water.

- Infectious synovitis caused by Muyoplasma synoviae
     200 to 400 mg/gal
     Which is 1 to 2 teaspoonfuls of powder/gallon

- Chronic respiratory diseases (CRD) and air sac infection caused by mycoplasma gallisepticum, Escherichia coli
     400 to 800 mg/gal
     Which is 2 to 4 teaspoonfuls of powder/gallon

- Foul cholera in growing chickens caused by Pasteurella multocida
     400 to 800 mg/gal
     Which is 2 to 4 teaspoonfuls of powder/gallon

Labeling states, "Treatment is not to be used for more than 3 weeks and has no withdraw time". But again, these antibiotics are not labeled or approved for use in chickens.

Sulfadimethoxine

This is a broad-spectrum, inexpensive drug used for outbreaks of cocidiosis and fowl cholera. Do not give to chicks under 4 months of age. Sulfadimethoxine dosage is 2 teaspoonful per gallon for treatment or as directed on the label.

Colloidal Silver

When any of my birds get an infection, I use colloidal silver for both topical (skin) and systemic (internal) treatment. After internal treatment my poultry patients receive probiotics to keep the good bacteria established in their GI tract. Colloidal silver is a liquid suspension of microscopic particles of silver. A colloid is when tiny particles of a material are small enough to remain in suspension in liquid form.

Silver has been used medicinally to treat conditions for centuries. Even so, colloidal silver is considered an alternative therapy. It's included in the Poultry's Pharmacy because of the excellent results we've experienced treating infected conditions in our animals. These include treating eye infections, cleaning wounds, and as a dressing on wounds and burns.

Colloidal silver is frequently promoted as a natural infection fighter, capable of killing both bacteria and viruses. The FDA has not approved colloidal silver and does not recognize it as safe or effective. Long-term use can result in silver salts being deposited in the skin, eyes, and internal organs. The skin and tissues can turn a bluish-gray.

A few prescription drugs containing silver are routinely used; for example, silver nitrate to prevent conjunctivitis (an eye infection) in newborn babies. Silver nitrate is also used to treat some skin conditions, such as corns and warts. A prescription-only burn cream is silver sulfadizine (Silvadine®).

Make Your Own Colloidal Silver

Colloidal silver is often available in natural foods stores and can be expensive. It can be easily made at home by purchasing a colloidal silver generator; although some of the generators make solutions that have dangerously large particles of silver. The best generators take two to three days to get the correct size and concentration of silver in colloid form.

Kits provide everything you need to make small batches of colloidal silver. YouTube.com has videos of how to make the colloidal silver solutions. The Internet can give you lots more information about using colloidal silver as an alternative treatment.

Never remove the electrodes from

Colloidal Silver Generator

A colloidal silver generator is easy to set up and use. The laser pointer light reflects off the silver in solution and gives an indication that the silver is in colloidal state.

the water if the unit is on. The silver strips should be at least ".999 fine" in purity. Do not use sterling silver (.9275) because sterling silver contains copper and nickel. Nickel can be toxic.

A laser pointer shining through the solution will reflect the particles and give you an indication of the colloidal silver in solution. A Hanna Pure Water Tester (PWT) meter is a way to test your colloidal silver to measure with reasonable accuracy the concentration you are making.

Eye Treatments

Colloidal Silver Eye Wash

Either purchase or make your colloidal silver solution. Use a 5 cc oral syringe to draw up a dose and administer using the chicken eye wash method described in the chapter on Poultry Primary Health Care.

Colloidal silver is especially effective for eyes that have froth (bubbles) around the eye, or drainage from an eye. Also give about 5 cc in an oral drench. Continue the treatment until the eye returns to normal, usually about 2 to 5 eye washes.

Normal Saline Eye Wash

You can buy a normal saline wash or use contact lens solution. You can make your own saline solution very cheaply, as the ingredients are simply distilled water and salt. Buy distilled (or deionized, or demineralized) water. The ideal source is laboratory grade sodium chloride. Next best would be a natural salt such as sea salt without any additives. Table salt usually contains silica-type minerals added to make it free-flowing. Silica is insoluble and can be gritty so you don't want it in your eye solution.

To make one liter of saline, mix in 8 grams of salt. If you don't have a way to weigh grams, then use a teaspoon measure that is used for cooking. A very slightly heaped teaspoon is about 8 grams and close enough for the eye wash purposes. A quart is close enough to a liter to measure the water. Just make sure it's a very full quart.

These measurements are approximate, but the solution doesn't have to be exact for use in eyes (or noses). It's better to have a slightly a weaker solution, because too much salt will make eyes sting slightly. Too little salt in the solution can produce a vaguely uncomfortable suction until the osmotic pressures equalise.

Topical Wound Treatments

Comfrey is probably one of the best known healing herbs in history. The medicinal use of comfrey dates back to 400 BC. Comfrey was used by the Greeks to stop heavy bleeding and treat bronchial problems. Dioscorides, a Greek physician of the first century, prescribed comfrey for healing wounds and mending broken bones. Comfrey has been used internally by the Native American Cherokee for any number of ailments. Comfrey leaves and roots have an ingredient called allantoin that is credited with its healing properties. Here are a few of its folk medicine uses.

• Comfrey is especially known for its ability to encourage tissue repair in the body.

• In folk medicine, comfrey is called the bone knitter because it helps bones to mend.

• Comfrey, when applied as a poultice, is effective for treating burns and swelling.

• Comfrey can stop bleeding and aid in soothing inflammation.

• Don't use comfrey on puncture wounds because cell growth stimulating factors in the comfrey can cause the wound to seal over before it has drained.

I've seen comfrey work miracles on wounds. As a case study, one of my hens was brutally mauled by a dog, most of the skin ripped off her back and upper thigh. The muscles were exposed, and the tear was so big that stitches were not an option. It was the worse wound on a chicken I'd ever seen. I didn't expect her to live through the night and was about to put her down. But she was one of my favorite hens and I wanted her to have a chance, and if she died, it would be peaceful.

I brought her inside and admitted her to my intensive care unit (a medium size pet carrier). I rinsed the wounds with colloidal silver using a 20 cc syringe (without a needle) to squirt all round and soak the tissues. I also gave her 5 cc's of colloidal silver as a drench. Her water had collodial silver and oral rehydration solution.

I then made and applied a fresh comfrey poultice on the exposed and torn muscles, just laying the green mass on top of her back with a large non-stick pad secured by paper tape. On the rest of her wounds I very gently spread a layer of comfrey salve so that most of the injured areas were covered.

She was on the edge of death for the first two days, barely moving. I kept treating her wounds as best as I could. The third day she ate a little oatmeal and drank some water mixed with colloidal silver and oral rehydration salts. By the end of the week she was outside walking around in the sunlight. I knew she was going to live. She finally rejoined the flock about 4 weeks later. Amazingly, all her feathers grew back and you couldn't tell she had been so horribly wounded.

Comfrey is a perennial herb that belongs to the Boraginaceae family. It's a valuable addition to every garden.

How to Make a Comfrey Healing Salve or Balm

A salve or balm is a soothing, healing, or comforting medicinal ointment. The two terms are often used interchangeably.

If you can read, and even better, if you can cook, then you can make a comfrey ointment in your kitchen. It's a simple two part process. First make a comfrey infused oil, and then use that oil to make an ointment, salve or balm. Here's how.

Making Comfrey Infused Oil

Making an oil infused with herbs is very easy and is the same principle as when making herbal oils for salads or other culinary delights. Here's how to make an infused herbal oil.

1. If the comfrey is fresh, rinse it in cold water, pat dry, soaking up as much moisture as you can. Leave the herbs to continue to dry in the sun or overnight if you can. Bacteria cannot grow in the olive oil, but it can grow on any residual water from the herb. If the comfrey is dehydrated, chop or crush into small pieces.

2. Put the leaves in a glass container and pour in extra virgin olive oil so that the top layer of leaves is covered.

3. Let soak for about 3 days at 100 degrees F, or for 5 days at room temperature.

4. Strain and press the soaked comfrey to squeeze as much of the oil out as you can. Seal in a light-resistant, air tight container. There will be some water and sediment in the bottom. Discard this layer.

5. Add about 1% vitamin E oil which is healing, and doubles as a preservative. You can get vitamin E oil from piercing capsules and squeezing the contents into the mixture.

6. Store the oil infusion in a glass bottle in the refrigerator. It will keep for some time. Smell for rancidity before using.

Making Comfrey Balm or Salve

You will use the comfrey infused oil to make the healing salve. It is not a long or complicated procedure. You can substitute different oils or just use olive oil if you prefer. Castor and apricot oils have been used in making the finer healing ointments for centuries. Most hobby stores carry beeswax.

The recipe below makes about 9 ounces of ointment.

1. Melt 25 grams beeswax over low flame or double pan boiler.
2. When melted add:

    60 ml (4 tablespoonfuls) comfrey infusion oils (see above)

    40 ml castor oil

    200 ml apricot oil or extra virgin olive oil

    1600 IU vitamin E

3. Do not overheat the beeswax such that it smokes. It has a low melting point.
4. Test the consistency of the batch by putting a small amount in a spoon and placing it in the freezer to cool quickly. Once it's cold, spread some on your hand or arm. It should soften quickly at skin temperature and blend in without lumps.

There are always tricks to every trade. Here are some tips for getting a professional product.

• Make sure the batch is hot enough so you don't have lumps in the final salve, but not so hot that it smokes or turns dark brown.

• Do the freezer test to see if the final product will be too hard. You want the balm to be soft and easy to spread, ideally melting at body temperature.

• Use a large syringe (without a needle) to draw up the liquid mixture and put into jars or lip tubes.

• Mix the hot oil mixture well before drawing it up in the syringe. Make sure color is consistent.

• Shake the tube or jar gently go get air bubbles out. Pouring slowly down the side will also keep air bubbles from forming.

• Watch for color consistency. If it's not consistent, put it back into the batch, stir and re-draw.

• Discard any sediment and water in the bottom.

• Store in air-tight containers and date. The balm will last several months depending on how it is stored. Always check for rancidity before using the preparation. If it smells like old nuts, it's rancid and should be thrown away.

How to Make Sulfur Ointment

A 5% to 10% sulfur ointment is remarkably effective as a topical treatment for insect bites, skin infections, infected wounds, skin irritations, fungi, and to kill mites (including burrowing mites such as scabies) and lice, on poultry.

Sulfur ointment can also be an antiseptic and an anti-fungal for conditions such as ringworm. A 5% to 10% mixture is soothing to the skin. A 10% sulfur product is as high a concentration as you want to use. Higher concentrations than 10% can turn to sulfuric acid and burn the skin.

How to Make a Sulfur Ointment

There are only 2 ingredients and they are both inexpensive.

1. Flowers of Sulfur powder, precipitated sulfur, sublimed sulfur, milk of sulfur and even brimstone are all names for the pure, elemental

sulfur you want to use. It is a fine crystalline powder, light lime color. It does not stink. Keep tightly sealed as it attracts moisture. The powder, undiluted, will burn your skin and irritate nasal passages. Be careful with it.

2. Use a base such as petroleum jelly (Vaseline®). You want a base that is really thick and sticks well to surfaces. You can add mineral oil to make the ointment less sticky. However, this also makes it less effective and you might need to use it longer.

You can also use the comfrey balm you made and add 5% sulfur. This gives a very effective treatment for skin that has scratches, lesions, or sores from a mite infection. It serves double duty as it also kills the mites both on and under the skin.

Here's the compounding procedure.
Measure out:
  4 ounces of base ointment
  6 grams (about 2.5 teaspoonfuls) of sulfur (for a 5% mixture)
  or 12 grams (about 5 teaspoonfuls) for a 10% mixture
The sulfur powder can be lumpy. Use the back of a spoon to crush any lumps into a very fine powder.

You'll need a flat surface to mix the ingredients together. A cutting board covered with wax paper, or a very smooth surface that you can easily clean will do. With a frosting knife or stiff spatula, mix in a little of the sulfur into the ointment. When this partial batch is well mixed, blend in, little by little, more of the ointment base and the powder until the mixture's color is even and the ointment is without lumps. Blending it this way avoids lumping. You can use the same container the ointment came in to store your finished product. Label and date.

Directions for Using Sulfur Ointment

For topical treatment, apply 1 to 3 times daily, or as needed. Apply to feathers, under wings, and on legs and feet. Discontinue if there are any signs that the sulfur might be irritating the skin.

For scaly leg, apply a thick coating daily, or every other day for 10 days. Repeat weekly until the legs return to normal. Lotions and creams will be easier to use on feathers and skin. Ointments are more effective on scaly leg, because it really sticks to the leg and works under the scales to kill the mites.

Treating Internal Worm Infestations

Worms can show up in hens from many sources. Just as urban dogs and cats get worms from the environment so also can hens.

Round worms are in the gut and can be about 2 inches long. A bird can tolerate some worms, but if there are too many they can get anemia and their eggs will have pale yokes. Most poultry wormers are simply added to the water to treat the entire flock, such as piperazine (Wazine®). Another wormer added to water is Wormout that contains praziquantel and oxfendazone. Note that these are not approved for use while a hen is laying because some of the metabolites pass into the eggs.

Ivermectin

Ivermectin is sold under several brand names including: Stromectol®, Mectizan®, Ivexterm®, Ivomec®, and Iver-On®. It is a broad spectrum wormer that is indicated in just about all species of livestock for roundworms, lung worms, grubs, horn flies, sucking and biting lice, and sarcophtic mange. Ivermectin labeling clearly states it is not to be used in food animals including dairy cows. There are not any products specifically approved for poultry but Ivermectin is widely used by small-scale poultry fanciers as wormers.

Some folks keep Ivermectin drench in their poultry first aid kits. Whenever they buy a new bird, they give it a few drops so that they know the bird has been recently wormed.

Ivermectin is oil based and doesn't mix well in water. The off-label and unapproved dosage suggestions used by some poultry keepers are below.

> • *In Waterers.* Ivermec Drench® put 6 to 8 cc/gallon for 1 day, repeat in 10 to 12 days.

> • *Oral Administration.* Use Ivermectin drench 0.5% to treat individual birds, see Hen Health Chapter 15 on how to give oral doses (drenches).

> Small bird (bantam) – 2 to 3 drops
> Medium bird – 3 to 4 drops
> Large bird – 5 drops (about 0.25 cc)

• *Topical Administration.* Ivermectin pour on 5mg/ml. Use an eye dropper and apply it to the chicken's skin on the neck or back. It's easiest to treat at night when the birds are on their roost and you can go down the line knowing which ones you've treated.

> Small bird (bantam) – 3 to 4 drops
> Medium bird – 5 to 7 drops
> Large bird – 8 drops

• *Injection.* 1% Ivermectin Injectable, give subcutaneously under the skin on the chest or upper thigh. See Hen Health Chapter 15 on how to give injection.

> Small birds 0.25 to 0.5 cc.
> Medium birds 1 cc.
> Large Standards: 1 to 1.2 cc
> Extra Large (Jersey Giants) 1.5 to 2 cc

Symptomatic Support

Symptomatic support for a sick bird is like admitting it to a hospital to give it a chance to heal itself. This means making sure it is in a clean, warm, draft free environment that has fresh water and high-quality, easily digestible feed available. It also makes the bird easily available for medical treatment (cleaning wounds or medical dosing). Additional vitamins and/or oral rehydration is often helpful in the water. Sometimes, just giving a sick bird a quiet and safe environment is enough for it to heal itself.

A dog carrier works nicely as an infirmary if you need to remove her from the flock.

Respiratory Conditions and Poultry Flu

Vet Rx is one of those old-time treatments with claims for a lot of ailments. It's similar to the camphor/menthol based Vick's Vapo Rub for people, which has a lot of off-label uses. The active ingredients include Canada balsam, camphor, origanum oil, rosemary oil, blended in a corn oil base.

The Vet Rx instructions say to use at the first signs of a respiratory disease when a bird is gaping, sneezing, gasping, with watery eyes, droopy appearance, and an unusual tendency to shake or turn the head.

In my experience, thieves' oil has been effective against respiratory diseases in chickens. Thieves' oil formulations contain varying amounts of essential oils, primarily lemon (citrus lemon), cinnamon bark (cinamonum verum), eucalyptus (eucalyptus radiata), clove bud oil (Syzygium aromaticum), and rosemary (rosemary officialis). The oil mixture can be mixed in a massage oil base for topical use. Thieves® Oil Blend is a proprietary blend of these essentials oils.

The aromatic oil formulation gets its name from 19th century, French perfumers, turned thieves. One story is that the thieves used the essential oils (in a base of white wine vinegar) and rubbed it on themselves to get protection from the plague. Thus, they could rob victims of Black Death plague and not contract the insect-vector disease. Perfumers had a reputation of being immune against the deadly disease.

Measurements & Conversions

Every Poultry's Pharmacy needs to calculate dosages and convert between the different measuring systems. This can be confusing. Some practical equivalents are below.

Liquid equivalents
    1 teaspoonful = 5 ml
    1 tablespoonful = 15 ml

    1 fluid ounce = 30 milliliters = 6 teaspoonfuls
    1 pint = 473 milliliters = 16 ounces = 2 cups
    ½ Pint = 236 milliliters = 8 oz = 1 cup
    1 gallon = 3,785 milliliters
    1 fluid ounce = 6 teaspoonfuls

Dry equivalents
    1 teaspoonful = 3.2 grams
    1 dry ounce = 28.35 grams
    1 pound = 454 grams
    1 kilogram = 2.2 pounds

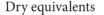

The fact that zoning in towns allows residents to raise a barking, crapping dog the size of a small elephant, but not four hens for a steady, fresh egg supply, shows just how lacking in common sense we have become as a society.

— GENE LOGSDON

18 Get City Chicks Legal in Your Town

As cohost of the Chicken Whisperer Backyard Poultry and Sustainable Lifestyles Talk Show, I've heard every opinion and belief about family flocks. The objections are just about all based on belief and misinformation, not fact. The seven most common myths about city chickens are below.

Myths vs Facts about City Chicks

1. Cleanliness and Odors. If not taken care of properly any animal (including humans) will cause offensive odors. A family flock of 10 chickens is not a factory farm with 100,000 birds. Proper flock care and the existing pet control laws are all that is needed to prevent problems. Unlike dog or cat poo, chicken manure can be composted and is valuable, rich fertilizer. Let's put this in context. One 40 pound dog generates .75 pounds/day of doggie doo. A flock of 10 chickens generates .66 pounds/day of poultry poo — that is valuable fertilizer for your garden and compost.

2. Noise. Hens are quiet. At their absolute loudest, they have a 70 decibel level which is only 10 decibels above normal human conversation (60 decibels). Compare this with barking dogs and lawn mowers that have decibel levels of 90. Roosters have decibel levels lower than barking dogs, but their crow is so shrill that it can be irritating and the sound carries far.

3. Attract Predator and Nuisance Animals. Wild bird feeders, pet food, gardens, fish ponds and trash all attract raccoon, foxes, mice, rats and snakes. A backyard chicken feeder is no more of an attraction than bird feeders and a chicken waterer is the same as an outside potted plant or pool. Predators and nuisance animals are already there because of the food supply that already exists is there. On the contrary, chickens have voracious appetites for ticks, fleas, flies, mosquitoes and just about any crawly, jumpy slimy thing they can get their beaks around; including mice and baby rats. Chickens can be part of a pest solution.

4. Real Estate Values Will Decline. We have NEVER seen one documented case of property values dropping because of a family flock of chickens next door or in the neighborhood. Not one! On the contrary, some home sellers and developers — as a sales incentive — are offering a free chicken coop with every house or lot sale. GreenWay Neighborhood in Buena Vista, Virginia is an example of using this strategy (www.GreenWayNews.com).

5. Disease, Avian Flu & Salmonella. It is the huge multimillion-bird, fragile-flock factory farms that raise immune-compromised, homogeneously-bred, mono-cropped chickens that are a far, far greater threat (and risk) of causing a disease epidemic. On the contrary, keeping backyard flocks can help prevent the spread of the Avian Flu H5N1 virus. Salmonella is the result of poor food handling practices, and insufficient refrigeration. Backyard flocks are not the source, nor the cause of salmonella or Avian Flu infections.

The Centers for Disease Control has stated: "In the United States there is no need to remove a flock of chickens because of concerns regarding avian influenza." And the 2006 GRAIN Report states: "When is comes to bird flu, diverse small-scale poultry farming is the solution, not the problem."

6. Neighbor Consent. Having to get a neighbors consent is an insult to our civil liberties. It's not required for dogs, cats, parties or guns. What if you had to get your neighbor's permission to have the noisiest and potentially most difficult of all: kids? There would be an outrage among the general public. Not in any part of the world is neighbor's consent required for just about anything — so why should it be for chickens? It's not the American way. The existing animal control codes are enough.

7. Appearances. Urban coops can be whimsical, charming, upscale and downright delightful. For examples, look in the Housing Chapter and see how coops can be architecturally blended with the surrounding homes. Coop tours are expanding nationwide and are a source of eco-tourism revenue. Cities that host coop tours include: Austin, Salt Lake City, Raleigh, Los Angeles, Seattle, and many, many others. The Shenandoah Valley Poultry and Garden Club is planning to host its first coop tour next year.

City Chick Laws in Your Town

The best way to learn about your local laws and ordinances is to contact your city's legal advisor. If you contact the administrative offices you will probably get a lot of opinions about the legality of chickens but not necessarily any answers. Don't let people's opinions stop you. Get a copy of the actual law and if city chicks are not allowed, ask what has to happen to get the code changed. Getting the laws changed is getting easier as the popularity of keeping chickens, and understanding how valuable they are in local food production is sweeping across America. Below is the story of our saga to get family flocks legal in Lexington, Virginia.

The Saga of Getting City Chicks Legal in Lexington, Virginia

Here's our story, complete with governmental red tape, the drama of special interest groups, and the power of personal politics. It began with Monika Eaton's birthday dinner. As she blew out the candles on her cake, she wished for a small flock of chickens to compliment her large yard and gardens. But Monika lives within the city limits of Lexington, Virginia, where "farm animals" are illegal.

"Then let's get it legal," said I, thinking this would be as easy as a piece of birthday cake. "Let's ask City Council to change the code". With that brash ambition, this real-life chicken saga began and it is still not resolved. Chickens are still illegal, contraband outlaws and fugitives within city limits.

Our persistence will ultimately pay off, as it's hard to keep a movement squelched when there are increasing numbers of outlaw flocks taking up residence within the city limits. We've detailed our experiences, setbacks, and strategies to help others convince their city government to legalize chickens.

Round 1. The Initial Proposal to City Council, September 2005

During the few weeks after Monika's birthday we gathered information, contacted key city officials, and wrote a City Chicks proposal paper. Monika was nominated to do the introductory presentation because her birthday wish started it all. Below is our written proposal, as Monika presented it to the City Council.

Draft City Chicks Code

Lexington city residents may keep up to 8 chickens on their property, provided that:

i. The principal use is for a single-family dwelling.

ii. No person shall keep any rooster without special permit.

iii. The chickens shall be kept in a covered enclosure with an appropriately fenced area.

iv. No enclosure shall be located closer than ten (10) feet to the boundary of an adjacent lot.

This proposal is sponsored by many Lexington city residents who are willing to sign a petition urging the adoption of the measure.

The REAL Reasons for Keeping Urban Chickens

• Food and yard bio-recycling keeping household and yard waste" out of the solid waste management streams (trash collection).

• Local fertilizer production, compost and top soil creation for gardens.

• Locally produced, fresh, heart-healthy, high-protein eggs.

• Hens serve as garden helpers, organic pesticiders, herbiciders, and fertilizers.

And Yet More GOOD Reasons for Urban Chickens:

• **Quiet Pets.** Hens are personable, easy to keep, and entertaining. Hens are far quieter than any barking dog and don't leave droppings on the curb.

• **Low Allergy Pets.** Often, folks with allergies to dogs or cats do not have allergies to chickens.

• **Kid & School Projects.** Hens provide an excellent way for kids to learn about pet responsibility, recycling biomass, composting, and food sources. The 4-H poultry club could expand to include City Chick projects.

Major Cities Encouraging City Chicks

The chicken savvy cities listed below have full nests, a local food supply and clucking civic workers that divert food and yard residues away from solid waste management streams. This list gets longer every day, and with every election of new city council members who are truly dedicated to creating "green" and sustainable municipalities.

The "Chicken Underground" has emerged all across North America. People are increasingly understanding their "Declaration of Local Foods Rights" and are pursuing their inalienable right to grow one's food in their own yard — and this includes a family flock of chickens!

Anaheim, California	Flagstaff, Arizona	New Orleans, Louisiana
Ann Arbor, Michigan	Fort Collins, Colorado	New York City, New York
Albuquerque, New Mexico	Gig Harbor, Washington	Oakland, California
Asheville, North Carolina	Gilbert, Arizona	Olympia, Washington
Atlanta, Georgia	Grand Forks, North Dakota	Omaha, Nebraska
Austin, Texas	Hamstead, New Hampshire	Portland, Maine
Baltimore City, Maryland	Houston, Texas	Portland, Oregon
Baton Rouge, Louisiana	Honolulu, Hawaii	Raleigh, North Carolina
Belmont, Massachusetts	Irvine, California	Richmond, Virginia
Berkely, California	Lakewood, California	Sacramento, California
Boise, Idaho	Lansing, Michigan	San Francisco, California
Brockton, Massachusetts	Lawrence, Kansas	Santa Fe, New Mexico
Burlington, Vermont	Laramie, Wyoming	Saint Louis, Missouri
Cambridge, Massachusetts	Laredo, Texas	Saint Paul, Minnesota
Camden, Maine	Las Vegas, Nevada	Salt Lake City, Utah
Casper, Wyoming	Little Rock, Arkansas	San Antonio, Texas
Chapel Hill, South Carolina	Long Beach, California	San Diego, California
Charlottesville, Virginia	Los Altos, California	Seattle, Washington
Chicago, Illinois	Los Angeles, California	Sioux City, Iowa
Colorado Springs, Colorado	Louisville, Kentucky	Spokane, Washington
Dallas, Texas	Madison, Wisconsin	Syracuse, New York
Davis, California	Miami, Florida	Tacoma, Washington
District of Columbia	Minneapolis, Minnesota	Tampa, Florida
Denver, Colorado	Mission Viejo, California	Topeka, Kansas
Des Moines, Iowa	Missoula, Montana	Syracuse, New York
Evansville, Indiana	Mobile, Alabama	Vancouver, Washington
Fayetteville, Arkansas	Nashville, Tennessee	Wichita, Kansas

• **Heritage Breed Preservation.** Just as some species of wildlife are in danger of becoming extinct, so are some breeds of chickens. The American Livestock Breeds Conservancy has credited backyard flocks with helping to stimulate interest in the rare and endangered chicken breeds. While breeding would not be done within the city limits, heritage chick sales could be a source of income for breeders.

• **Local Commerce and Food Production.** Other benefits to the community include: increased interest in gardening, more representation at Farmers' Markets, and an increase of poultry-related supply and feed sales from the Farmers' Co-op and Tractor Supply Company.

Our City Chicks request was well received and, at the end of the meeting, City Council members seemed positive and open to the proposal. Per protocol, it was referred to a subcommittee for further review and recommendation. Lexington's local papers (*The News Gazette, The Rockbridge Weekly,* and *The Advocate*) ran the story. Here is the story from *The News Gazette.*

Proposal Would Allow Chickens in City

By Roberta Anderson

A group of Lexington residents would like to see a change in the city ordinances to allow individuals to keep a limited number of chickens within the city limits.

Monika Eaton last week told Lexington City Council that she represents a contingent of about half-dozen chicken fanciers who would like to see Lexington join a growing number of cities nationwide that allow residents to keep certain poultry species. In making her request, Eaton emphasized her desire to remain sensitive to the impact that chickens could have on other city residents. Eaton had prepared a packet of information on her proposal that she distributed to Council members.

Eaton said her request did not include keeping roosters. She suggested that a new ordinance could specify that the chickens be kept in enclosed areas and could include other restrictions such as a limit on the number of chickens allowed.

Among the reasons for Eaton's request to keep chickens are the health benefits of having home-raised eggs. "Chickens just make wonderful pets". She added. "They are so much fun for children".

The request was referred to the social and economic services committee meeting. In the meantime, Eaton was circulating a petition in support of the city chick proposal.

Round 2 October 2005 — City Council Subcommittee Review

We knew there was dissent among the subcommittee members about our proposal. Some viewed chickens as dirty farm animals. What else to do but take a live hen to the subcommittee meeting so they could see how clean and amicable a hen can be?

I enlisted one of my favorite ladies, Attila the Hen, for this public service task. Attila is one of the most friendly girls in my flock and a true character. I put her in a cat carrier and — with my version of "chicken in a basket" — we went through City Hall security with no problem.

I slid into the meeting room and sat demurely in the back row with Attila beside me in the carrier. Her curiosity quickly got the best of her

Chicken Visit To Council Successful

Lexington Councilman Tim Golden gets up close and personal with Attila the Hen, who, along with her owner, appeared before Council last week to ask that chickens be allowed in the city. An ordinance is now being drafted to that effect. For details, see page 5. (Roberta Anderson photo)

Chicken Ambassador Attends City Council

Most of the City Council members had never touched a chicken. Here Attila the Hen makes an excellent first impression to the Council members, not to mention a great photo opportunity and she began making a low, gargling "*gurrrrrr*" noise, which is Chicken for, "What's going on?"

Attila was just loud enough to cause giggles among the audience. Finally one of the Council members asked, "Is that a chicken?" I recognized my performance cue and said, "Yes, would you like to meet her?"

I took their stunned looks and open mouths as a "yes" and took Attila the Hen out of the carrier. The city council members stroked her downy feathers and fondled her wattles. Ever her diplomatic self, Attila continued with her vibrant low semi-purring "*gerrrrah*?" while making eye contact with the council members. It was the first time many of them had touched a live chicken. They could see, feel, and smell how clean, fluffy, and gentle chickens can be. I genuinely think the audience and Council members had fun. After the meeting, one Council member — who had been totally opposed to chickens — passed me in the hall way and said, "I admit that Attila softened me a bit".

The press was delighted to see a live chicken attending the Council meeting. They whipped out their cameras faster than cowboys reaching for pistols and flashes began to fly. *The News Gazette* had a front page photo of Attila being stroked by Council member Tim Golden. Here's the article.

The News Gazette: Chickens in The City One Cluck Closer

By Roberta Anderson

Attila the Hen clucked and cooed sociably through her introductions to members of the Lexington City Council social and economic services committee last week. Her charm was such that by the end of the meeting, the committee directed City Manager Jon Ellestad and attorney Larry Mann to draft an ordinance to allow chickens to be kept within the city limits.

Attila arrived at the meeting on the arm of her owner, fowl fancier Patricia Foreman, along with Monika Eaton. Last month Eaton had requested City Council to consider amending the city ordinance that prohibits the keeping of fowl and livestock within the city. Eaton requested that city residence be permitted to keep a limited number of chickens on their property. She told Council that hens make good pets and are not noisy like roosters or other types of fowl. There are also health benefits to having eggs from a noncommercial source.

Attila apparently had not worked her charm on the city animal control officer Captain Roger Clark, whose feathers seemed a bit ruffled by the proposal. Clark expressed his opposition to the proposal in a letter to Council members.

"If we allow the chickens and fowl, what will happen when someone wants to bring in pygmy goats, miniature horses, and of course pot-bellied pigs and other fad type of exotic animals?" Clark wrote: "Fowl would be an irresistible draw for such undesirables as foxes, raccoons and snakes". Clark argued: "and the city pound lacks proper chicken accommodations. Councilwoman Mimi Elrod, who heads the social and economic services committee, wasn't so sure. "We all think of dogs and cats being city animals". Elrod said. "But we all know dogs can be a nuisance. Maybe chickens are less of a nuisance".

Of greatest concern to the committee is a minimum distance between the chicken house and a neighboring home to be specified in a proposed ordinance. Ellestad said the minimum distance in Virginia cities and towns that allow livestock appears to be 125 feet. (But chickens can be considered pets, not livestock). A similar specification in Lexington would limit the number of properties on which chickens could be kept because of the small size of most lots within the city.

Eaton noted that most ordinances include all types of fowl and livestock, not just laying hens. An ordinance currently in force in Madison, Wisconsin requires only 25 feet between the hen house and neighboring homes. Eaton also said she does not believe that there will be an exponential explosion in the local chicken populations if an ordinance is approved.

Eaton said she has surveyed all of her neighbors and none have objected to her proposal. She produced a petition of signatures in support of her request. She also addressed Clark's concerns about predators, saying suitable chicken houses would eliminate that risk. County Extension agent Jon Repair has indicated that the presence of chickens would not contribute to the possibility of an outbreak of bird flu, Eaton added.

Councilman Tim Golden asked however, if allowing laying hens in the city would not somehow represent "a step backwards".

Eaton responded that keeping chickens in an urban environment is more common in Europe and becoming increasingly common in this country as more farmland is lost. "It is not necessarily a step backwards when people can work together, especially to produce food", she said.

A public hearing, followed by final approval by Lexington City Council is require to enact a new ordinance.

Round 3: Second Subcommittee Review City Chicks

December, 2005 (2 months later), the City Council, City Attorney Larry Mann, and City Manager Ellestad had researched and prepared a draft city ordinance for review by the Council. After discussing some revisions to the ordinance, the committee took a vote. Our proposal did not have the majority vote needed to send it to a Public Hearing. The motion failed by two votes. Dang!

After the meeting we were feeling rather low. On the way out, the owner and editor of The Advocate, Doug Harwood, confided in us that he had run the City Chicks story and had received calls in favor of the proposal. He had one caller from out of state asking, "What's going on with chickens? Is Lexington getting progressive?" That gave us a good laugh. Chuckling about Doug's comments, we plodded over to the Southern Inn to have a drink and console each other. It was time to let the City Chicks movement brood for a while.

Round 4: September 2007 —
City Chicks Two Years Later

Time passed and two things changed:

    1. The local foods movement was gaining popularity and

    2. The county landfill was almost full and must close.

Since we first presented our proposal, there had been increasing international interest about the possibility of using chickens for recycling yard and food waste, as chickens have the potential to divert tons of biomass away from landfills and into compost. A city in Belgium gave 2,000 residents 3 hens each, not for eggs, but to decrease the amount of biomass in trash collection.

    Rockbridge County's landfill was almost at capacity and was

Press Photo Opportunity

A large, over-stuffed chicken served as a prop to give the press an entertaining photo opportunity during the Lexington City Council meeting.

NEWS/Page A2
CHICKENS IN THE CITY?
Supporters Of Idea Say Chickens Could Serve As Bio-recyclers

scheduled to close soon, at which time the refuse would be transported out of the county. It was a big issue and an even bigger budget item.

We felt that the "diverting-waste-to-compost" line of reasoning might help get City Chicks approved, if not for long-term residency, at least for a special study. We brought a gigantic stuffed chicken as a prop for photo opportunities at the next City Council meeting. Yes, it was embarrassing to stand in the court room, before the distinguished members of City Council, holding a stuffed chicken. But this sacrifice was beyond myself. I did it for the cause and for future generations of City Chicks.

After all, small flocks of chickens in hundreds or thousands of backyards can recycle millions of tons of yard and food waste into compost, fertilizer, garden soil, and protein-rich eggs. The stakes are high, both for the city budget, and for local food production. The write up in the *The News Gazette* says it all.

Chickens in the City Advocates Try Again

By Roberta Anderson

Chickens in the city advocate Patricia Foreman brought a large stuffed chicken with her to Lexington City Council last week as she tried to convince Council to authorize a study to determine how much money could be saved in landfill fees if household food and yard waste was recycled by hens with the added bonus of getting fresh eggs and good compost.

Advocates of city chicks keep pecking away at Council to allow city residents to keep hens as pets for their eggs at their homes. Patricia Foreman, author of *City Chicks* and *Chicken Tractor,* and past supporter of allowing hens in the city, last week approached Lexington City Council to request a study that would determine how much food waste can be diverted from the landfill and converted into compost and eggs. The hens would take on the role of bio-recyclers.

Foreman said the participants would be restricted as to the number of hens allowed per household and the housing of the chickens would be closely monitored. Foreman had a list of volunteers ready and eager to sign up for the project.

A request put forward by Foreman and other chicken fanciers almost two years ago to legalize chickens in the city failed to win Council approval. With the exception of Councilwoman Mimi Elrod, Council was not more receptive this time to the idea of chickens, bio-recyclers.

"The take on the proposal is different this time," argued Elrod. "The take isn't different," countered Golden. "We already voted on it".

Foreman asked how the city could turn down the proposal if the study could prove that the hens could save the city money. "How can you turn it down?" she asked.

"Very easy," responded Golden. "We don't want chickens in the city. That's why we live in the city".

Foreman countered that there were a lot of people who would like the privilege of keeping a small flock of laying hens. Foreman's proposal was turned over to the Social and Economic Services Committee for further study.

The discussion was testy, and we were surprised at the hostility expressed by one especially vocal council member. In spite of his ill feelings, per protocol, City Chicks was referred to subcommittee for review. Here is a copy of our second proposal:

Biomass Recycling Research Proposal Presented to the Lexington City Council

Key focus: documenting how much food and yard waste can be diverted from the Rockbridge County, Virginia landfill and converted into compost, fertilizer, and eggs, using chickens as bio-recyclers.

The Problem: landfills of Rockbridge County and other cities in the area are filling up fast. Some cities have run out of landfill space and are exporting their trash to other locations. Rockbridge County's landfills have only a couple years of service left; transfer stations are already being planned. The county has declared a "Zero Waste Resolution" which eliminates the concept of waste. Under this approach, yard and food waste become "residuals" and are raw matter assets to create compost, potting soil, and topsoil.

This research project investigates a low-tech, high-output methodology to divert food and yard residuals from entering the county landfill system by employing small flocks of chickens to help transform the residuals into compost, fertilizer, and eggs.

If this pilot study is successful, it can be expanded to a nationwide study involving major cities. This project is supported by the Gossamer Foundation, a non-profit organization dedicated to ecological preservation and restoration.

Research Design Using Chickens as Biomass Converters

A small number of city residents will keep micro-flocks of 6 to 8 chickens. Participating households will estimate how much food waste they give the hens and will keep a journal documenting their experience with the micro-flocks. Worksheets will keep track of food waste, egg production, and the household's experience with the hens.

Hen housing and management will be approved and monitored by the research team for the hen's safety, sanitation, and well-being.

The research team will collect data and report quarterly. A summary review of the project will be presented to the council after one year. Results will be submitted for publication in national poultry journals and posted on the web site www.ChickensInTheCity.com.

Questionnaires and feedback from the establishments and individuals involved will be collected and will address the questions presented in this study, including:

- The participating household's experience with hens.
- Neighbor's experience of households with hens.
- How city police and animal wardens are affected by the presence of hens.
- The impact (if any) hens in the city have on the animal shelter
- Input from the Rockbridge Area Conservation Council (RACC).
- Input from the Agriculture Extension Service.

Indicators in the study design will be:
- Approximation of food waste diverted from the local landfill.
- Average bio-conversion per hen, including the composting of yard waste.
- Number of eggs produced for local consumption.
- An estimation of how much fuel is saved by not transporting food waste.

This project addresses the feasibility of keeping household food and yard waste out of landfills and using chickens to transform that waste into valuable, locally produced compost, fertilizer, and eggs.

Lexington residents who have committed to participate in the study include 16 adults and 17 kids. There are others interested if this proposal is approved.

Frequently Asked City Chick Proposal Questions:

Question: Why is this research important?

Answer 1: America has a problem with trash disposal. One hen can biorecycle about 7 pounds of food waste a month. A few hens — in hundreds or thousands of backyards can transform tons of biomass kitchen and yard "waste" into compost. If 100 families had 6 hens, and composted/recycled their food and yard waste (and that of their neighbors) they could keep about 8,400 pounds (4.2 tons) of food waste and double that of yard waste out of landfills.

Answer 2: There is a national movement encouraging local production and consumption of food. This research examines the decreasing fuel costs of transporting food and helping to decrease the carbon dioxide levels associated with global warming. It is in line with the "thinking globally and acting locally" credo.

Question: What about the risk of Avian Flu?

Answer: There has not been a single case of human illness due to family flocks causing avian flu in the United States. The incidence of avian flu in small flocks is virtually zero. If this disease were to be pandemic, it would be more hazardous to commercial flocks raised intensively in stressful conditions.

Question: How will the results of this research be used?

Answer: The results of the Lexington chicken project will be submitted for publication in national poultry magazines and be posted on the website www.ChickensInTheCity.com. We expect to collaborate with environmental studies students in universities to expand the research in other cities.

Question: How is this project funded?

Answer: The non-profit Gossamer Foundation is collaborating with Good Earth Publications, Inc. to sponsor the research with a grant, in-kind contributions, donations, and volunteer support.

Question: What will it cost to participate in the study?

Answer: Each participating household will be responsible for the cost of:

• Purchasing 6 to 8 chickens at a cost of $15 to $25 each.

• Feed, scratch, and grit (oyster shell).

• Building or buying suitable hen housing, nest boxes, feeders, and waterers. There are companies that sell stylish coops for suburban chickens.

• Every household will be given data sheets and questionnaires.

• A course on chicken keeping will made available to participating households.

Question: Who is on the research team?

Answer: The principle researcher and organizer is Patricia Foreman. She will be working with local volunteers and environmental students from various colleges and universities.

Question: What do you expect from this pilot study in the long-term?

Answer: There are several potential long-term ramifications from this study.

1. If the research shows that chickens have a significant potential to divert household food and yard waste from the landfill, then we will pursue expanding the practice throughout the country.

2. The research will document the participant's experience with micro-flocks and is in the spirit of the Rockbridge Area Conservation Council's efforts to increase local food production. This is also in accordance with the efforts of countless organizations to decrease global warming.

The research proposal was referred to the Subcommittee for consideration. They would meet in one month.

Round 5: The Third Subcommittee Consideration of City Chicks

Twenty five months after our initial request we met with the review committee again. This time our advocate group included some of Lexington's most respected citizens, including two professors from Washington and Lee University: Professor David Harbor, Head of the Geology Department, and Professor Matt Tuchler, Head of the Chemistry Department.

These professors gave our proposal scientific validity. It showed that we were not just a group of wacked-out, crazy people who had nothing better to do than promote the husbandry of our fine feathered friends.

We presented the pilot study. Our presentation fell on deaf ears. Some members of the Council were adamant that they didn't want to hear anything about chickens or the landfill. In fact, some Council members were quite angry that we petitioned the Council again. They said, in various ways, that we were wasting their time.

A letter was read out loud by Mr. Kludy's. He clearly illustrates why many city dwellers oppose urban chicken residency. His views would be valid if the old way of keeping chickens — in stationary coops and non-rotating, toxic runs — IF this system were still practiced today. But it isn't. It's true the old stationary coop and run system stank. It was unsanitary. It was fly ridden and it reeked of ammonia so badly that you could hardly breathe without gagging. The chickens living in this environment were filthy and vermin was a constant concern. This is not the way City Chicks are kept.

The books: *City Chicks* and *Chicken Tractor* offer new ways to keep small flocks of hens. When managed properly, a few chickens don't create the problems he describes. My chicken coop is next to my home; the run abuts my house and the coop is 35 feet from my side door. I can see the run, coop, and gardens from my office window. Because of the way we manage the hens, we have none of the problems Mr. Kludy is concerned about.

Perhaps the Lexington City Council, Mr. Kludy, and others will read *City Chicks* and will understand that the old stinky chicken house paradigm has evolved to much higher standards and has more to offer than egg production.

It's not just the hens. How they are managed and employed as garden and compost workers is just as important.

Our hope is that the advantages of producing local foods, creating topsoil, and the economics of diverting biomass waste from landfills will tip the personal poultry prejudice scale towards legally allowing laying chickens in urban backyards. Add to that our inalienable right to grow our own food, including a family flock of chickens and the justice scales have to eventually tip to city administrators encouraging folks to keep chickens.

All of us promoting the City Chicks project want to thank the Lexington City Council for their service and attention. They are a highly dedicated group of our finest residents who put in long hours and deal with many difficult and complicated issues. The Council members, Mayor, and city officers serve our city well. We want to especially acknowledge and thank Council member Dr. Mimi Elrod for her support and guidance during the city chick quest for approval. Dr. Elrod has been a special friend and advisor. She was recently elected the first woman mayor of Lexington in 167 years!

Thanks to all the council members: Lawrence Broomall, Frank Friedman, Jim Gianniny, Tim Golden, Jack Page, and Ron Smith. We are also grateful to John Knapp, former Mayor, Larry Mann, City Attorney, T. Jon Ellestad, City Manager, Bill Blatter, Director of Building, Planning and Zoning and Debbie Desjardins, Clerk of Council.

Deep thanks to our local newspapers and their editors: Matt Paxton, Publisher, Darryl Woodson, Editor of *The News Gazette,* and reporter Roberta Anderson for her coverage and clear, witty writing. Thanks to Jerry Clark, Editor of *The Rockbridge Weekly,* and reporter Patty Wood for their dedicated service. Special thanks to Doug Harwood, Owner and Editor of *The Advocate* and Elizabeth Sauder for their special support.

It's the beginning of a new phase, and with recent elections, some city council members are retiring. Bestowing humorous parting gifts at their final meeting was a memorable way to end the year. It's amazing how a legacy can be left behind. and what one is remembered for after years of service. Jack Page gave the council member who was staunchly against City Chicks:

> *"A toy chicken to forever commemorate opposition to advocates of locally grown foods who lobbied to keep a limited number of hens in the city for local food production."* — THE NEWS GAZETTE

Mimi Elrod, former council member and newly elected mayor, was given an egg timer that was used by the former mayor to "time comments during public hearings on controversial issues".

Our city chicks advocates are convinced that the "egg timer" is an auspicious metaphor symbolizing that it's only a matter of time before hens can live legally within the city limits to product local eggs.

Summary of Chicken Myths & Legal City Chicks Lessons Learned

Addressing the legality of keeping city chicks has taken twists and turns that surprised us all. Both sides of the fence hold remarkably strong and sometimes valid opinions. Below is a summary the facts, and what we've learned about the politics of city chicks.

1. Local Laws Pertaining to Chickens. If chickens are illegal then develop a strategy to change it. Get to know city officials and ask their positions. Present them with the facts that counter biased and false beliefs about city chickens. Explain how chickens can help save taxpayer dollars by bio-recycling kitchen and yard waste. Explain how important chicken skill sets are in local food supply and for emergency preparedness. Get family flocks as debated issues in the next elections. Include candidates positions printed in your newspaper. Educate folks about how family flocks can be a part of emergency preparedness and national defense strategies. City council members across the country have lost elections due to being against allowing folks the right to have a family flock and supporting green projects.

2. Form a Chicken Advocate Group; Local poultry clubs including 4-H are a good place to start. Reach out to other chicken advocates across the country who can help with strategy and facts. Generate eMail campaigns to your local city officials that favor micro-flocks of family chickens. At public hearings, have your group wear pro-chicken T-shirts, and hats, that help visually show how strong the poultry advocates are. There's power in numbers, votes and persistence.

3. Your Right to Grow Your Own Food. Invoke your right to have, and control, your own food supply. The Declaration of Local Foods Rights states:

> *We hold these truths to be self-evident, that all people*
> *have certain unalienable Rights, that among these are*
> *Life, Liberty, Pursuit of Happiness, and the right to grow*
> *one's food in their yards — including a family flock of*
> *chickens!*

4. Emphasize the Economics of Chickens as Biomass Recyclers. It's quite simple and that's why it's so powerful. Chickens eat food scraps and their manure can be used to compost leaf and yard waste. This decreases the amount of trash that has to be picked up, transported, possibly transferred and dumped into a land fill. Encouraging residents to keep city chicks is a no-cost, ecological and sustainable technology will save local tax payer dollars. Any elected official that "talks green" but doesn't support — and even encourage — micro-flocks of chickens and backyard composting is simply not walking their talk! It's that simple.

5. Emergency Preparedness Partners. Explain how chickens can be part of a household's emergency plan. At most there are 3 days of food supply in the local grocery store; and that assumes no hoarding. With a family flock of chickens, even if the food supply lines are severed, you can still have an omelette. This can also be part of a National Defense Strategy. The Pentagon has repeated stated that our food and water supplies are the most vulnerable links in terrorism attacks.

6. Chicken Ambassadors. To help change the prejudice and stereotypes of chickens, take a sociable chicken to City Council meetings and even to court. In a "Are Chickens Pets or Livestock" court case in Maryland I took a Buff Chantecler hen as an "expert witness" that chickens can be pets. She charmed everyone, even the guards and the press.

7. Chickens as Pets. Most chickens in family flocks have names. In micro-flocks chickens have a chance to express their personalities and their even senses of humour. They can do tricks to amuse guests and neighbors. For these with allergies to dogs or cats, chickens can be alternative pet

8. Design Chicken Tractors, Coops, and Arks to be Attractive, Worthy of Being in the City. Take into consideration the type of siding, roofing, fencing, size, and height of your systems. Plant shrubs, flowers, or install attractive fencing that will help your poultry system blend in.

9. Keep Flocks Small and Sized to Match the Chicken Housing. Just as with people, overcrowding causes behavior problems and filthy environments Match the flock size to the coop and available space.

10. Be a Local Food Activist. Point out that folks interested in keeping hens in the city are doing so primarily as an act of taking control of their food supply and reducing their carbon footprints. Educate the powers that be of the importance of local food supply to emergency preparedness, national defense and healthy food.

11. Document the Lessons You Learn and Share the Results. Utilizing urban hens is on the cutting edge, and there is a lot to learn and quantify. Document how much biomass is diverted from the trash and converted into compost. Track how much you save on feed by giving hens kitchen and restaurant residues. Estimate the government funds saved by not having to transport, transfer, and landfill biomass trash. Document other benefits to your household and the community, such as increases in local soil fertility and garden production. Get the facts on how many complaints there were in your city, and others about chickens compared to dogs or cats.

12. Be Diplomatic, Positive, and Persistent. These attributes make your City Chicks efforts unstoppable. The realities of finding economical and ecological waste management solutions and a sustainable, wholesome local food supply will create new paradigms and overcome poultry prejudice. Be persistent about your right to keep a family flock. It might even take a new petition submitted shortly after every election or appointment of new city officials. Know the voting track record of your elected officials and let voters know who's "Green" and in favor of local food security.

There is a pervasive underground "city chick" movement sweeping across North America. Cities that don't allow chickens have flocks residing illegally and quietly; and the numbers are exponentially growing.

Thousands of baby chicks are being donated by hatcheries collaborating with local sponsors to offer "Chicken Stimulus Packages" and "Chicks for Charity" fund raisers to support local food supply. The purpose of these programs are to help more folks gain experience with keeping family flocks. Sponsors include The Chicken Whisperer Talk Show, and Meetup poultry clubs such as the Shenandoah Valley Poultry and Garden Club.

The Chicken
Have — More Plan

Grass roots campaigns are leading municipalities to allow chickens within city limits. Mayor Dave Cieslewicz of Madison, Wisconsin stated in 2004:

> *"It's a serious issue . . . it's no yolk . . .Chickens are really bringing us together as a community. For too long they've been cooped up."*

— MAYOR DAVE CIESLEWICZ

MADISON, WISCONSIN

Mayors and city council members across the country are agreeing that it's time to "think outside the coop" and employ chickens, and their skill sets, to serve their communities.

May Truth and Justice prevail.

Eggs and hope spring eternal.

...and may the flock be legally with you

— Evermore!

Chickens are very popular....every carnivore loves a chicken dinner.

— CITY CHICKS

19 City Chick Predators

The loss of chickens to predators is heartbreaking, discouraging, and frustrating. Chickens are popular food for carnivorous animals, birds of prey, and, yes, people. Chicken nuggets for all.

Chickens are low on the food chain. You will probably lose a chicken to a predator eventually. It's awful to see the tell-tale pile of feathers in the yard.

The most common city predators are dogs, raccoons, opossums, and people. Birds of prey, feral cats, rats, and stray ferrets are less common, but still problematic where they exist. Each of these chicken stalkers is described below.

Dogs

It's amazing how a sweet little poochie can instantly morph into a crazed poultry killer. Neighbors' dogs are usually the biggest threat to City Chicks. Even if the dogs can't get into the coop to forcibly kill the chickens, they can run back and forth in front of the pen, jumping and barking. The frightened chickens will crowd in a corner and pile up, and the ones stuck on the bottom can suffocate or die of fright.

Our new neighbor's two Lhasa Apsos would charge the hen pen, yapping crazily. We quickly and firmly tutored the pooches in correct poultry etiquette. They stopped the annoying behavior, and even bonded with the chickens and would try to protect them by herding them away from the street when they were let out to free range.

Teaching a dog how to behave around chickens is critical. It usually doesn't take much training. Dogs must clearly understand that chickens are a NO. I start the canine-chicken introduction by putting the dog (or puppy) on a leash and walking to the coop. If the dog lunges at a chicken, snap the leash sharply and loudly say NO. Walk around and inside the run, repeating this procedure, until the dog shows no interest in harassing the chickens.

Next, get the chicken and dog up close and personal. Let the chicken ride on the dog's back and flutter in its face. The dog will eventually understand that chickens are not to be chased like squirrels.

After a dog kills a chicken and tastes its blood, training the dog to ignore chickens gets tougher. The only way to stop the carnage is to learn how to train your dog to obey all commands you give – the first time.

Luckily, dogs are highly trainable and will learn to guard your chickens instead of chasing or harming them. Having a chicken-killing dog in a chicken-keeping home causes too much tension and is an accident waiting to happen. It's worth the time to train dogs to be chicken-friendly poultry defenders. You have to be firm, clear, and consistent with the dog: barking at, chasing, and killing chickens is totally unacceptable behavior.

If a dog does kill a chicken, the time-tested punishment for the crime is to tie the dead chicken around the dog's neck. Leave the dog tied outside for an extended period of time to contemplate the deed. It has to be an unpleasant experience.

Neighbors' dog(s) that roam and kill is a civil matter. I heard about a family that had been fined 3 times and gone to court twice for a loose dog, and they still refused to confine the dogs to their property. For the nearly $1,000.00 paid in fines, they could have built a great fence.

If you have trouble with a stray dog, document everything and call the police or animal control officer. The law will be on your side. Here's the law in Virginia:

> *Code of Virginia ß 3.1-796.116. Dogs killing, injuring or chasing livestock or poultry. It shall be the duty of any animal warden or other officer who may find a dog in the act of killing or injuring livestock or poultry to kill such dog forthwith whether such dog bears a tag or not. Any person finding a dog committing any of the depredations mentioned in this section shall have the right to kill such dog on sight as shall any owner of livestock or his agent*

finding a dog chasing livestock on land utilized by the livestock when the circumstances show that such chasing is harmful to the livestock.

Raccoons

Raccoons are common in urban areas. A raccoon, with all its fastidious habits, can be a particularly gruesome murderer. If a hen is sleeping too close to the side of the pen, a raccoon will reach through the wire, grab her, and pull the live chicken apart through the wire. It's a horrible way to die.

Raccoons are strong and smart. They can pull chicken wire off a coop to get to the birds. They can manipulate latches and pry open windows and pop hole doors. They are good climbers and can go up and over the tops of fences. If a raccoon gets to the hens, he will usually eat the crops (throat area) and decapitate some of them. If given the chance, he will return every few days for another meal. Raccoons like eggs and will usually remove one end of the shell without crushing it.

Opossums

Opossums are North America's only marsupial (mammals whose young develop in a pouch), and are closely related to kangaroos and koalas. They range in weight from 4 to 14 pounds. A opossum visiting your hen house will leave eggs chewed into many small pieces. If it eats a chicken, the carcass will be mauled from the rear forward.

Birds of Prey

Owls and hawks usually kill birds at night. Remember that birds of prey mostly eat mice, rats, moles, and other potential pests. They help keep nature in balance. Owls are very adept at carrying off chickens. Hawks leave behind everything but the chicken's head and neck.

If birds of prey are a problem in your area, installing netting over the run might be the solution. Letting chickens roost in trees at night is an owl's invitation to dinner.

Feral Cats

Most cats will not chase or attack a full grown hen, but they will eat chicks and smaller bantams. Trap the feral feline with a Hava Heart live trap baited with hamburger or cat food.

Rats

Rats and mice are common in urban areas. Rats will kill baby chicks and drag them to their burrow. Mice don't bother chicks or chickens. In fact, a chicken will gulp down a mouse or baby rat. Rodents love living in spaces between walls and under floors. They also love poultry feed and will chew holes in feed bags if they have access to them. Keep feed in metal cans secured with bungee cords, and feed will stay rodent-free.

Minks & Ferrets

Mink, weasels, and ferrets can get through unbelievably small holes. They steal eggs by rolling them away. I had a ferret or weasel wipe out my flock in one night. Only one hen survived; twelve were brutally slaughtered. I thought the fence was predator-proof but obviously it was not. Invaders will continue to find and create different ways to get to the chickens. The calling card of these smaller rodents is small bites around the chicken's head and neck. Weasels, minks, and ferrets usually crush an egg while eating it.

The best solution is to buy or borrow a live wire trap. Ask your local vet, Humane Society, Animal Control Unit, or wildlife trust if they know anyone who could loan you one. Bait it with cat food in gravy; the smell is irresistible to them!

Foxes

Foxes probably won't be a problem in larger cities, but they can thrive in rural towns and suburbs. Foxes like eggs and will carry them away to eat.

Foxes are in a league all of their own when it comes to stealing eggs and killing chickens. Their boldness and cunning is legendary, and they hunt both day and night. Even though my hen house is only about 20 feet from my home, in sight of my office, patrolled by 2 mixed-terrier watchdogs and 4 cats, and surrounded by a 4 foot fence, a fox still got to my hens. At about 10:30 in the morning I heard the chicken distress call and ran out to see a fox in my backyard, cornering a hen.

Another time my neighbor saw one of my chickens moving fast across her lawn...in the mouth of a fox. That hen wasn't so lucky, but at least the kits (baby foxes) got a tasty meal. A good rule of thumb is if a squirrel can get in, so can a fox.

You will know there was a fox attack if the coop is unusually quiet and the hens that normally greet you aren't there. There will be a trail of feathers across the yard, in fact, there might be feathers everywhere. A few of the hens will be huddled on their perches with fearful eyes, making low, scared sounds. On the floor and in the run there will be lifeless carcasses but not much blood. A couple of the birds might be missing. Even if you search the coop to find how the foxes got in, you might not find any clue at all. Such is the way of a fox.

Protecting Your Flock from Predators

The best defense against predators is prevention. Here are some ways to protect your chickens (and their eggs) from becoming someone else's dinner:

- Build tightly constructed coops. Ferrets, rats, and weasels can get through remarkably small openings. Cover any openings that are larger than 2 inches.

- Close the door and pop holes at night to keep nocturnal possums and raccoons out.

- Design the doors and windows so they can't be pried open.

- Keep the birds inside until after the dawn "killing hour." The early bird gets eaten.

- Install secure perimeter fencing (preferably at least 4 feet high) to keep foxes and dogs out.

- Use fence skirts, especially around the perimeter fence, to keep critters from digging under.

- Put netting over the runs if birds of prey are a persistent problem.

- To keep predators from climbing over a wire fence, install a tightly stretched electric wire about 3" above the fence.

- Install an electric fence around the hen house or run.

Make your property unattractive to potential predators so that they prefer to go somewhere else instead of to your land. Removing predator food sources and making potential shelter not appealing as housing make the real estate less desirable even to visit. There are several ways you can do this.

- Put trash out in morning instead of evening.

- Raccoon and opossum-proof garbage cans with tight fitting lids, and/or secure them with straps or a bungee cord.

- Don't leave pet food out at night.

- Enclose compost piles in a framed box using hardware cloth, in a sturdy container (such as a 55-gallon drum) or in a commercial composter.

- Keep the area under bird feeders clean.

- Remove brush piles and debris.

- Close garage doors at night.

- Put L-shaped wire skirts around the bottom of fences and stationary coops and pens so critters can't dig under fences.

- Keep a dog. Besides barking and chasing off predators, dogs will mark the territory which can serve as an olfactory "no trespassing" sign to intruder predators.

If you lose hens to predators, console yourself by acknowledging that every being has a right to live and needs to eat. Predators that eat chickens also keep the rodent, squirrel, and rabbit populations down. Some of the predators will be migratory, just passing through.

Finally, remind yourself that chickens usually have short lives. Compared to the billions of chickens that live their entire lives without ever seeing sunshine or touching the earth, know that your hen enjoyed a blessed and remarkable life while in your care.

20 City Roosters

Cock-a-doodle do — or Cocka-a-doodle don't? The issue of urban roosters is important enough to include it as a separate chapter. Roosters are beautiful. They can be charismatic and charming. They are spunky, flashy, and handsome. Some breeds are awesomely colorful and have plumage that would make a peacock jealous.

Sometimes called roos or gallos, roosters can be likeable. Their cock-a-doodle-do-ing in the morning and throughout the day can be a pleasant natural sound. All roosters seem to think they are Romeos and their crowing is an expression of their masculine desirability and in-your-face self esteem.

Crow-ologists found that crowing has harmonic overtones. Each breed tends to have a unique pattern, number of notes, tempo, accents, clarity and pitch. Some breeds have 2 note to 4 note crows with different inflections and emphasis on the different notes of the scale. One breed, the Tomaru, is known as the black crower and has a rich two-tone deep call.

Some county fairs hold rooster crowing contests and give a prize to the cock who crows the most in a certain time frame, usually 30 minutes to an hour. Three breeds from Japan are known as "longcrowers". These males can sing for over 15 seconds in one breath.

In the wild, roosters protect their flock by being aggressive toward predators. I've heard heroic stories of roosters battling hawks and foxes to protect their flocks. Like a noble knight in the days of old, a rooster will sacrifice his life so his ladies may live.

Roosters serve other important roles, such as food finders. They will show their hens where a feast is waiting and will share their own food. A rooster will explore new territory, boldly going where no chicken has gone before. Sometimes that new territory is over the top of the fence into the yard. Roosters have their diplomatic side and often mediate pecking order disputes.

But roosters have their dark side. The two qualities that make roosters indispensable in the wild also make them undesirable in the city: crowing and aggression/protection.

Crowing is Noise Pollution to Some Ears. Like a dog barking incessantly, roosters can be high-pitched noise-makers all day and night; they will crow and crow and crow. Every morning they announce to the world that they deserve credit for the sunrise. Their shrill tones can invoke a negative physical response in humans, like fingernails on a blackboard. Roosters can interrupt the sleep of almost everyone in the neighborhood.

Many municipalities that allow chickens forbid roosters, and crowing is on most city lists of noise pollution violations. Even the most tolerant neighbors are usually quick to report a clandestine cock. Once found, fines for contraband roosters run up to $1,000, so it's not worth the risk or loss of neighborly goodwill to house a fugitive rooster.

Theories abound regarding why roosters crow. Some research concludes that crowing is a sign of fighting ability. A loud, long, fierce crow implies that the cock must have the physical ability to back up his verbal challenges. By listening to each other, the crower and the hearer of the crow (crowee?) can judge their chances of winning a fight. Having a louder, longer crow might bluff a guy's way out of a fight by intimidating a competitor that is bigger/meaner/older/healthier. If crowing can avoid serious fights and the accompanying injury and death, then bellow it out.

Some researchers think that a rooster crows to call his hens and to mark territory. A dog marks his territory with urine. Since roosters don't have a hind leg to raise, why not stake their claim with sound?

They also crow to signal danger to the flock. A rooster will sound the alarm if startled during the night. Oddly, this trait doesn't help much because chickens are night blind and couldn't see to escape if something were stalking them.

There's volume in numbers; more roosters means more crowing. When we had broiler breeding flocks, there would be a daily crow-off competition between the roosters. They would crow for hours; it was as though they were relieving the burden of un-given speeches. One would crow because another crowed, and none of them wanted to be the first to stop; so they just kept on crowing.

One bawdy comment on an Internet chat group suggested bantams (smaller roosters) crow more because they suffer small cock envy. This could be the subject of an inter-species episode of City Chicks Sex in the City.

Should Roosters be Allowed in City Limits?

Sometimes roos can be intown. There are special cases where roosters have been allowed to be kept within the city limits. In these cases, special permits were issued and the permit was subject to approval by the neighbors. These roosters had become pets and in a couple of cases were being used to raise endangered heritage breeds of chickens.

Can Roosters be Kept Quiet?

People have tried stiffle their roos — with varying degrees of success. Here's what seems to decrease crowing. Three of main means are:

- Blocking light from the coop at night.

- Sound-proofing the coop.

- Bring the roo inside at night to a garage or basement.

Chickens are night blind and can't see anything in the dark so roosters tend to be quiet when they are in the dark — until something startles them, then they sound their alarm; that's their job. The darker and quieter the better for all the flock to sleep.

Roosters tend to be quieter when they are the only male; however, being the only cock doesn't always keep a rooster quiet. This was evidenced recently by an online chat room comment about the remarkable volume produced by a lone little bantam Silver Laced Sebright. The roosters of this breed are considered the Pit Bulls of the bantam world and are known for being loud, flighty, and ready to fight – even when there is no other rooster to challenge their dominion.

Are there breeds that have quiet roosters? As far as I know, little research has been done on which chicken breeds crow loudly or quietly. Smaller breeds (bantams) might crow more softly. Some of the quieter breeds (according to discussion groups) are Modern Game Bantams, Asilis, Orloffs, Seramas, Silkies, and Brahmas. Larger breeds might have more volume accompanying their size. If someone were to breed a silent variety of rooster, they might be popular birds for urban flocks, but that would be like breeding a silent dog – not very practical. On the contrary, crowing ability has been genetically selected. It's a guy thing.

Surgically removing the roo's voice box probably isn't practical either, given the cost of veterinary procedures. The bottom line is, unless the roo is an extremely special, valuable guy, it probably isn't worth it.

To conclude, can a rooster be kept from crowing? I think the answer is generally no, but there are some strategies to try if you absolutely, positively have to keep a roo.

Rooster Agression. Aggression is the second biggest problem with some roosters; they can morph into clueless pit bulls-with-feathers. Some breeds have more agressive roosters then others. Injuries to children, pets, and strangers are likely if your rooster is aggressive. Some people have horrific memories of being mauled by roosters. Friends have told me they have used trash can lids as shields and baseball bats as weapons to protect their legs and ankles from a rooster attack. Some rooster owners routinely wear gloves while collecting eggs to protect their hands from the gashing spurs of a rooster's feet.

Children are even more vulnerable to rooster attacks. Because their size is closer to that of a rooster, kids can be viewed as threats and viciously assaulted. You do not want to have to explain to the authorities — or the child's parents — why your rooster attacked a child.

A fully grown rooster can weigh up to 15 pounds, and it's not all feathers. The spurs on a rooster are pointed — like daggers — and he skillfully uses them as weapons. A friend of mine got a terrible gash on her palm from the rooster's spur. "Spike" became "stew" that evening.

Roosters can be aggressively hard on hens, especially in micro-flocks. The ratio of rooster to hen should be about 10; which is larger than some jurisdictions allow. I've seen hens with their backs shredded by the sharp spurs and toenails of an over-amorous rooster. It looks like they received 100 undeserved lashes. Roosters bite the hen's neck to hang on and can inflict bruises and bleeding while mating. There is no foreplay when some cocks mate. No Tantric intimacy for these guys. Some poultry breeders buy "mating saddles" to put on hens so the rooster can't do as much damage with the numerous mating attempts.

Just like compost, roosters happen. A male will be mis-sexed and put in your female-only chick order. Or you might order chicks that were not sorted by sex at the hatchery and half the brood will be male. Science projects or Easter hatchings will also produce little baby roos.

If you do end up with a rooster and your neighbors complain, the police will legally side with the folks who are complaining. You might be putting your entire flock – and the flocks of other urban poultry keepers – at risk. Stick with hens and enjoy them.

No matter how cute the cockerel is or how much you enjoy hearing him cock-a-doodle-do all day and night, the bottom line is roosters, in general, should not be kept in urban environments with out special permit and requirements.

What to do with a rooster if you end up with one? You have options.

- Find him a new home.
- Donate him to the local zoo.
- Apply, or petition, for a special rooster permit.
- Have him for dinner.

This last option has advantages. Coq au Vin (roo in wine) is a classic French dish of slow simmered rooster cooked in red wine. There's a saying among chefs: The older the bird, the richer the sauce. With access to exercise and fresh air, mature roosters have stronger and denser bones than indoor birds and make a delicious sauce. Roosters are tough. The trick to cooking roosters is to cook them slowly for a long time so the meat will be tender and the broth rich.

In summary, in my opinion, roosters don't belong in urban areas without special conditions which would probably be in the form of a specific permit. Roosters are banned because of their crowing noise pollution. They can be aggressive and are a legal liability if they attack anyone. Their presence is not necessary for hens to lay eggs, and a rooster in a micro-flock can harm the hens by repeated mating.

If your local laws don't allow, or have special permits for them, then don't keep a rooster within city limits. Irregardless of how handsome, personable and gentle he is, no amount of moxy a rooster might have can compete with legal fines and losing the goodwill of your neighbors.

Local food supply systems are growing in popularity, importance, and necessity. Changing weather patterns are forcing us to consider, and prepare for, food shortage prevention and local food self-sufficiency.

We must prepare for food shortages not only for lack of calories, but also for want of nutritional value.

We can do this by strengthening and expanding local food production systems that use biologically sustainable methods — such as described in City Chicks.

Epilogue: The Chicken Have-More Plan Revisited

About 60 years ago, Ed and Carolyn Robinson wrote a classic book called: *"The Have-More" Plan: A Little Land — A Lot of Living*. Their book inspired millions of people, including myself, to be more self-sufficient. It showed folks, in Ed's words, *"How to do things simply and well"*.

Like the Robinson's book, *City Chicks* is also about how to do things simply and well — by employing poultry power and chicken skill sets. This book began with briefly outlining four intertwined areas that are all influenced by a chicken have-more plan. These 4 areas are:

1. *Enhancing Local Agriculture.*
2. *Diverting Food and Yard "Waste" Out of Landfills.*
3. *Decrease Oil Dependence and Lower Carbon Footprints.*
4. *National Defense & Emergency Preparedness.*

Now let's look at the unintended consequences and benefits of keeping family flocks. On a small scale, *City Chicks* describes how to integrate small flocks in with gardening, and how you can have a good meal of eggs and garden goods that only travel the short distance from your backyard or community garden, instead of the 1,500 miles from a factory farm to your grocery store. Let's explore the multi-layered ramifications large-scale food production systems.

Food Production and Topsoil

Without topsoil, it is difficult to grow good food easily. *City Chicks* outlines resourceful techniques for yard and kitchen waste management that morph the "waste" into topsoil and increased soil fertility.

But for food production to be biologically and economically sustainable, drastic changes must be made to large-scale production processes. These processes begin in the soil...or not. It is alarming that most commercial food is grown on biologically dead soils. Decades of applying toxic fertilizers, pesticides, herbicides, and fungicides to fields have killed the vast majority of beneficial soil dwellers.

Additionally, commercial fields have been alternatively deeply plowed and left winter bare for so long, that they have suffered serious wind and water erosion. The "dirt" in most mono-cropped fields is more like clay or sand subsoils. The topsoil structures are nearly devoid of humus, the rich organic matter that supports soil life and retains moisture. Not only is there inadequate topsoil to support root structure, but the remaining soil has become toxic due to agricultural "plant support" chemicals, or by salinization and calcification from irrigation.

Then there is the issue of genuinely nutritional foods. Think about this. The USDA definition of "Organic" tells you absolutely nothing about the nutritional value of food. "Organic" only means that certain toxic chemicals (poisons) have NOT been added to the soils, sprayed on the plants, or fed to livestock. In other words, UDSA Organically Certified only means certain specific chemicals have not been used. It tells you nothing of the antioxident, vitamin, mineral, protein or lipid quality or quantities present in the food.

The question remains: can healthy crops be produced from unhealthy soils? The answer is no. Even worse, topsoil destruction by erosion and chemical contamination carries serious long-term consequences for the health of all beings on our planet.

Local creation and enrichment of living topsoils is where small flocks of chickens can serve as world class heros. Their manure, added to local carbon-based materials, can create compost to grow your local foods.

Hidden Costs of Food Production

New agricultural techniques, costs, and economies of scale continually push food production towards the easier, quicker, cheaper, and larger. The hidden economic and environmental costs are being ignored, and even worse, it is creating a debt that future generations will have to reconcile. Examples of these hidden costs are:

• Loss of Topsoil. Massive amounts of topsoil have been lost because of over-plowed and under-protected fields.

• Waste Pollution. Massive amounts of manure waste are produced by beef, hog, and poultry operations. This concentrated waste can drain into rivers and seep into ground water, polluting ecosystems in the process. The methane gas from the concentrated manure contributes to global warming and global weather pattern changes.

• Chemical Pollution from Agriculture. Run-off of fertilizers, pesticides, and herbicides drain and leach from fields into streams, rivers, lakes and bays causing "dead zones" (oxygen-poor areas created by excessive algae) that kill most aquatic life.

Plowing and Erosion Cause Topsoil Loss

Decades of over-plowing and under-protecting fields have resulted in wind and water erosion causing severe topsoil loss. Topsoil is necessary for healthy crop production, supporting soil life, nutrients and moisture retention.

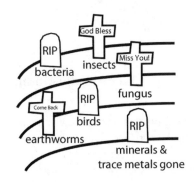

Fields Without Topsoil Become Soil Dweller Graveyards

Applications of oil-based fertilizers, pesticides, fungicides, and herbicides kill soil dwellers and the surrounding wildlife. Runoff poisons the streams and rivers.

Genetically Modified Crop Nutrition and Safety

The nutritional and safety issues of genetically modified organisms (GM or GMOs) are another area of concern. GMOs are common in animal feed and are increasingly showing up in human food. The nutritional and metabolic effects of GM foodstuffs in humans have not been extensively studied. GM plants are selected for size, shape, appearance, and shelf life, and bred for disease, insect, and pesticide resistance. GMOs are popular among commercial growers because they simplify production and lower costs — not because they have superior nutrition.

Nutrients such as antioxidants and vitamins have been shown to be lower in GM plants. Some genetically modified produce are more like empty nutritional shells, but are able to endure bumpy trips during transport, or

Genetically Modified Super Seeds & Plants

The only plants that can grow in depleted soils and survive applications of herbicides, pesticides and fungicides are genetically engineered crops. The plants are bred for their long shelf-life, ability to withstand transportation, and how pretty they look when they arrive at the supermarket. They are not selected for optimal nutritional value.

The soils they grow in often lack trace minerals, finer oils, and amino acids; so these don't pass up the food chain. The result can be a nutritionally deficient harvest, even if it's labeled "USDA organic".

last for extended periods of time on a shelf and still look picture perfect. The mass-producers of food have very different objectives than those of the nutrition conscious consumer.

A study published in *The Lancet*, vol 354 in October 1999 entitled "Effect of Diet Containing Genetically Modified Potatoes Expressing Galanthus Nivalis Lectin on Rat Small Intestine" concluded:

> *"...there is intestinal damage caused by affecting the mucosa of the gastrointestinal tact and this exerts powerful biological effects that may also apply to Genetically Modified (GM) plants, such as soya beans or any plant expressing lectin genes or trans-genes.*
>
> *"*

Although this study was performed on rats, its importance lies in understanding the link between the GI tract and your immune system. As GMO food consumption has increased over the last several decades, there has been a corresponding rise in allergies and autoimmune diseases to epidemic proportions.

Although the link has not been conclusively established, I wonder if these diseases are linked to a food chain laced with GM plants and other synthetic chemicals and molecules that the human body has never metabolized before. Dr. Jeffrey Bland of the Institute for Functional Medicine calls these synthetic molecules "New to Nature". They are in plants, preservatives, artificial colorings, fake flavorings, and the plastic packaging prevalent in our grocery stores.

"Leaky Gut Syndrome" is a medical condition you will be hearing more about in the news, and hopefully from your doctor. Leaky gut is what its name implies: it's a condition where the normal protective barrier in the gut lining has been breached, allowing increased permeability of toxins, bacteria, viruses, undigested food, and inappropriate molecules.

The breach in the gut lining can cause an immune reaction and systemic responses to the foreign invading substances, thus invoking allergic reactions. Leaky gut syndrome is increasingly associated with many conditions including the autoimmune diseases of: rheumatoid arthritis, lupus, and scleroderma. In many cases, treating leaky gut greatly helps patients with colitis, chronic fatigue syndrome, multiple sclerosis, and Crohn's disease.

Evidence suggests that this gut damage can be caused by antibiotics, toxins, fast food diets, parasites, infection, and synthetic "New to Nature" molecules. Many of these diseases take years to develop, and may even be passed on to future generations.

Add to the hidden costs of food production, the lives of stress-whacked meat animals. These "production units" are force-fed unnatural diets made from "rendered" by products, and from the cheapest GM feeds possible that are grown on eroded, dead soils.

Thus we have the recipe for a toxic food production system that has dramatic ramifications throughout our culture ranging from the "health care crisis", to dropping scholastic rates of our progeny, and environmental contamination below, on, in and above our planet.

Commercial Meat Production

As we work up the food chain, most factory farm produced meats and dairy products contain antibiotics, growth-enhancing hormones and chemical residues. These chemicals can wreak havoc on our endocrine systems and can slowly accumulate and poison us as we consume them over the decades of our lives. Nutritionally depleted soils and contaminated protein eventually affect us in adverse ways we are only beginning to understand.

> **If the axiom "you are what you eat" is true, then it follows that "you are what your food eats."**

We Feel What We Eat: Transferred Stress

On a moral level, much of large scale commercial agriculture treats meat animals (and especially poultry) inhumanely; and that affects you. Scientific studies backed by research done at Harvard and Stanford conclude that food carries more than just nutrition. These studies show that stress can be imported into our bodies by consuming products from stressed sources. That's right, it's possible that

Stressed Livestock

Stress might follow up the food chain. Is it possible that you also feel what you eat? This transferred body-state signaling is called xenohormesis.

eating stressed animals can cause second-hand stress, as if you didn't have enough of your own.

This phenomenon, called xenohormesis, is the hypothesis of how body-state signaling (fight or flight stress) produced by one species (xeno), and then eaten by another species, can set in motion a stress response (hormesis). A paper titled, "Are We Eating More Than We Think? was published in *Medical Hypotheses,* (Volume 67, Issue 1, 2006, by Yun, JA, Lee, PY and Doux, JD of Stanford University). The abstract states:

> *"Modern techniques of husbandry and agriculture can produce stress in the food chain, such that food itself can act as an illegitimate signal of chronic stress. Obese livestock and unusual fat profiles in farmed fish, meat, and eggs may reflect stress phenotypes. Consumers of stressed foods may sense those signals — a phenomenon known as xenohormesis — and assume the stressed phenotype. This maladaptive process may promote obesity."*

In other words, when a highly stressed cow, chicken, pig, or fish becomes meat, some of its hormonal stress and anxiety markers are biochemically passed up the food chain to you, the consumer, producing the same biochemical response in your body. That ancient saying: "you are what you eat" looks back at you from the overcrowded, filthy feedlots and stressful factory farms.

What We Eat is Our Health

My degrees from Purdue University in Animal Science (genetics and nutrition) and Pharmacy (Clinical) have given me a lifelong interest in nutrition and the role it plays in obtaining and maintaining optimal health. My belief about nutrition is that it starts with healthy soils, and travels up the food chain to the end consumer...or not. Disease (dis-ease) starts with the soils, and bears ill tidings up the food chain, making each level increasingly dis-eased and further from optimal health.

Let us not forget how we are bombarded by advertisements of top athletes downing caffeinated, carbonated soft drinks, and loving Mom serving mouth-watering processed dinners (wrapped in layers of endocrine-disrupting plastic packaging) to seemingly healthy and happy families.

The adverse health affects of these processed foods, and their packaging, are never mentioned and rarely researched.

Our current health care system down plays the relationships between food and health. It's easier (and more profitable for drug companies) for consumers to pop a pill than it is to take responsibility for their lifestyle and food preferences.

I continue to work as a pharmacist to stay current with new developments in medicine. The advantage of working at a pharmacy in a large-chain grocery store was that I could see how food and health are linked directly. As I got to know customers better, I learn more about their lives and lifestyles. I could see what they ate by what was in their grocery carts. I also get to know the physicians' drug prescribing patterns. As a customer comes to the pharmacy with their grocery cart, I can pretty accurately tell you what diseases they have (or will have) and what drugs they are (or will be) taking. The sequence goes like this:

- What is in their grocery carts (what they eat),
    results in …
- What a customer's body looks like (phenotype),
    which is an indicator of, or results in …
- What chronic diseases they manifest,
    which results in …
- What prescription drugs they will be prescribed.

A direct, connect-the-dots pathway exists between what a person habitually eats and what chronic diseases they will, or won't, develop. The following description of the four archetypes of eaters and shoppers describes the general food-disease connections.

Four Archetypes of Eaters and Grocery Shoppers

1. The Interior Eaters

These consumers primarily eat highly processed and canned foods; they eat few fresh vegetables, fruits or unprocessed meats. Interior eaters tend to be overweight and many of them have what's called "metabolic syndrome". Metabolic syndrome is a combination of several diseased states that include insulin resistance, high blood pressure, elevated triglycerides, and central obesity with fat deposits around the waist and hips. Metabolic syndrome currently affects about 1 in 4 Americans. The incidence is rising rapidly — especially among young people — along with increased risk of developing cardiovascular disease and diabetes.

2. The Soda Drinkers & Sugar Eaters

The soda drinkers and high sugar eaters are interior shoppers, but they have a serious sugar (sweet tooth) addiction. These are the shoppers with six-pack sodas hanging from their grocery carts. They buy significant amounts of sweet foods, especially those items containing high fructose corn syrup. The soda/sugar consumers have more pronounced metabolic syndrome tendencies, and the tendency to be obese. They are well on the road to diabetes, heart disease, and dental problems.

3. The Perimeter Eaters

These consumers generally buy foods on the perimeter of the grocery store — the vegetables, fresh meats, fish, and dairy. They purchase some, but not a lot, of the processed foods in the middle of the store. The Perimeter Eaters are generally slimmer and healthier looking than the Interior Eaters.

4. The Locavores

These are consumers who prefer to eat locally grown food. They are part of a broader stainability movement and tend to be more aware of how their food sources and selections affect their health. They are more tuned-in to how they feel after eating. Because of this awareness, they tend have slimmer, fitter phenotypes. They probably buy foods from local food networks that include farmers' markets, food co-ops, and community farms. They might have a kitchen garden and grow some of their own vegetables. They might keep a micro-flock of chickens for eggs and fertilizer. I see fewer of these people at the pharmacy.

The next time you are in a grocery store, look for the four different grocery cart archetypes and notice who is pushing the cart. The bottom line is you have choices about what you eat. Your choices make a difference in your health — or disease — states. No one knows the extent of how nutritionally depleted and genetically manipulated our food supply has become. Nor do we know the potentially grim, long term consequences of our schizophrenic food culture and our Standard American Diet (SAD). What we do know is that the choices we make as a society, and as individuals, are reflected in our grocery carts, what we see in the mirror, and how frisky or drained we feel.

Oil In Food Production

Commercial food production systems cannot work without oil. The hidden costs of getting food to your table significantly exacerbates the already massive pollution, economic, and foreign entanglement problems we are facing as a society.

A brief look at our oil-related pollution gives us a sense of the scope of the hidden costs and problems discussed earlier. It has been estimated that over 17 percent of America's oil is used for agriculture production. This includes fuel for tractors and other farm tools, petroleum based fertilizers, pesticides, herbicides, greenhouse heating, irrigation, temperature control, processing, packaging, refrigeration, and transportation to your dinner table.

Over a quarter of this oil is used for fertilizer. Because commercial soils have lost their natural fertility, they cannot produce crops without continual fertilizer fixes. The bottom line is, commercial agriculture is addicted to, and cannot produce food without, massive amounts of petroleum-based products. Every man, woman, and child in this country requires about one gallon of oil per day, just to bring food to their table.

You may already be aware of this commonly cited statistic: the average food item travels more than 1,500 miles from source to your table. But here is the real eye-opener: the total energy input of producing, processing, packaging, transporting, and storing food is greater than the calories consumed.

In other words, our current food production systems require more energy to produce — and get food to you— than the energy you will get from eating the food. How sustainable is that? The real price our planet, and future generations, will pay for our current food practices is incalculably, unimaginably and exorbitantly high. It is high not only in terms of environment damage, but also in terms poor health and the associated medical costs for chronic diseases caused by dietary choices.

Food Supply and National Security

Published in the October 12, 2008 *New York Times Magazine*, is a letter written to Barack Obama by Michael Pollan, author of the *Omnivore's Dilemma*. It's entitled "The Farmer in Chief". Pollan contends that national food policy is not something American presidents have given much attention. Current federal policies promote maximum production of corn, soybeans, wheat, and rice. These crops have been subsidized by the government so they can be produced in huge quantities.

Pollan's letter makes the point that the health of a nation's food system is linked to national security. He believes food supply is about to take center stage among the crisis issues. Here's what he wrote:

> *"You will need not simply to address food prices but to make the reform of the entire food system one of the highest priorities of your administration: unless you do, you will not be able to make significant progress on the health care crisis, energy independence or climate change."*   — MICHAEL POLLAN

He concludes that *"cheap food is food dishonestly priced — it is in fact unconscionably expensive"*. According to Pollan, if the world is to move past the oil era, improve the health of the masses, restore the environment, and lessen climate changes, then we must do so by *"turning food consumers into part-time producers"*. This brings the focus to local, nutritional food production, national defense and the health care (disease management) crisis.

The Poultry National Defense League

It's an ancient truth that whoever controls your food supply controls you. So how does this relate to family flocks of chickens? It's because of the value chickens have as protein producers and the roles and skills sets they bring toward urban food self-sufficiency. Throughout *City Chicks* we have talked about chicken skill sets and how valuable they are in food production systems. We also talk about how chickens can help us be less dependent on oil-derived "crop enhancement products" (as called by Big Ag). You can almost think of chickens as clucking oil-wells right in your yard. Acres of feathered diamonds!

The Cluck Heard Around the World

Can chickens in family flocks serve as part of national defense and emergency preparedness systems and strategies? YES!

Ponder on this. Your local grocery store has, at most, 2 to 4 days of food supply at any one time. This assumes people are not hoarding. You've seen how quickly stores empty with approaching storms or

Industrial Food Production Systems

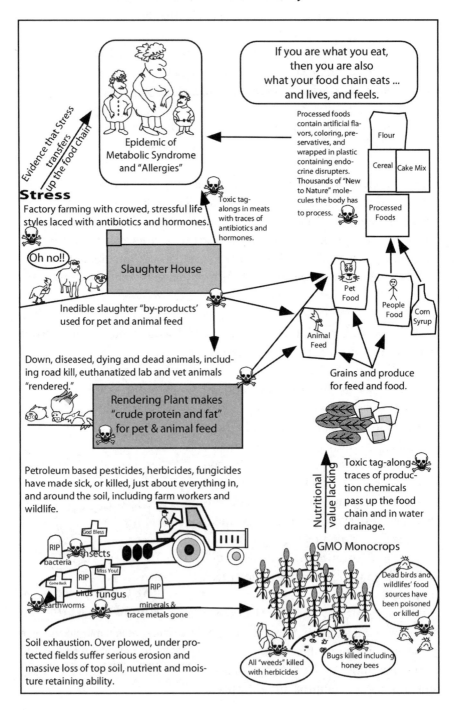

when supply lines are disrupted. We hear in the news, time after time, that when a disaster happens, the lack of food and water can become a matter of life and death and the highest priroty.

Even without a disaster, there is the matter of locally available protein. Kitchen gardens and urban homesteads can provide vegetables, nuts, fruit and small grains. But your diet could still be malnourished without the protein provided by eggs and chicken meat.

Local Food Production

Barbara Kingsolver's book: *Animal, Vegetable, Miracle* is a must-read to raise awareness of where food comes from and how your daily food choices make a huge difference for the planet. She writes:

> *"If every U.S. citizen ate just one meal a week (any meal) composed of locally and organically raised meat and produce, we would reduce our country's oil consumption by over 1.1 million barrels of oil every week. Small changes in buying habits can make big differences. Becoming a less energy-dependent nation may just need to start with a good breakfast".*
>
> — BARBARA KINGSOLVER

Authors like Joel Salatin are providing valuable how-to resources for anyone with some acreage wanting to produce wholesome food. Farmers' markets are providing customers for independent growers and becoming ever more popular thanks to the "Buy Local" programs and other programs like Women, Infants and Children (WIC) assistance programs that allow food coupons to be used at these local markets.

But small family farms have almost disappeared in America. The segment that is growing consists of smaller, specialty food producers, who use chinks of land niched between housing developments. Also becoming more prevalent are Community Supported Agriculture farms.

These producers are raising crops like trout, catfish, free-range poultry, vegetables, berries, fruits, nuts, mushrooms, herbs, and value added products like jellies and home-baked breads.

Communities are recognizing that independent food producers are not only important for local food production, they also increase local employment. They promote barter and currency exchanges that keep local resources in the community.

Tending Your Own Garden

Here's what Michael Pollan says about planting a garden.

> *"Planting a garden sounds pretty benign, but in fact*
> *it's one of the most powerful things an individual can*
> *do — to reduce your carbon footprint, sure, but more*
> *important, to reduce your sense of dependence and*
> *dividedness: to change the cheap-energy mind."*
> — Michael Pollan

Think about that. Planting a garden is one of the most powerful acts you can do. Michelle Obama did it on the South Lawn of the White House. It's amazing how that 20-by-50 foot plot of vegetables had such a powerful impact. Obama's simple act of connecting the dots between food and health has inspired kitchen gardens across America. Now national poultry advocates, including myself, are encouraging the Obamas to include a family flock of first chickens.

Why are we so passionate about gardening? Because gardening gets you up close and personal with your food sources. You know the quality of what comes from your garden. You help your community and play a part in solving the issues we are facing, such as dependence on oil, global warming, pollution of our environment and sustaining healthy lifestyles.

Gardening is also one of the healthiest hobbies. There's evidence that working in a garden acts as an anti-depressant, and working with living, wholesome soils can enhance your immune system. Families can garden together as a way to teach children (and adults) the value of work, patience, love, and learning a life-long skill.

How many gardeners are there in America? The National Gardening Association did a survey in 2005 and estimated that there were over 90 million households with gardens, and this number grows significantly every year and is fertilized by the high unemployment.

An article in the *British Guardian* entitled "Seeds of Change: Cabbages and Carrots Could Replace Flowers in Royal Parks" explains how London's Royal Parks will have flower beds converted into garden plots to demonstrate how city folks can grow fruits and vegetables. Chicago's Grant Park recently demonstrated that edible landscapes containing fruit and nut trees and vegetable beds can be as attractive as flowers.

Small is beautiful in food production. Combining aesthetics with the practical might be the new paradigm we are looking for in local food production today. The "past is prolog" metaphor, coupled with the World War II Victory Gardens model could serve as a partial food supply solution for our times.

During World War II, twenty million Americans produced about 40 percent of the nation's produce from their backyards. That's a wow! Integrating chickens with modern technology and intensive garden systems can make wholesome, healthy, local food production move well beyond the levels of the Victory Gardens.

What does all this have to do with *City Chicks* and the Chicken Have-More Plan? A lot! Like the Robinson's *Have-More Plan* of fifty years ago, keeping chicken micro-flocks can help each of us, our families, and our communities to have-more of so many things and benefits including:

• Health by using locally produced produce grown in living soils and high quality protein from eggs. Better nutrition will help address the "health care crisis".

• Chickens give us locally produced, non-oil based fertilizer to grow foods.

• Use poultry power in your garden as pestaciders, herbiciders insecticiders and fungiciders.

• Recycle kitchen and yard waste to make compost — keeping biomass out of landfills while at the same time creating topsoil to grow food.

• Save local government funds by NOT having to transport as much garbage and waste biomass from your trash can to the landfill.

• Breathe cleaner air, due to less methane gas production from anaerobic biomass breakdown in landfills and fuels used to produce, process and transport foods

• Establish connections with, and an understanding of, your food sources. Gardens can be an endless topic for good conversations, seed swaps, and crop sharing.

• Learn about growing foods as you study seed catalogues, read gardening books, get involved with garden and poultry clubs, go to farmers' markets and, attend gardening workshops.

• Enhance garden beds using chickens to glean, clean and till.

• Promote chicken welfare. Backyard micro-flocks get to live in the sunshine, scratch in the dirt, stretch, run and flap their wings, and have a place to nest. Simple quality-of-life things that battery cage hens never get a chance to experience.

• Preserve rare chicken breeds. Like crops, poultry has also been commercially mono-cropped. Most commercial layer and broiler chickens originate from a very small gene pool. Keeping small flocks of chickens can help preserve the genetic pools, raise interest in endangered heritage breeds of poultry and support their breeders.

• Barter and have another income stream with eggs and chickens.

• Make a lot of friends through poultry. I'm amazed at how many people I have met and contacts I've made because of my interests in poultry and local food systems. It has enriched my life, and my health.

Keeping chickens isn't for everyone. It doesn't have to be. But, a few households keeping a few chickens in every neighborhood, and a few households composting for themselves and their neighbors, can make a difference to the community and the environment.

David Gershon's book: *Low Carbon Diet: A 30 Day Program to Lose 5,000 Pounds* describes how individuals, neighbors and support groups can make small changes to promote sustainable life style choices. He writes about forming eco-teams. With an eco-team approach, neighbors collaborate collecting kitchen, yard and garden waste for composting. They can take turns looking after flocks and sharing eggs. They can barter services for garden vegetables and fruits.

In closing, *City Chicks* might be more descriptively entitled: *How Chickens Can Help Save You, and the Planet, in Times of Commercial Food Supply Contamination, Food Nutritional Deficiencies, Topsoil Erosion,*

Pollution of Water Supplies, Energy Shortages, Waste Management Difficulties, Landfill Closings, Weather Pattern Changes and Global Warming. But *City Chicks* is easier to remember.

It is my intention and prayer that *City Chicks* might help and inspire you to explore new ways to feed yourself, your family, your friends, and support your community local food production.

If you do that — with the power of one — you will be simultaneously be conserving our valuable resources and repairing, restoring, and preserving our global environment for all living beings — for generations to come.

Many blessings, love and hope.

...and may the flock be with YOU!

Quoth the Chicken, "Evermore".

Glossary

A

Alektorophobia. Fear of chickens.

American Standard of Perfection. A book published by the American Poultry Association describing the characteristics of each breed recognized by that organization.

B

Bantam. A miniature chicken variety that is about one-fourth to one-half the size of a regular chicken. They are kept mainly for ornamental purposes. Some chickens come in both Standard and Bantam varieties.

Banty (plural, banties). Affectionate word for bantam.

Barnyard chicken. A chicken of mixed breed.

Beak. The hard, protruding portion of a bird's mouth, consisting of an upper beak and a lower beak.

Bedding. Straw, wood shavings, shredded paper, or anything else scattered on the floor of a chicken coop to absorb moisture and manure.

Biddy. Affectionate word for a hen.

Bloom. The moist, protective coating on a freshly laid egg that dries so fast you rarely see it.

Bran. The outer coating of a kernel of grain. It is extremely high in silicon, which slows down its decomposing in the soil and is a cheap by-product of milling. Often given away free by large mills.

Breed. 1. (noun). A group of chickens that are like each other and different from other groups. 2. (verb). Pairing a rooster and hen for the purpose of obtaining fertile eggs.

Breeder. 1. A mature chicken from which fertile eggs are collected. 2. A person who breeds chickens.

Broiler. A young chicken destined to be eaten; also called a "fryer".

Brood. 1. (verb). The act of a person or hen caring for a batch of chicks. 2. (noun). The chicks themselves.

Brooder. A heated enclosure used to imitate the warmth and protection a mother hen gives her chicks during the brooding process.

Broody. (adj.) Describes a hen's desire to sit on eggs for an extended period of time to incubate and hatch them.

C

Candle. (verb). To examine the contents of an intact egg with a strong light source.

Candler. (noun). A device that uses strong light to examine the contents of the egg.

Cannibalism. The bad habit of chickens eating each other's flesh, feathers, or eggs.

Capon. A castrated rooster.

Carrier. 1. An apparently healthy individual that transmits disease to other individuals. 2. A container used to transport chickens.

Chicken Diapers. Yes, you can get diapers for poultry. Bring your birds inside — or to meetings — without being embarrassed by their bowl movements at sensitive moments.

Chooks. Popular Australian term for chickens.

Clean legged. Having no feathers growing down the shanks.

Cloaca. The chamber just inside the vent where the digestive, reproductive, and excretory tracts come together.

Clutch. A batch of eggs that are hatched together, either in a nest or in an incubator (from the Old Norse word "klekja," meaning to hatch), also called a "setting".

Coccidiasis. Infection with coccidial protozoa without showing any signs.

Coccidiosis. A parasitic protozoal infestation, usually occurring in damp, unclean housing conditions.

Coccidiostat. A drug used to keep chickens from getting coccidiosis.

Cock. A male chicken; also called a "rooster".

Cockerel. A male chicken under 1 year old.

Comb. The fleshy, usually red, crown on top of a chicken's head.

Conformation. A chicken's body structure.

Contagious. 1. Description of a disease that's readily transmitted from one individual or flock to another. 2. Refers to an animal who is infected with a disease and can infect others.

Coop. The house or cage in which a chicken lives.

Crop. 1. (noun). A pouch at the base of a chicken's neck that bulges after the bird has eaten. 2. (verb). To trim a bird's wattles.

Crossbreed. The offspring of a hen and a rooster of two different breeds.

Cull. 1. To eliminate (kill) or remove a non-productive or inferior chicken from a flock. 2. The non-productive or inferior chicken itself.

D

Debeak. To remove a portion of a bird's top beak to prevent cannibalism or self-pecking.

Down. The soft, fur-like fluff covering a newly hatched chick; also, the fluffy part near the bottom of any feather.

Droppings. Chicken manure or "poop".

Droppings Tray. A collection unit located underneath roosting poles that collects droppings for easy disposal.

Dub. To trim the comb.

Dust Bathing. A behavior pattern in which chickens dig themselves a hole in the ground and immerse themselves in the loosened earth, rolling around and trying to get as dirty as possible. Dust bathing is an important defense against mites and lice, and if they don't have access to a dust bath, they need an artificial dust bath set up indoors.

E

Egg tooth. A horny cap on a chick's upper beak that helps the chick pip through the shell.

Embryo. A fertilized egg at any stage of development prior to hatching.

Exhibition breeds. Chickens kept and shown for their beauty rather than their ability to lay eggs or produce meat.

F

Feather legged. Breeds with feathers growing down the shanks, i.e., Cochins and Brahmas.

Fecal. Pertaining to feces.

Feces. Droppings or body waste – chicken manure or "poop".

Fertile. Capable of producing a chick.

Fertilized. Containing sperm.

Flock. A group of chickens living together.

Fodder. Also called animal feed. Any foodstuff that is used specifically to feed domesticated livestock, such as cattle, goats, sheep, horses, chickens and pigs. Most animal feed is from plants but some is of animal origin. Fodder refers particularly to food given to the animals (including plants cut and carried to them), rather than that which they forage for themselves (see Forage). It includes hay, straw, silage, compressed and pelleted feeds, oils and mixed rations, and also sprouted grains and legumes.

Forage. 1. (verb). The act of an animal looking for and eating food as they free-range. 2. (noun). The food that animals find and consume of their own accord, as opposed to that which is collected and fed to them (see Fodder).

Forced-air Incubator. A mechanical device for hatching fertile eggs that has a fan to circulate warm air.

Fowl. Domesticated birds raised for food.

Free range. To allow chickens to roam a yard or pasture at will.

Fryer. A chicken destined to be eaten; also see "Broiler".

G

Gizzard. An organ that contains grit for grinding up the food a chicken consumes.

Grade. To sort eggs according to their interior and exterior qualities.

Grain. Small, hard seeds, particularly those in the grass family. Grains are rich in carbohydrates, B-vitamins, and phosphorus. Whole grains are also a fair source of protein.

Grit. Sand, pebbles, oyster shell, or any hard matter a chicken can ingest and use in its gizzard to grind up other food.

H

Hackles. A rooster's cape feathers.

Hatch. 1. (verb). The process by which a chick emerges from an egg. 2. (noun). A group of chicks that pip out of their shells at approximately the same time.

Hatchability. How likely fertilized eggs are to hatch after incubation.

Hen. An adult female chicken.

Hen Apron. An apron or saddle that protects hens' backs from being scratched by a rooster's toenails while he is mating (it also helps to keep his toenails trimmed and blunted). Also protects from pecking, especially while molting, covers injury to prevent further damage and enables prompt healing.

Host. Any animal (including birds) a parasite or infectious agent lives in or on.

Hybrid. The offspring of a hen and rooster of different breeds. The term is often mistakenly used when referring to the spawn of a hen and rooster of different strains within a breed.

I

Immunity. Ability to resist infection.

Impaction. Obstruction of a body orifice, such as the crop or cloaca. Also see "Pasting".

Incubate. To maintain the ideal environment for hatching fertile eggs.

Incubation period. How long it takes for fertilized eggs to hatch, usually about 21 days.

Incubator. A mechanical device for hatching fertile eggs.

K

Keel. A chicken's breastbone, which resembles the keel of a boat.

Leaky Gut. A condition in which spaces form between the gut wall's cells which allow bacteria, toxins, and food to leak into the bloodstream. This condition can affect all animal species, from humans to chickens. It results in an overworked liver, compromised immune system, and widespread inflammation, and pain.

L

Litter. Straw, wood shavings, shredded paper, or any material scattered on the floor of a chicken coop, run, or brooder to absorb moisture and manure.

M

Mate. The pairing of a rooster with one or more hens.

Mite. A tiny parasite with jointed legs that lives on chickens.

Molt. The annual shedding and renewing of a bird's feathers.

Morbidity. Percentage of a population affected by a disease.

Mortality. Percentage of a population killed by a disease.

N

Nest. A secluded place where a hen feels she may safely leave her eggs.

Nest Box. A human-made box designed to encourage hens to lay eggs in it. One nest box is required for every 4-5 hens in a flock.

Nest Egg (fake). A wooden or plastic egg placed in a nest to encourage hens to lay there.

O

Oocyst. The egg forms of animal parasites, including protozoa.

Oviduct. The tube inside a hen through which an egg travels to be laid.

P

Parasite. An organism that lives on or in a host animal and uses it for food or protection. It does not benefit its host.

Pasting. Loose droppings sticking to vent area, also known as "pasting up" and "pasty bottoms".

Pecking order. The social rank of chickens.

Pen. The fenced outside area around a coop.

Perch. 1. (noun). The place where chickens sleep at night. 2. (verb). The act of resting on a perch (See "Roost".)

Permaculture. The design and preservation of sustainable agriculture systems that mimic natural ecosystems.

Pinfeathers. The tips of newly emerging feathers.

Pip. 1. (noun). The hole a newly formed chick makes in its shell when it is ready to hatch. 2. (verb). The act of making the hole.

Plumage. All feathers covering a chicken.

Poultry. A term for chickens and other birds cared for by humans and used for food, eggs, or companionship.

Prebiotics. Non-digestible food ingredients that help an animal by stimulating the functionality and growth of "good" bacteria in the colon, and thus improving overall health of the animal.

Predator. An animal that hunts other animals and eats them.

Probiotics. Dietary supplements containing live microorganisms and "good" bacteria to help the body's naturally occurring gut flora microbes to re-establish themselves.

Pullet. A female chicken one year old or younger.

Purebred. The offspring of a hen and rooster of the same breed.

R

Range fed. Describes chickens that are allowed to graze freely.

Ration. The combination of all feed consumed in a day.

Resistance. Immunity to infection.

Roost. 1. (noun). The place where chickens spend the night. 2. (verb) The act of resting on a roost. (See "Perch".)

Rooster. A male chicken; also called a "cock" or a "roo".

S

Saddle. The part of a chicken's back between the wings and tail.

Scales. The small, hard, overlapping plates covering a chicken's shanks and toes.

Scratch. 1. (verb). The habit chickens have of scraping their claws against the ground to dig up things to eat. 2. (noun). Any grain fed to chickens.

Set. To keep eggs warm so they will hatch.

Sexed. Newly hatched chicks that are sorted into pullets and cockerels.

Sexing. A process by which the sex of a baby chick is determined.

Sex Link. A relatively new breed of chicken whose sex is indicated as soon as they hatch by the color of their feathers. Sex Link females are known for excellent egg production, and males are good "fryers".

Shank. The part of a chicken's leg between the claw and the first joint.

Spurs. The sharp pointed protrusions on a rooster's shanks.

Stacking. A method of farming, gardening, or chicken keeping in which each system or implement serves several functions or supports other functions.

Standard. The description of an ideal specimen for its breed; also, a chicken that conforms to the description of its breed in the *American Standard of Perfection.*

Started pullets. Young female chickens almost old enough to lay eggs.

Starter. A feed ration for newly hatched chicks, also called "crumbles".

Straight run. Newly hatched chicks that have not been sexed; also called "unsexed" or "as hatched".

Strain. A flock of related chickens selectively bred by one person or organization for such a long period of time that the offspring become uniform in appearance or production.

Stress. Any physical or mental tension that reduces resistance to disease or infection.

V

Variety. Subdivision of a breed according to color, comb style, beard, or leg feathering.

Vent. The outside opening of the cloaca, through which a chicken emits eggs and droppings from separate channels.

W – Z

Wattles. The two red or purplish flaps of flesh that dangle under a chicken's chin. Sometimes the wattles waddle when chickens walk.

Zoning Laws. Laws that regulate or restrict the use of land for a particular purpose, such as raising chickens.

Resources & Bibliography

Hatcheries

Increasingly there are small, heritage breed hatcheries that specialize on a few breeds. There might be one close to you. Check the internet for hatcheries and contact the American Livestock Breeds Conservancy who maintains breeders contact information.

Cackle Hatchery
www.cacklehatchery.com
 PO Box 529, Lebanon, MO 65536, 417-532-4581
Foy's Pigeon Supplies, 877-355-7727
www.foyspigeonsupplies.com

Hoffman Hatchery
Box 128, Gratz, PA 17030, 717-365-3694. www.hoffmanhatchery.com

Marti Poultry Farm
PO Box 27, Windsor, MO 65360, 660-647-3156. www.martipoultry.com

Meyer Hatchery
626 State Route 89, Ohio, 44866 888-568-9755
In business since 1985. Over 115 varieties of poultry. www.MeyerHatchery.com

Murray McMurray Hatchery
P.O. Box 458/191 Closz Drive Webster City, IA 50595
800.456.3280. These folks have a beautiful color catalogue.
 www.mcmurrayhatchery.com

Mt. Healthy Hatchery, 9839 Winton Rd. Mt. Healthy Ohio, 45231. 800-451-5603

Privette Hatchery
PO Box 176, Portales, NM 88130, 877-PRIVETT. 877-774-8388, www.privette-hatchery.com

Ridgway Hatchery
PO Box 306, LaRue, OH 43332, 800-323-3825. www.ridgwayhatchery.com

Sand Hill Preservation Center
 1878 230th St, Calamus, IA 52729, 563-246-2299
Neat company that sells heritage chicken breeds and heirloom seeds. www.
sandhillpreservation.com

Schrock's Mother Goose Hatchery
www.schrocks.com
14928 CR28 Goshen, IN 46528, 574-825-1325

Equipment Dealers and Gardening Suppliers

Clausing Company
Nocatee, FL 34268, 941-993-2542. clausing@desoto.net

Countryside Natural Products
PO Box 997, Fishersville, VA 22939, 888-699-7088 or 540-932-8534
They stock most natural feed supplements and soil treatments. They also have
an organically certified dealership for grains. www.countrysidenatural.com

Egganic Industries
Creative coop designer and custom builder including the egglu and henspa.
800-783-6344, www.henspa.com.

Foy's Pet Supplies
3185 Bennett's Run Road, Beaver Falls, PA 11510 877-355-7727
foyspigeonsupplies.com

Fertrell Company
PO Box 265, Bainbridge, Pennsylvania 17502, 717-367-1566
This is a major supplier of nutritional supplements and soil amendments.
www.fertrell.com.

Gardener's Supply Company
128 Intervale Road, Burlington, VT, 05401
800-427-3363. Innovate company supporting sustainable and local gardening. Catalogues are chucked full of handy products and delightful garden
gift ideas. GSC's research gardens are being used to conduct trials using
garden chicks in food production systems. www.gardners.com.

Hoegger Supply Company. Quality products for goat and small farm owners
since 1935. Color catalogue. Fayetteville, GA 30214. 770-461-6926.
www.HoeggerGoatSupply.com.

North Country Organics
PO box 372, Bradford, Vermont 802-222-9661
All natural land care products. www.norganics.com.

Premier 1 Supplies
800-282-6631. Offer a comprehensive catalog of fencing and related accessories. www.premier1supplies.com

Randall Burkey Company
Provides quality products since 11947. Offers an extensive inventory of poultry and game bird supplies. Excellent catalogue and customer service. 800-531-1097, www.RandallBurkey.com

Real Goods:
Solar Living since 1976.Helping people live a greener lifestyle. Real Goods provides products and education on topics such as solar electric, wind power, hydro-electric turbines, energy-efficient appliances, organic cotton, and much more! Really great catalogue. 888-507-2651 in CA, 888-212-5643 in CO www.RealGoods.com and www.RealGoodsSolar.com

Redmond Minerals, Inc.
PO Box 219 Redmond, Utah 84652, 800-367-7258
These folks mine the natural, good-for-you and your livestock, rock salt. www.redmondinc.com.

Seeds of Change, Certified Organic Herloom Seeds. Preserving biodiversity and supporting sustainable organic agriculture. Mission: to preserve biodiversity and promote sustainable, organic agriculture. By cultivating and disseminating an extensive range of organically grown vegetable, flower, herb and cover crop seeds, we have honored that mission for 20 years. 1-888-762-7333, www.SeedsofChange.com.

Seedway: Vegetable Seed for the Organic Grower
99 Industrial Road, Elizabethtown, P 17022
800-952-7333 phone, www.seedway.com.

Seven Springs Farm
426 Jerry Lane NE. Check, Virginia 24072, 540-651-3228
Organic farming and gardening products. www.7springsfarm.com.

Southern Exposure Seed Exchange: Saving the Past for the Future.
Their mission is to ensure that people retain control of their food supply, that genetic resources and conserved, and that gardeners have the option of saving their own seed. PO Box 460, Mineral, VA 23117, www.SouthernExposure.com.

Territorial Seed Company: All the Pieces to Your Gardening Puzzle
www.territorialseed.com.
800-626-0866 phone, 888-657-3131 fax

Tripple Brook Farm
37 Middle Road, Southampton, MA, 01073, 413-527-4624.
www.tripplebrookfarm.com.

Poultry & Urban Agriculture Magazines

Acres USA Magazine, A Voice for Eco-Agriculture. Acres U.S.A. is North America's oldest, largest magazine covering commercial-scale organic and sustainable farming. Has an excellent book catalog. PO Box 91299, Austin, TX 78735, phone 512-892-4400. www.AcresUSA.com.

BackHome Magazine, Your Hands-On Guide to Sustainable Living. PO Box 70, Hendersonville, NC 28793. Phone 800-992-2546. An excellent resource for sustainable living with articles by folks who know, and live, what they are writing about. www.BackHomeMagazine.com.

Backyard Poultry Magazine, Dedicated to More and Better Small-flock Poultry. 145 Industrial Dr., Medford, WI, 54451, 800-551-5691. www.Back-YardPoultryMag.com.

Hobby Farms Magazine. Colorful and informative magazine. P.O. Box 6050, Mission Viejo, CA, 92690-6050, 800-627-6157 or www.hobbyfarms.com.

Practical Poultry: the UK's Best-selling Poultry Magazine. Cudham Tithe Barn, Berry's Hill, Chadham, Kent, TN16 3AG, www.practicalpoultry.com.

Mother Earth News, The Original Guide to Living Wisely.
 www.motherearthnews.com.

Small Farm Today Magazine. The Original How-to Magazine of Alternative and Traditional Crops and Livestock, Direct Marketing, and Rural Living — Established. 1984 Regularly carries articles on pasture poultry and small-scale sustainable agriculture. 3903 West Ridge Trail Road, Clark, MO 65243, phone 800-633-2535. www.SmallFarmToday.com.

Other Resources
American Livestock Breeds Conservancy
PO Box 477, Pittsboro NC 27312, 919-542-5704
This is a non-profit association dedicated to preserving America's heritage livestock. Membership is a good way to network with like-minded poultry people. www.albc-usa.org.

For information about Canadian hatcheries, contact:

Rare Breeds Canada, Trent University, Program in Environment and Resources Studies, Box 4800, Peterboro, Ontario K95 7B8, 705-748-1634. www.rarebreedscanada.ca.

AGRICulture OnLine Access (AGRICOLA) – the Department of Agriculture's Library search at: INK http://www.nal.usda.gov/ag98/english/catalog-basic. html www.nal.usda.gov/ag98/english/catalog-basic.html

www.ATTRA.org/search.html
Contains lots of useful poultry information.
www.msstate.edu/dept/poultry
Great resource for information about chicken health.

BackyardChickens.com. Established in 1999, BackyardChickens has information you can use to raise, keep, and appreciate chickens.

Eggbid.com
An auction site for purchasing eggs.

FeatherSite.com maintained by Barry Koffler.
An on-line zoological garden of domestic poultry, including photos, video and information about various breeds of fowl. Barry Koffler.

Living Eggs; Education Through Life program. Provides complete kits and learning objectives for a 10 day school project. www.LivingEggs.co.uk

MyPetChicken.com. Chicken suppliers of baby chicks, chicken coops, accessories, books and free "how to" information.

National Gardening Assoication, 1100 Dorset Street, South Burlington, VT 05403, Phone: (802) 863-5251. Website has one of the largest and most respected array of gardening content for consumers and educators, ranging from general information and publications to lessons and grants.

Seattle Tilth Association, 4649 Sunnyside Avenue N, Room 120, Seattle, WA 98103, Phone (206) 633-0451
Website has loads of useful information. Hosts a coop tour every year. email tilth@seattletilth.org, www.seattletilth.org

Society for the Preservation of Poultry Antiquities (SPPA), 1057 Nick Watts Rd., Lugoff, SC 29078, http://www.feathersite.com/Poultry/SPPA/SPPA.html

Bibliography

The ABC and XYZ of Bee Culture, by A.I. Root, A.I. Root Company, 1945.

The Accounting Game: Basic Accounting Fresh From the Lemonade Stand, by Darrell Mullis and Judith Orloff, Sourcebook, Inc., 1998.

The American Poulterer's Companion: A Practical Treatise on the Breeding, Rearing, Fattening, and General Management of the Various Species of Domestic Poultry, by C.N. Bennet, Harper & Brothers, 1853.

The American Poultry Yard; Comprising the Origin, History, and Description of the Different Breeds of Domestic Poultry, by D.J. Browne, 1850.

Animals in Translation, by Temple Grandin, A Harvest Book, 2005.

Animals Make Us Human: Creating the Best Life for Animals, by Temple Grandin, Houghton Mifflin Harcourt Publishing Company, 2009.

Animal, Vegetable, Miracle: A Year of Food Life, by Barbara Kingsolver, HarperCollins Publishers, 2007.

Avoid the Vet: How to Keep Your Birds Healthy and Happy, by Practical Poultry, Kelsey Publishing Ltd..

Backyard Livestock: Raising Good, Natural Food for Your Family, by Steven Thomas, revised by George P. Looby, DVM, Countryman Press, 1990.

Backyard Market Gardening: The Entrepreneur's Guide to Selling What You Grow, by Andy Lee and Patricia Foreman, Good Earth Publications, 1993.

Backyard Poultry Keeping, by G.T. Klein, Everybodys Book Publishing Company, 1943.

Backyard Poultry Naturally: A Complete Guide to Raising Chickens Naturally, by Alanna Moore, Acres U.S.A., 2007.

Chicken Coops: 45 Building Plans for Housing Your Flock, by Judy Pangman, Storey Publishing, 2006.

The Chicken Health Handbook, Gail Damerow, Storey Publishing, 1994.

Chickens in Your Backyard: A Beginner's Guide, by Rick and Gail Luttmann, Rodale Press, 1976.

Chicken Tractor: The Permaculture Guide to Happy Hens and Healthy Soil, by Andy Lee and Patricia Foreman, Good Earth Publications, 1994.

Collapse: How Societies Choose to Fail or Succeed, by Jared Diamond, Penguin Books, 2006. A *New York Times* best seller and a must read for everyone who wants our society to survive.

Comfrey Report: The Story of the World's Fastest Protein Builder and Herbal Healer, by Lawrence D. Hills, The Rateavers, 1975.

Day Range Poultry: Every Chicken Owner's Guide to Grazing Gardens and Improving Pastures, by Andy Lee and Patricia Foreman, Good Earth Publications, 2002.

The Development of the Chick: An Introduction of Embryology, by Frank R. Lillie, Henry Holt and Company, 1908.

Diseases of Poultry, by Leonard Pearson, Clarence M. Busch. State Printer of Pennsylvania, 1897.

The Dollar Hen: The Classic Guide to American Free-Range Egg Farming, by Milo M. Hastings, Norton Creek Press, 2003.

Eggs All the Year Round at Fourpence Per Dozen, and Chickens at Fourpence Per Pound: Containing Full and Complete Information for the Successful and Profitable Keeping of Poultry (Second Edition), James Maclehose, 1876.

Eggs and Egg Farms, Third Edition, Reliable Poultry Journal Publishing Co., 1907.

Empty Harvest: Understanding the Link Between Our Food, Our Immunity, and Our Planet, by Dr. Benard Jensen and Mark Anderson, Avery Publishing Group, Inc., 1989. This book is a masterpiece. It starts by saying: "If they get you asking the wrong questions, they don't have to worry about the answers". It's time to ask the right questions.

Essentials of Medical Geology: Impacts of the Natural Environment on Public Health, edited by Olle Selinus et. al, Elsevier Academic Press, 2005.

Everything I Want to Do IS Illegal: War Stories from the Local Food Front, by Joel Salatin, Polyface, Inc, 2007.

Farm Poultry: A Popular Sketch of Domestic Fowls for the Farm and Amateur, by George Catchpole Watson, The Macmillan Company, 1919.

Farm Poultry Production, by Leslie E. Card and Melvin Henderson Interstate Printing Co., 1935.

The Feeding of Chickens: Farmers Bulletin No. 1841, by Harry W. Titus, U.S. Department of Agriculture, 1939.

Feeding Poultry: The Classic Guide to Poultry Nutrition, by G.F. Heuser, Norton Creek Press, 2003.

Feeds and Feeding, by Frank B. Morrison, The Morrison Publishing Company, 1954.

Free-Range Poultry, by Katie Thear, Farming Press, 1997.

Fresh-Air Poultry Houses: The Classic Guide to Open-Front Chicken Coops for Healthier Poultry, Norton Creek Press, 2008.

Fruit Trees for the Home Gardener, by Allan A. Swenson, Lyons & Burford, 1994.

The Garden Design Primer, by Barbara Ashmun, Lyons and Burford, 2003.

Gardening for the Future of the Earth, by Howard-Yana Shapiro, and John Harrisson, Bantam Books, 2000.

A Guide to Better Hatching, by Janet Stromberg, Stromberg Publishing, 1975.

How to Raise Chickens: Everything You Need to Know by Christine Heinrichs. Voyager Press, 2007.

The Henwife: Her Own Experience, in Her Own Poultry-Yard, by Blair Fergusson, Hamilton, Adams, and Co.

The "Have-More" Plan, by Ed Robinson and Carole Robinson, Storey Publications, Inc. 1973,

The Herbal Medicine-Maker's Handbook: A Home Manual, by James Green, The Crossing Press, 2000.

The Herb Gardener: A Guide for All Seasons, by Susan McClure, Storey Publications, 1996.

Holistic Herbal: A Safe and Practical Guide to Making and Using Herbal Remedies, by David Hoffman, Time-Life Books, 1996.

In Defense of Food: An Eater's Manifesto, by Michael Pollan, The Penguin Press, 2008. *Incubation: A Guide to Hatching and Rearing*, by Katie Thear, Broad Leys Publishing Co, 1997.

Keep Chickens! Tending Small Flock in Cities, Suburbs, and Other Small Spaces, by Barbara Kilarski, Storey Publications, 2003.

Keeping Chickens: The Essential Guide to Enjoying and Getting the Best From Chickens, by Jeremy Hobson and Celia Lewis, David & Charles, 2007.

Keeping Pet Chickens, by Johannes Paul and William Windham, Barron's Educational Series, 2005.

Livestock and Complete Stock Doctor: A Cyclopedia, by A.H. Baker, The Thompson Publishing Company, 1913.

Living with Chickens: Everything You Need to Know to Raise Your Own Backyard Flock, by Jay Rossier, The Lyons Press, 2004.

Low Carbon Diet: A 30 Day Program to Lose 5000 Pounds, by David Gershon, Empowerment Institute, 2006. www.EmpowermentInstitute.net

Making Plant Medicine, by Richo Cech, Horizon Herbs Publication, 2000.

The Merck Manual: Ninth Edition, Merck & Co., 2005.

The Merck Manual: Third Edition, Merck & Co., 1967.

Nutrition and Physical Degeneration, by Weston Price, Keats Publishing, 1997.

The Omnivore's Dilemma: A Natural History of Four Meals, by Michael Pollan, Penguin Press, 2006. Named one of the ten best books of 2006 by the *New York Times* and *the Washington Post*.

Pasture Poultry Profit$: Net $25,000 in 6 Months on 20 Acres, by Joel Salatin, Polyface, Inc, 1993.

Permaculture: A Practical Guide for a Sustainable Future, by Bill Mollison, Island Press, 1990.

The Pleasures and Profits of Our Poultry Farm, Chapman & Hall, 1879.

The Popular Edition of Wright's Book of Poultry, by Lewis Wright, Cassell & Cmpany, LTD., 1885.

Poultry Ailments and Their Treatment. For the Use of Amateurs, by D.J. Gray, James P. Mathew & Co., 1885.

Poultry and Poultry-Keeping, by Alice Stern, Merehurst Press, 1988.

Poultry: A Practical Guide to the Choice, Breeding, Rearing, and Management of All Descriptions of Fowls, Turkeys, Guinea-Fowls, Ducks and Geese, For Profit and Exhibition (Second Edition), by Hugh Piper, Groombridge & Sons.

Poultry Culture: How to Raise, Manage, Mate and Judge Thoroughbred Fowls, by I. K. Felch, A. A. Donohue & Co., 1902.

Poultry Diseases: Causes, Symptoms, and Treatment, With Notes on The Poultry Handbook, edited by Rudolph Seiden, D. Van Nostrand Company, Inc., 1949.

Poultry Keeping in a Nutshell, by Henry Trafford, Poultry Success Co.

The Poultry Yard; Comprising Management of All Kinds of Fowls (Including the Cochin-China), by W. C. L. Martin, George Routledge & Co., 1853.

Practical Incubation, by Rob Harvey, Hancock House Publishers, 1990.

The Practical Poultry Keeper, by Lewis Wright, Cassell and Company.

Practical Poultry Keeping, by David Bland, The Crowwood Press, 1996.

Practical Experience in Breeding, Rearing, and Fattening the Common

Kinds of Domestic Poultry, With an Account of Experiments in Artificial Incubation, by Joseph Newton, Ickwell: Published by the Author, 1852.

A Practical Treatise on Breeding, Rearing and Fattening All Kinds of Domestic Poultry, Pheasants, Pigeons and Rabbits: Also, the Management of Swine, Milch Cows and Bees; With Instructions for the Private Brewery on Cider, Perry, and British Wine Making, by Bonington Moubray, Sherwood, Gilbert and Piper, 1842.

Profitable Poultry; Their Management in Health and Disease, by W. B. Tegetmeier, Darton & Co., 1854.

Principals and Practice of Poultry Culture, by John H. Robinson, Ginn and Company, 1912.

Principles of Poultry Science, by S.P. Rose, CAB International, 1997.

Productive Poultry Husbandry, by Harry Lewis, J.B. Lippincott Co., 1933.

Reproduction in Poultry, by Robert Etches, University of Guelph, 1996.

Rodale's Illustrated Encyclopedia of Herbs, edited by Claire Kowalchik and William H. Hylton, Rodale Press, 1987.

Small-Scale Poultry Keeping: A Guide to Free-Range Poultry Production, by Ray Feltwell, Faber and Faber, 1980 & 1992.

Soil Fertility and Permanent Agriculture, by Cyril G. Hopkins, Ginn and Company, 1910.

Sprout Garden: Indoor Grower's Guide to Gourmet Sprouts, Revised Edition, by Mark M. Braunstein, Book Publishing Company, 1999.

Standard of Perfection, by The American Poultry Association, The American Poultry Association, 1926.

Standard American Perfection Poultry Book: Describing All of the Different Varieties of Fowls, Their Points of Beauty and Their Merits as Setters, by Felch, I.K., M.A. Donohue & Co., 1903.

Success With Baby Chicks: A Complete Guide to Hatchery Selection, Mail-Order Chicks, Day-Old Chick Care, Brooding, Brooder Plans, Feeding and Housing, by Robert Plamondon, Norton Creek Press, 2003.

Textbook of Functional Medicine, edited by David S. Jones, MD, The Institute for Functional Medicine, 2005.

A Veterinary Materia Medica and Clinical Repertory, by George Macleod, The C.W. Daniel Company Ltd., 1997.

Water-Conserving Gardens and Landscapes, by John M. O'Keefe, Storey Publishing, 1992.

Water for Every Farm: Using the Keyline Plan, by P.A. Yeomans, Second Row Press, 1981.

You Can Farm: The Entrepreneur's Guide to Start and Succeed in a Farming Enterprise, by Joel Salatin, Polyface, Inc., 1998.

Zen and the Art of Chicken Maintenance: Reflections on Keeping Chickens, by Martin Gurdon, Lyons Press, 2004.

Index

Symbols

Acknowledgements

The saying: "If I can see so far, it's because I stand on the shoulders of giants" is how I feel about so many people who have influenced my life. Two very awesome people have been my parents, Marie and Richard Foreman, who gave me a loving wholesome upbringing, and served as positive influences. My mother, in her 90s continues to be a role model on how to go through life gracefully, gathering bunches of close friends along the way. Thanks to my dear brother, Richard Foreman, for always being there. Thanks to Pat, Michael, Robert, Amanda, Cody and Austin Foreman for sharing holidays & family times.

Long-term thanks to Andy Lee for being a talented and original guy. Without Andy, I would not have had so many experiences with poultry, organic gardening, and green construction. Without Andy, none of the books we co-authored would have been written. Andy Lee, you deserve a standing ovation!

Gratitude to other members of my extended clan and community including: Scott Mitchell, Betty Gough, Tamara Santa Ana, Katie Letcher-Lyle, Nick Charles, Barbara Lane, Rosie McKnight, April Getz, Tasha Walsh, Teresa, Chris & Sarah Holler, Barbara Lane, Michel duBois, Paula Martin, Nancy (Scooter) McMoneagle, Beth Johnson, Manal Laurion Stulgaitis, Kip Brooks, Kate Robinson, Laure, Tim, Jacob and Leo Stevens-Lubin, Laurie Macrae, Monika Eaton, Alex and MaryLynn Lipscomb, Mimi Elrod, Nancy (Tweedy) McNeil, Paige Cooper, Wendy and Lee Duke, Rosie & Jim McMillan, Ansis, Lisa, Jack and Buddy Helmanis, Marsha Heatwole, Elizabeth Sauder, Cathy Wells, Penny Holland, Bev and Don Burks, Emily Gale Eubank, Nick Charles, Howie Mitchell, Sandy Masamoto, Andy Marsinko and Luke Sevison. Hugs to David, Cathryn, Erin, Kaya, and Thomas Harbor for so many entertaining dinners, discussions and antics. Top marks to Tasha, David, Brenna, Brian & Dylan for hot tub evenings, football games and camp cooking. Thank you P. Duaine Fitzgerald of the Bank of Botetourt for your support and valuable advice through some really hard times.

To Don Burks, my yoga teacher. Thanks for your friendship and for stretching my physical and mental edges, and to Lisa Burks for sharing the "toys".

"Splashes back at you" and "high-five paddle slaps" to my kayak buddies for the wonderful river trips and camping adventures: Andy Lee, Jann Ross, Karen McKnight, Bob Foote, Alicia Jahsmann, Barb Franko, Terri Bsullak, Dev Malore, Joey Dimegio, and all the other Virginia Paddlers.

Respect and gratitude to all the staff of the Rockbridge County SPCA for the service they provide to so many homeless and helpless animals.

Applause to all those who helped evolve the *City Chicks* manuscript including Fiona Raven, manuscript design consultant (www.FionaRaven.com), Wendy

Duke, Barbara Lachman, Alice Lynn, Marie Foreman, Don Burks, Laure Stevens-Lubin, Tim, Jacob and Leo Stevens-Lubin Stevens, Cathryn Harbor, Jennifer Bodnar, and Tasha Walsh for their hours of help with production, research, editing, proofing, re-proofing, and doing the countless tasks that are behind bringing a book to print. Thanks to Phil Laughlin for his creative covers.

To Andy Schneider, The Chicken Whisperer. It's been an incredible adventure co-hosting with him on the Chicken Whisperer Backyard Poultry and Sustainable Lifestyles Talk Show.

Special appreciation and respect to Will Raap, founder of Gardener's Supply Company. Gratitude to Frank Oliver and Kathy LaLiberty.

Hats off to Karl Hammer, owner of Vermont Compost Company, for truly being a poultry pioneer and a leader in urban compost creation. Well done Karl!

Appreciation to my pharmacy colleagues including: Charlie Kahle, Bonnie Lodge, Tonia Holley, Kathy Johnson, Carla Davis, Sandra Lewis, Cherity Smith, Katie McClung, Kirby Peters, Debbie Arthur, and Tim Ferguson.

Thanks to past, current and future residents of GreenWay Neighborhood in Buena Vista, Virginia for sharing the dream of creating a green community that respects the environment, preserves food production land and maintains edible landscape and forests for both people and wildlife.

The content in *City Chicks* is a result of the inspirations, ideas, writings, lectures, discussion groups and activities of so many people. I'm deeply grateful to the heritage poultry breeders for keeping the valuable poultry gene pools viable to pass along us from the past and to future generations.

My gratefulness to many author mentors, most of whom I've never met. These amazing writers and doers include (not in any particular order, nor is the list complete): Robert Rodale, George DeVault, Wendell Berry, Wes Jackson, Weston A. Price, Bill Mollison, Frank B. Morrison, Joel Salatin, Michael Pollan, Barbara Kingsolver, Temple Grandin, Jeffery Bland, Buckminster Fuller, Jean Houston, Ray Bradbury, Gail Damerow, Dave and Elaine Belanger (*Backyard Poultry Magazine*), Chris and Rachel Graham (*Practical Poultry Magazine*) and so many others whose works I've studied and benefited from.

This might be over the top, but I want to express my appreciation of the many poultry flocks I've been associated with over the decades. They have been my teachers, selfless food producers, entertainers, and a few of them my feathered family, especially the unflappable and unforgettable: Attila the Hen and Oprah Henfrey.

Most of all, acknowledgement with deepest respect and love for this very special living planet, Gaia, that supports us all.

— PF

About Good Earth Publications, Inc.

At Good Earth Publications we have a vision of the world as it could be. We want to promote a world that values nutritious foods grown on living soils. A world where livestock is humanely raised, respected, and honored for the food they pass on to us. A world where people protect and regenerate the environment. A world where we all respect our planet as we are part of a sacred living web of life. A planet where neighbors and nations are guided by the principles of goodwill, cooperation, and ever are striving to be better. Granted, this is all idealistic. But these ideals inspire us to do our best. We know that a dedicated group of people can form a force strong enough, a fulcrum long enough, and a consciousness aware enough to help move the world.

Having begun business in 1989 in Vermont, we are inspired by, and agree with, the socially responsible and linked prosperity business philosophy developed by Ben & Jerry's. We do our best to base our decisions and directions on actions that we feel are socially just, economically fair, and environmentally responsible. To us, business is much more than just a short-term profit or bottom line motivation. To this end, we have a three part mission objective.

• Product Mission: We view our books, training materials and courses as leveraged change agents that are directed towards life-affirming models. These models are regenerative, sustainable, and life-supporting for all beings of our planet.

• Social Mission: To operate our business in a way that actively recognizes the central role that business plays in the structure of society. Our goal is to initiate innovative ways to improve the quality-of-life of our community, at all levels: locally, nationally, and globally.

• Economic Mission: To operate the company on a sound financial basis of profitable growth coupled with creating career opportunities and financial rewards for our employees.

We conduct all our affairs with the highest levels of honesty, goodwill, and integrity. If an initiative is not mutually beneficial to all involved, we won't do it.

Other books by the authors Patricia Foreman and Andy Lee

Chicken Tractor: The Permaculture Guide to Happy Hens & Healthy Soil
The all new Straw Bale Edition. A chicken tractor is a portable pen that fits over your garden beds! Just set it wherever you need help in your garden. The chickens peck and scratch the soil to clean your beds, eat pest bugs and weed seeds, and fertilize the beds with their manure. Best of all, they provide eggs and meat with that old-fashioned flavor and homegrown goodness.

Day Range Poultry: Every Chicken Owner's Guide to Grazing Gardens and Improving Pastures, including the management of breeder flocks, egg handling, incubating, hatchery management, building shelters, marketing, advertising, soils regeneration, compost creation, processing poultry humanely and effectively, and much, much more! Foreword by everybody's favorite contrary farmer, Gene Logsdon.

Backyard Market Gardening: The Entrepreneur's Guide to Selling What You Grow
Discover how easy and profitable it is to grow and sell vegetables, fruits, flower, herbs, and small livestock from your backyard market garden.

A Tiny Home to Call Your Own: Living Well in Just Right Houses. Small is in! You can live in an attractive, aesthetically appealing, upscale tiny house AND you can do it so that it is quality built, architecturally beautiful, highly marketable, and profitable. This book shows you how small is beautiful, functional, and economical in housing. Smaller scale homes can be more livable, cozy, homey, enjoyable, affordable and socially responsible than the current trend of MacMansionization that is chewing up America. Why not impress your neighbors with style and freedom rather than a large-scale house?

Ask About Wholesale and Bulk Discount Pricing
Good Earth Publications, Inc.
20 GreenWay Place
Buena Vista, VA 24416
Phone & Fax. 540-261-8775
Email: info@GoodEarthPublications.com
www.GoodEarthPublications.com